ENVIRONMENTAL ECONOMICS
and Natural Resource Management

The tools of environmental economics guide policymakers as they weigh development against nature, present against future, and certain benefits against uncertain consequences. The policies and research findings explained in this textbook are relevant to decisions made daily by individuals, firms, and governments. This textbook offers instructors and students a user-friendly, relevant, and up-to-date introduction to these topics while covering recent advancements in the field and significant political and economic changes.

The book has been thoroughly updated while retaining the story-based narratives and visual emphasis of previous editions, capturing students' attention with full-color photos, graphs, and illustrations. This sixth edition includes:

- Updated coverage of international environmental regulations, the effects of the COVID-19 pandemic on the environment, the effects of war on the environment, recent environmental summits and agreements, the evolving energy and transportation sectors, and trailblazing policies and research
- Expanded coverage of environmental issues and approaches in underrepresented countries and continents
- New discussions of EV batteries, populist leaders, carbon leakage, food waste, and ecological resilience
- Revised digital supplements, including a solutions guide, PowerPoints, and sample tests.

Environmental Economics and Natural Resource Management promotes environmental and economic literacy with policy-oriented, application-based content delivered in concise, accessible discussions. Through its engaging approach, the text brings the economic way of thinking into discussions of personal, community, corporate, and government activities that affect environmental assets and the quality of life.

David A. Anderson is the Paul G. Blazer Professor of Economics at Centre College, Kentucky, USA. He received a BA from the University of Michigan, an MA and PhD from Duke University, and a Graduate Certificate in Innovation and Entrepreneurship from the Harvard Extension School. Dr Anderson's research focuses on the economics of the environment, law, crime, dispute resolution, and public policy.

"This book provides a nice backdrop to the material I choose to present in lectures. It also provides a nice roadmap for choosing which topics to cover and when to cover them. It is well-written and well-researched. Highly recommended."

Arthur J. Caplan, *Utah State University, USA*

"I just wanted to let you know how much I enjoy using this text in my environmental economics class. The book works very well for my students, who are a mix of economics and environmental science majors. The material is accessible enough for the non-majors yet challenging enough for the majors. I also appreciate the way the author addresses the normative issues central to environmental economics using the framework of economic analysis."

Margaret A. Ray, *Instructional Associate Professor, Department of Economics, Texas A&M University, USA*

"Anderson's book is just enough of a challenge for my students as it does not overwhelm them with models or policy prescriptions. Instead, this book treats environmental economics as an interdisciplinary endeavor which makes it a perfect fit for my community college students."

Paul Briggs, *Professor (Economics), Windward Community College, USA*

ENVIRONMENTAL ECONOMICS
and Natural Resource Management

Sixth Edition

David A. Anderson

Routledge
Taylor & Francis Group

LONDON AND NEW YORK

Designed cover image: Legna69 / Getty Images

Sixth edition published 2025
by Routledge
4 Park Square, Milton Park, Abingdon, Oxon, OX14 4RN

and by Routledge
605 Third Avenue, New York, NY 10158

Routledge is an imprint of the Taylor & Francis Group, an informa business

First edition published by South-Western 2004
Fifth edition published by Routledge 2019

British Library Cataloguing-in-Publication Data
A catalogue record for this book is available from the British Library

ISBN: 978-1-032-55044-2 (hbk)
ISBN: 978-1-032-55039-8 (pbk)
ISBN: 978-1-003-42873-2 (ebk)

DOI: 10.4324/9781003428732

Original Text Design by David A. Anderson

Typeset in Century Schoolbook
by Apex CoVantage, LLC

Access the Support Material: www.Routledge.com/9781032550398

For Donna, Ava, and Ally

Contents

CONTENTS

Chapter 6

Environmental Quality 141

Chapter 7

Energy 177

Chapter 8

Sustainability 207

Chapter 9

Population, Poverty, and Economic Growth   237

Chapter 10

Biodiversity and Valuation   259

Chapter 11

International and Global Issues   287

Chapter 12

Perspectives on Environmental Policy   315

Chapter 13

Natural Resource Management: Renewable Resources   341

Chapter 14

Natural Resource Management: Depletable and Replenishable Resources   367

CONTENTS

Preface

Courses in environmental economics and natural resource management are unique among college offerings. Like the arts, the environment serves as a basis for cultural identity and a fount for social welfare. As in the natural sciences, students of environmental economics and natural resource management seek a better understanding of the natural world. Yet this course holds special importance because it directly informs decisions that determine our fate. Economic analysis addresses the specifics of what products best serve society, what regulations provide net benefits, what incentives are optimal, and what resources should be prioritized for conservation.

As both the world's most dangerous predator and its last great hope, humans have the capacity to protect, alter, or destroy natural resources on a grand scale. Policymakers confront trade-offs between lives and profits and can be guided by greed or emotion in the absence of compelling information. Economic analysis reveals that the value of human life is not infinite and the optimal amount of pollution is not zero but that ignorance of economic insights results in undervalued lives and excessive pollution. The purpose of this book is to make economic tools readily available to college students interested in the environment or natural resources.

Most people are aware of debates involving the environment and natural resources. Relatively few are familiar with the economic way of thinking. Even fewer know the means by which to weigh short-term costs against long-term benefits or the costs and benefits of alternative fuels. When information is lacking, critical environmental policy is more easily swayed by questionable arguments. It may be inhuman to be dispassionate, but economic theory provides opportunities to displace emotions with concrete criteria. The challenge, then, is to apply the most valid methodology earnestly and honestly. This book explains relevant techniques and points out likely missteps.

The sixth edition of *Environmental Economics and Natural Resource Management* retains the story-based narratives of the previous editions, discusses the latest policy initiatives and developments in the field, and simplifies the most challenging topics. Plentiful full-color photographs, diagrams, and graphs offer visual perspectives on global environmental and resource issues. "Reality checks" in each chapter delve more deeply into the application of economic principles in the real world. Review problems and "websurfing challenges" reinforce understanding, and suggested Internet links and additional readings serve students whose interests have been stirred. Some of the more challenging models appear in appendixes to grant instructors the flexibility to cover them or not. Above all, this textbook addresses the critical objectives of environmental and economic literacy with policy-oriented, application-based content that is straightforward but compelling.

Ethical dilemmas surround environmental economics, but criteria for deciding right from wrong receive little coverage in many textbooks. The need for education on ethical considerations is punctuated by daily headlines about corruption and severe abuses of the environment. The allocation of scarce resources involves moral quandaries over the treatment of humans, wildlife, and future generations of the same. Chapter 16 of this textbook explains secular ethical theories and highlights the role of ethics in environmental and resource policy.

Alarm about resource scarcity earned economics a reputation as the "dismal science," yet there is hope for a marriage between growing consumer demands and progress on environmental fronts. The navigation of economic growth through sensitive environmental waters requires deliberate practices and a firm understanding of the relevant theory and evidence. May reading this book provide meaningful guidance along that journey.

General Overview

This textbook is divided into three parts. Part I introduces environmental economics and provides a review of the more useful tools in the field. Part II lays out current areas of interest and concern and explains alternative approaches to problem solving and the attainment of efficiency. Although discussions of environmental policy appear throughout the text, Part III emphasizes policy and public-sector oversight. Decisions regarding natural resources cannot escape the realm of ethics, so the final chapter provides a foundation in environmental ethics.

One cannot discuss or apply environmental economics appropriately without adequate knowledge of the underlying concepts and definitions. With no understanding of the food chain, one cannot appreciate the economic value of plankton. Not knowing the meaning of hedonic pricing, one cannot speak intelligently about estimating the value of biodiversity. For these reasons, the opening sections of most chapters contain definitions and perhaps a taste of chemistry, biology, or political science. The alternative would be to assume that readers have taken and remember all of those classes that complement environmental and natural resource economics—an expectation I would not want applied to myself!

Chapter 1 The Big Picture

This chapter illuminates the path ahead for readers. Each of the nine sections introduce a compelling environmental and natural resource issue and its ties to economics. The highlighted topics include market failure, waste and recycling, environmental ethics, sustainable development, biological diversity, environmental degradation, alternative energy sources, population and economic growth, and natural resource management. The chapter sketches out the importance of each challenge and the promise that economic tools will contribute to a beneficial resolution.

Chapter 2 Efficiency and Choice

This chapter covers the primary tools of economic analysis, explaining marginal analysis, expected-value calculations, supply and demand, and consumer choice. It serves as a review for students who have seen the basic concepts in an introductory economics course and as a reference for students who encounter applications of this material later in the text and want to re-read explanations of the underlying concepts.

Appendix: Efficiency Criteria in Greater Detail

This appendix provides a mathematically rigorous explanation of efficiency criteria.

Chapter 3 Market Failure

This chapter explains how market outcomes diverge from socially optimal outcomes. Discussions cover the sources of market failure—externalities, public goods, imperfect information, and imperfect competition—in detail, providing graphical analysis and real-world examples. This chapter also presents the Coase theorem using numerical examples, and it foreshadows the policy solutions in Part III of the textbook.

Chapter 4 The Role of Government

Chapter 4 analyzes the role of government in efforts to avoid market failure. Discussions address the need for government, the solutions government brings, and some pitfalls of both public and private approaches to externalities. The chapter also identifies opportunities to gain by substituting regulation for liability risks. Outlines of key environmental agencies and legislation are provided.

Chapter 5 Trade-Offs and the Economy

The most difficult challenges in environmental and resource economics involve painful trade-offs. Decision-makers struggle with choices between short-run and long-run benefits and between financial and environmental gains. This chapter explains the tools of discounting and their applications to tough decisions. The chapter then covers methods for weighing economic growth against environmental degradation and explores prospects for economic growth that are consistent with environmental goals.

Chapter 6 Environmental Quality

This chapter explains measures and determinants of environmental quality, including air quality, water quality, light pollution, and noise pollution. Case studies of solutions include policy, education, technology, product substitution, and market-based incentives. Tradable emissions permits are introduced and receive more thorough coverage in Chapter 12.

Chapter 7 Energy

This chapter addresses traditional and alternative sources of energy. The trade-offs between various fuels, political and economic barriers, and future prospects all receive attention. An examination of price ceilings under various market structures provides a backdrop for discussions of energy policy. The influence of politics and big players in energy-policy debates is not overlooked.

Chapter 8 Sustainability

Decision-makers with the goal of sustainability have a guiding question for every relevant activity: For how long can this activity continue? This chapter discusses multiple interpretations of this question, appropriate applications of sustainability criteria, and promising opportunities for sustainable development.

Chapter 9 Population, Poverty, and Economic Growth

Demographic trends have close ties to the environment. Past theories, including those of Malthus and Kuznets, are coupled with more recent perspectives on municipal waste generation and the determinants of resource use. The chapter concludes with a discussion of how current and proposed government policies affecting poverty, population growth, and economic growth are likely to affect the environment.

Chapter 10   Biodiversity and Valuation

This chapter addresses optimal levels of biodiversity, issues of species prioritization, and the valuation of natural resources. Readers will learn methods for estimating the marginal value of specific species. Topics include the interpretation of market prices, contingent valuation, hedonic pricing, and the travel cost method.

Chapter 11   International and Global Issues

Many aspects of environmental and natural resource economics transcend national boundaries. This chapter covers initiatives for international cooperation and the associated organizations and agreements. Topics include the CITES and Paris agreements, global warming, acid deposition, natural disasters, global scarcity, poaching, and the strengths and weaknesses of international law.

Chapter 12   Perspectives on Environmental Policy

Building on the review of marginal analysis in Chapter 2, this chapter explains the application of cost-benefit analysis to major environmental policy initiatives. The chapter explains the pros and cons of command-and-control regulations and incentive-based solutions. Case studies include congestion pricing, tradable emissions permits, and excessive deterrence.

Chapter 13   Natural Resource Management: Renewable Resources

Although many of the chapters in this text pertain to natural resource management, Chapters 13 and 14 have a narrower focus. Chapter 13 introduces a model of renewable resource use that serves as a basis for policy discussions in this and the following chapter. The chapter highlights case studies of fisheries and forest management.

Chapter 14   Natural Resource Management: Depletable and Replenishable Resources

Discussions of fuels and water underscore the particular challenges of managing depletable and replenishable resources. Topics include transitions between resources, the allocation of resources, water rights, and pricing plans.

Appendix: Intertemporal Allocation and Hotelling's Rule

The Appendix offers a rigorous explanation of optimal allocation across time periods and a derivation of Hotelling's rule to supplement the rule's introduction in the chapter.

Chapter 15   Environmental Dispute Resolution

The controversial trade-offs inherent in decisions related to the environment create many opportunities for disagreement. Liberals and conservatives battle over policy. Businesses and communities battle over growth. Owners of natural resources battle with multiple parties over use restrictions, liability, and conflicting ownership claims. The resolution of these disputes determines the allocation of natural resources and the success of environmental preservation. This chapter emphasizes efficient mechanisms for dispute resolution, including "cake-cutting" techniques, mediation, arbitration, and offer-of-settlement devices.

Chapter 16   Morals and Motivation

At the intersection of economics and the environment are moral issues involving the appropriate treatment of flora, fauna, fellow humans, and future generations. This chapter considers the motives behind our behavior: in essence, the elements of our utility functions. Topics include ethical theories such as ethical egoism, utilitarianism, deep ecology, social ecology, and ecofeminism. The chapter concludes with several alternative "tests" for whether particular actions that affect the environment and natural resources are acceptable.

Acknowledgments

It was my pleasure to work with Chloe Herbert and Michelle Gallagher on the development of the sixth edition. I am indebted to Eric Dodge, Paul Briggs, David Martin, Mark Smith, and Anne Lubbers for valuable discussions. Sarah Howard, Nathan Olsen, John Takach, and Ashley Vinsel provided research assistance. My parents, wife, and children are to thank for providing inspiration and practicing patience. For thoughtful comments and checks of accuracy, I am grateful to the following reviewers:

Johari Amara
Bond University, Australia

Kathleen P. Bell
University of Maine

Arthur J. Caplan
Utah State University

Allan Collins
West Virginia University

Jay R. Corrigan
Kenyon College

Bob Cunningham
Alma College

Lotanna E. Emediegwu
Manchester Metropolitan University

Molly Espey
Clemson University

Christina Fader
University of Waterloo

Sue E. Hayes
Sonoma State University

Lilly P. Harvey
Nottingham Trent University

S. Aaron Hegde
California State University, Bakersfield

Andrew T. Hill
Washington College

Joe Kerkvliet
Oregon State University

Rajaram Krishnan
Earlham College

Charles Krusekopf
Austin College

Roland Lewin
UC Santa Barbara

John B. Loomis
Colorado State University

Allan MacNeill
Webster University

Frederic Menz
Clarkson University

Gretchen Mester
University of Oregon

Jeffrey A. Michael
Towson University

Diane K. Monaco
Manchester College

Craig Morley
Toi Ohomai Institute of Technology, New Zealand

Brian Peterson
Manchester College

Margaret A. Ray
Mary Washington College

George D. Santopietro
Radford University

Eric C. Schuck
Colorado State University

Davis F. Taylor
College of the Atlantic

Kenneth N. Townsend
Hampden-Sydney College

Cees Withagen
VU University Amsterdam

Anonymous reviewers from
Bates College
Hanover College
Southern Oregon University

University of Applied Sciences
Neu-Ulm, Germany
University of Copenhagen,
Denmark
University of California
University of Durham (UK)
University of Nevada
University of Texas
University of Wisconsin—
Green Bay
Webster College

ACKNOWLEDGMENTS

ENVIRONMENTAL ECONOMICS
and Natural Resource Management

Reto Stöckli, Nazmi El Saleous, and Marit Jentoft-Nilsen, NASA GSFC

The Big Picture

*W*hy study environmental economics? For starters, the oxygen you breathe comes from plants. It takes 300 to 400 plants of average size to produce enough oxygen to keep you alive. Out a window you might see trees and grasses that require moderate temperatures and clean water to survive. Your body requires the same, as do the sources of the natural fibers in your clothing, the wood in your desk, the pages in this book, and the food you ate for breakfast. While the environment sustains your life, you enjoy manufactured goods, electricity, housing, and travel at the environment's expense. Environmental economics is about making wise decisions about trade-offs such as these.

Natural resources are those components of nature that humans find useful. Some natural resources, such as trees and fish, can be harvested repeatedly with proper management. Others are available for extraction only once. The minerals used to make your cell phone, bicycle, and washing machine come from nonrenewable sources, as do the petroleum-based synthetics in your backpack and shoes. The tools of **natural resource management** address critical decisions of whether, when, and how to tap supplies of the Earth's raw materials.

Economists pursue the goal of efficiency by asking the question: What would maximize the net benefits for society? Properly conducted comparisons of costs and benefits could guide anyone toward efficiency but are often absent or flawed. Pollution costs seldom sway private decisions to drive another mile or build another factory. The benefits of habitat protection can be neglected

DOI: 10.4324/9781003428732-1

3

in cost-benefit analyses of development projects. Decisions about some environmental regulations are made without weighing costs against benefits. All of this means there is much society could gain from better decisions. And that makes the study of environmental economics and natural resource management important and exciting. This chapter highlights nine major issues to whet your appetite for the discussions that follow.

Market Failure: Can We Trust the Free Market?

In 1776, Scottish economist Adam Smith wrote that self-interested individuals operating in a free market could achieve efficiency as if guided by an "invisible hand." Given these words from one of the founders of economics, why would anyone want to meddle with the market? Unfortunately, as Smith himself understood, the conditions required for free-market efficiency are seldom fully met. This section introduces four reasons markets can fail to be efficient when not assisted by the not-so-invisible hands of policymakers: *imperfect information, imperfect competition, externalities*, and *public goods*.

What You Don't Know Can Hurt You
When producers hold inside information about product-related dangers, consumers may use risky products excessively or unsafely. A lack of information can also lead consumers to under-consume products with unrecognized benefits. That's why government agencies in countries ranging from Armenia to Zambia step in to require hazard warning labels on pesticides and to teach consumers the benefits of eating vegetables. For such products, consumption closer to the efficient level can result from warnings, education programs, and other forms of information sharing that tend not to arise in a free market.

Imperfect information about product safety led the United Nations to develop internationally recognizable labels for hazardous materials. This one indicates toxicity to aquatic wildlife.

Source: https://unece.org/transport/dangerous-goods/ghs-pictograms.

Competition Underpins Efficiency

The European Court of Justice fined Google €4.1 billion for alleged anti-competitive behavior. Why the large fine? Because a lack of competition generally leads to higher prices, lower quality, and smaller quantities. When it is difficult for competitors to enter a market, a small number of firms may become powerful and threaten efficiency. The challenge is for the government to limit market power while promoting innovation and permitting adequate incentives for entrepreneurs.

Side Effects Matter

Chapter 3 explains that *externalities* are effects felt beyond or "external to" the people causing them. When individuals decide how many cigarettes to smoke or how many trees to plant in their yard, they may not consider the costs or benefits to others. This common form of neglect results in too many purchases of goods like cigarettes that cause detrimental *negative externalities* and too few purchases of goods like trees that generate beneficial *positive externalities*. The environment often bears the burden of inefficiencies that arise due to externalities. This book explains how governments use taxes, subsidies, and property rights to address externality problems by helping people feel for themselves—or *internalize*—the effects of their own behavior.

Free Riders Cause the Underprovision of Public Goods

Public goods are goods whose benefits (1) can be received by more than one person at a time and (2) cannot be kept from people who did not pay for them. Streetlights, TV signals, and military protection are classic examples. Public goods present challenges because individuals can *free ride* on the purchases of other people by receiving benefits from goods they did not pay for. As a result, too few public goods are purchased. Many goods and services related to the environment face the free rider problem. Consider efforts to remove toxins from rivers and lakes. Everyone who uses the water benefits from the cleanup efforts, whether or not they help pay for the cleanup. Free riders would rather have someone else pay to make the water safe, so too few people pitch in for cleanups. In such cases, the government can tax beneficiaries and use the revenue to fund cleanups, among other public goods. This textbook elaborates on the threats to free-market efficiency and explains the pros and cons of intervention to address market failure.

If someone pays to protect the environment, everyone benefits. But who wants to be the one who pays?

Orgánico

Waste and Recycling: Where Can We Put It All?

People buy a lot of stuff, so sooner or later there's a lot to throw away. In a typical year, the average child in Australia receives $725 worth of toys and the average Canadian buys $3,500 worth of clothing. Worldwide, consumers purchase $600 billion worth of home furnishings and appliances each year. In many countries, including the United States, the number of *landfill* disposal sites is decreasing while the volume of waste is increasing. As disposal sites near cities fill up, waste must be transported greater distances at a higher environmental cost. Beyond that are problems with groundwater contamination from landfills and ash toxicity from waste incineration.

Approaches to growing waste problems include *zero waste* initiatives that encourage households and firms to imitate sustainable ecosystems in which old plants become new soil and waste becomes fertilizer. Likewise, the linear path of consumer goods that begins with resource extraction and ends with disposal can become circular, with old goods becoming new resources. Some solutions come from firms, such as Nike's use of billions of recycled plastic bottles to make soccer uniforms. Households can compost, reduce, reuse, and recycle, and governments can provide needed incentives. For instance, thousands of U.S. communities have adopted *pay-as-you-throw* programs that encourage waste reduction by having people pay a set amount for each bag of waste sent to the landfill. Chapter 9 explains more solutions to the mounting waste problem.

Trash bins in Mexico guide users to sort organic and inorganic items to cope with large volumes of waste.

Inorgánico

Sustainable Development:
How Long Can This Last?

Along with deciding how to dispose of things, we must decide how to make things. Some production processes can be continued long into the future; others cannot. Most of our energy comes from nonrenewable oil and coal reserves. We use stocks of old-growth forests, groundwater, marine life, and fertile topsoil at unsustainable rates. We release harmful chemicals faster than they can dissipate into the air or water, so they collect and increase in concentration. The damaging path of most modern development has inspired attempts to change course. For example, the Leadership in Energy and Environmental Design (LEED) rating system provides guidelines for socially responsible building design and acknowledges developers who act to protect natural resources.

Recent innovations in industries that include forestry, mining, cement, transportation, and energy represent progress toward sustainability, but most improvements come at a cost. Pressing questions complicate the issue: How much economic growth is enough? What resources should we protect for future generations? Should we act, as philosopher John Rawls suggested, so that we would be indifferent between living now or in the future?

Economists have developed ways to assess the values of animal species.

Biodiversity:
What Is a Flamingo Worth?

Biological diversity, or **biodiversity**, refers to the variety of ecosystems, species, and genetic differences within species. There are trade-offs between biodiversity and economic development, which comes at the cost of lost habitat for wildlife. The advancement of human civilization has ushered in a dramatic increase in the rate of extinctions, now estimated to occur at 100 to 1,000 times the natural or "background" rate of between 1 and 10 extinctions per year. This impels us to consider the value of biodiversity, the value of development, and the best way to balance our conflicting interests.

Wildlife provides benefits to humans, including natural beauty, recreational opportunities, medicinal cures, and the air we breathe. It is an **anthropocentric** approach to use human-centered benefits such as these as the basis for decisions. An **ecocentric** approach includes recognition that wildlife has value in and of itself. Proponents of this approach argue that, even in the absence of human life, plants and animals are worthy of preservation. Although economists understand both of these bases for decision-making, most concentrate on the sufficiently challenging question of what biodiversity is worth to humans, rather than the more daunting task of assessing the value of biodiversity apart from human interests.

Economists have developed primarily anthropocentric techniques for estimating the values of wilderness areas and animal species, including everything from slugs to *Homo sapiens*. While more or less imperfect, these methods provide superior alternatives to throwing up our hands and saying we cannot determine the values, which in the past has led policymakers to use values for wildlife ranging from zero to infinity. Chapter 10 explains several specific ways to place values on biodiversity.

Environmental Degradation: How Much Pollution Is Too Much?

If no pollution were allowed, we couldn't drive accident victims to the hospital, build homes, or produce most goods. In fact, to live is to pollute. Animals, including humans, emit hydrogen, methane, hydrogen sulfide, nitrogen, and carbon dioxide in the processes of breathing, eating, and digesting. Even setting these essential emissions aside, the optimal

Industries send toxic emissions into the atmosphere, contributing to pollution and global climate change. Controversy rages over what levels of pollution are acceptable and how tighter controls would affect workers and consumers. Economic tools help to answer these questions.

level of pollution is positive. Why? Because the benefits from the most important sources of pollution, such as hospitals, food production, and basic housing, outweigh the negligible effects of the first few puffs of smoke emitted from cars and smokestacks.

As communities promote economic development to create more jobs and raise the standard of living, the challenge from a societal standpoint is to know when further development is appropriate and when another tree should not fall and another product should not roll off an assembly line. Should economic development be limited in some areas? Should we conduct further research or take immediate action against global climate change and similar threats? Despite apparent confusion among politicians, the answers to these questions are clearer than you might think.

Alternative Energy Sources: Why Aren't They Here?

Over the last half century, the Internet has connected the world, spaceships have landed on Mars, and more than 85 percent of the world's citizens have obtained smartphones. Yet advancements in energy production and conservation have been relatively modest. Why aren't solar panels on more rooftops or windmills on more mountaintops? Why aren't hybrid electric cars in more garages? Why did scientists develop the perfect fat substitute before the perfect fossil fuel substitute? Economic principles help to explain the contrasting growth rates in these indus-

Despite the barriers of politics and inertia, the availability of clean energy technology feeds optimism for increased implementation.

tries. Complicating factors include politics, imperfect information, high start-up costs, and the profit-maximizing strategies of firms that compete with clean energy sources. At the same time, there is reason to be optimistic about the future. In the words of Karl R. Rábago, Principal of Rábago Energy, LLC,

> *The rapid growth of renewable energy is today an established feature of global energy systems. Solar energy has grown faster than all other generation sources for 18 years in a row. Clean energy investment is expected to soon lead fossil energy investment by a ratio of nearly two-to-one. While the transition to renewable energy seems assured,*

a deeper and more complex transformation of the ways we make, dis-tribute, and use energy must be realized. Nothing less than equitable access to affordable and sustainable clean energy resources for all is the challenge of the coming decades.[1]

Population and Economic Growth: Are We Doomed to Starvation?

Doomsayers have predicted our demise for centuries. In 1798, English economist Thomas Malthus heard that the population of the United States was doubling every 25 years. In contrast, he estimated that food supplies would increase by a fixed amount each year. Malthus's famous conclusion was that population growth would outpace growth in food supplies, meaning that starvation was ahead. Since then, advancements in agricultural and industrial productivity have spoiled predictions of doom, as the world has produced unprecedented quantities of food, fuels, and all other essentials.

New discoveries and technological improvements help ward off short-ages, as do price changes. The market forces of supply and demand increase prices when there is a shortage, and higher prices motivate suppliers to find ways to produce more. Higher prices also lead con-sumers to desire less. Both of these incentives decrease the likelihood a shortage will persist. Of course, sooner or later we must confront the many binding resource constraints of the Earth, and this will force dif-ficult questions on society: How can trends in consumption be reversed? If population control is part of the solution, should it be mandated or self-imposed? What is the ideal population size? What role, if any, should more-developed countries play in the population-control strate-gies of less developed countries? Economists have theories and opinions on each of these weighty questions, some of which might surprise you.

Natural Resource Management: When Should I Harvest My Elms?

Shakespeare said the world is a stage and we are just players. In some other respects, the world is a farm and we are just farmers. As con-sumers, voters, or cultivators ourselves, we all influence decisions about whether and when to harvest the Earth's crop of wildlife and its stock of minerals. When farming our vast forests and oceans, we must also

1 Written for this textbook.

decide how many trees and fish to reap, how to bring them to market, and how to reseed the land and waters. Indeed, these issues apply to water itself. Less than 6 percent of extractable groundwater is available on a renewable basis. Decisions about these resources must be made in light of their repercussions for current and future generations. This is the task of natural resource management.

Beyond questions of whether, when, and how to harvest natural resources is the issue of how to manage access to resources that are not privately controlled. Harvests of fish from the world's oceans started to decline in the 1990s due to past overfishing, pollution, and inadequate regulation. When those with access to natural resources do not feel the effects of their decisions because they do not own the resources they are depleting, their actions may be inefficient and harmful to society. The challenge is to establish guardrails or incentives for responsible use of natural resources. This can mean limiting access, requiring resource restoration, or creating rewards or punishments that are effective despite the difficulties of monitoring remote wilderness areas. This textbook presents models of natural resource management and methods for encouraging responsible resource use.

Environmental Ethics: What Can We Do? What Must We Do? What Should We Do?

Environmental and natural resource dilemmas are intertwined with ethical issues, especially when they pit the welfare of society against personal interests. While pollution-control efforts tug on our purse strings, the act of polluting pulls on our moral fiber. Is it ethical to build homes in the remaining wilderness areas? How much trash generation is acceptable? Do we have a moral responsibility to recycle? How much value should we place on sustainable development for the sake of future generations? What right do we have to eliminate species of wildlife? And what ethical issues should we consider when planning our contributions to the planet's population?

Some ethical quandaries stem from uncertainty over the appropriate goal for society. We seek happiness and satisfaction, but what trade-offs should be made between maximizing the total amount of satisfaction in society, equality in satisfaction across individuals, equality of satisfaction across generations, and merit-based distributions of satisfaction? Scholars have also suggested a number of approaches to individual ethical dilemmas, such as whether to purchase material goods that damage the environment and whether to sell products that may cause harm to consumers. Chapter 16 explains options for how you might address these and other dilemmas.

Summary

The summit of the dormant Mauna Kea volcano in Hawai'i is one of the most beautiful places on Earth. It is also one of the best places to build telescopes used to study the skies, which has sparked controversy over the appropriate balance between environmental preservation and scientific research. Despite common misconceptions, economics covers far more than money and applies to myriad environmental issues such as this. Economics provides tools to deal with trade-offs between, for example, the benefits of development and the loss of wilderness areas. In this textbook, you will learn how economists place values on wildlife, factor in elements of uncertainty, and seek efficient solutions to environmental and natural resource problems.

More broadly, **economics** is the study of the allocation of scarce resources among competing ends. With most environmental assets being scarce, economics applies to most cases of tension between economic growth and environmental health. The stakes are high, and mistakes can have profound repercussions.

There are several views on environmental and natural resource policy. Some people accept the status quo and feel it is unnecessary to use natural resources more efficiently. Other people seek lower environmental standards and fewer limits on natural resource depletion, asserting that the problems are exaggerated and that financial resources could be better spent creating jobs or lowering taxes. Still others push for progressive environmentalism. They believe that a lack of foresight could take us so far down a path of environmental neglect that the damage would be regrettable and irreversible. Who is right? This textbook and this course can help you make your own informed decisions about the proper allocation of environmental assets.

Some of the most beautiful places on Earth, such as the summit of Hawai'i's Mauna Kea volcano, are also the most valuable places for development. Economic analysis helps policymakers weigh the costs and benefits of environmental preservation.

This chapter introduced the intrigue and import of environmental and natural resource economics. As you read more, you will discover answers to many of the questions presented here, although these answers tend to spawn new questions: Given optimal pollution levels, what is the best way to monitor factories and prevent excessive releases? How can society implement improved techniques for natural resource management? How can we find the self-discipline required to do the right thing? Do not be afraid of new questions; be afraid of not knowing what the important questions are. Deliberate thought and debate will lead to new answers and still more questions, with each round bringing greater levels of understanding and new opportunities to improve the quality of life.

• • • • • • • • • • • • • • • • • • •

Problems for Review

1. Explain why each of the following is, or is not, a natural resource:

   a) A blueberry bush.

   b) A lake.

   c) A coronavirus.

2. To examine how natural resources and the environment connect with your daily life, list two consumer goods you purchased recently. For each, indicate where the main raw material for that good came from. For example, a bike tire is made from latex harvested from a rubber tree, and a plastic computer mouse is made using oil extracted from underground. If you don't know where a raw material came from, do a quick Internet search!

3. From your own perspective, when it comes to deciding on policies for natural resource preservation, does society take more of an anthropocentric view or an ecocentric view? Explain which view you think society *should* take and why.

4. What criterion did John Rawls suggest for decisions about how to act? Would you favor limits on gasoline use in the present under this criterion? Explain your answer.

5. What is the source of energy for the building where you live? What barriers prevent more energy from being obtained from clean sources?

6. Is the optimal level of pollution zero? Why or why not?

7. What four broad categories of problems can cause market failure?

8. What has prevented the starvation predicted by Thomas Malthus?

9. According to the chapter, in what way are we all farmers?

10. Beyond the examples of ethical issues mentioned in this chapter, list one more ethical issue related to the environment that you face on a regular basis.

websurfer's challenge

Find five websites devoted to the environment (not including those listed under Internet Resources in this book). For each website write down

1. The name or URL

2. An environmental insight you learned from the website

3. A connection between the insight you learned from the website and environmental economics as discussed in this chapter.

Key Terms

Anthropocentric
Biodiversity
Ecocentric

Economics
Natural resources
Natural resource management

Internet Resources

Australian Department of the Environment, Water, Heritage and the Arts:
www.environment.gov.au

Chinese Ministry of Ecology and Environment:
english.mee.gov.cn

Environmental and Natural Resource Economics News, Opinion, and Analysis:
www.env-econ.net

Environment Canada:
www.ec.gc.ca

European Environment Agency:
www.eea.europa.eu

Indian Ministry of Environment and Forests:
https://moef.gov.in/moef/

New Zealand Ministry for the Environment:
www.mfe.govt.nz

South African Department of Environmental Affairs and Tourism:
www.environment.gov.za/

U.K. Environment Agency:
www.gov.uk/government/organisations/environment-agency

U.S. Environmental Protection Agency:
www.epa.gov

Further Reading

Anderson, David A. *Survey of Economics.* New York: Worth, 2019. An overview of the fundamentals of economic theory that serve as building blocks for the economic analysis of environmental and natural resource issues.

Chow, Gregory C. *Economic Analysis of Environmental Problems.* Hackensack, NJ: World Scientific, 2015. Provides a rigorous approach to environmental economics for those who want to see the calculus.

Devine, Robert S. *The Sustainable Economy: The Hidden Costs of Climate Change and the Path to a Prosperous Future.* New York: Anchor Books, 2020. An insightful, readable description of market failure and the economic tools needed to address the environmental problems that result.

Smith, Adam. *The Wealth of Nations.* London: Timeless Publications, 2023. One of the books that started it all—the study of economics, that is. Smith discusses the efficiency of markets as mentioned in this chapter.

Thunberg, Greta. *The Climate Book: The Facts and Solutions.* London: Penguin Press, 2023. Science-based arguments that everyone should prioritize attention to environmental problems now, written by an influential climate activist the age of today's college students.

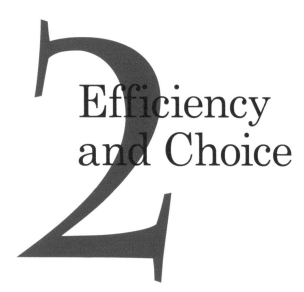

Efficiency and Choice

*E*ven the best things in life—hikes in the woods, chocolate consumption, kisses—end voluntarily when they do not end by necessity. Naturally, it's all a matter of economics. A golden rule of economics is that every action should continue until the additional benefit of more would fall below the additional cost. Adherence to this rule yields the largest net benefit from the activity. This rule is appropriate for decisions ranging from when to wake up in the morning to how many toxic industrial sites to clean up. This chapter applies this rule and other principles of economics to issues involving the environment and natural resources.

Surprisingly to some, economists see beyond money in their calculations of efficiency. The cost-benefit analysis that prevents hikers, chocolate eaters, and kissers from going on forever may have little to do with money. While soaking up the joys of acceptance and affection, kissers may also be thinking about the increasing likelihood of a roommate entering the dorm room, about their own fear of commitment, about not appearing too eager, or about the large and growing opportunity cost of not being able to spend the time studying environmental economics. Of course, they may also get caught up in the moment and neglect some of the costs. That's a problem in environmental economics, too!

Many economic models deal with how individuals maximize not money but **utility**—an abstract measure of happiness. Others consider **social welfare**, which is the collective well-being of society. These models can account for the importance of joy, sorrow,

DOI: 10.4324/9781003428732-2

love, guilt, spite, disease, free time, and anything else of interest or ill favor. In fact, to omit any significant influence on happiness, monetary or otherwise, would be to jeopardize the integrity of the models.

Scrutinizing Efficiency

Italian economist Vilfred Pareto provided a well-accepted criterion for efficiency: **Pareto efficiency** is achieved if no one can be made better off without making at least one person worse off. Suppose a national fish and wildlife service stocks ponds in two neighboring villages with fish. Each spring, Bassville's pond receives 150 bass and Catville's pond receives 200 catfish. If preferences for variety would make each community better off with a pond stocked with 75 bass and 100 catfish each year, the initial distribution is not Pareto efficient. An exchange of 75 bass from Bassville's allocation for 100 catfish from Catville's allocation would be mutually beneficial. If no additional change could make one community better off without making the other worse off, Pareto efficiency is achieved.

With a Pareto efficient allocation, there are no opportunities for cost-free benefits, but the division of benefits may not meet other goals, such as fairness. Suppose Bassville received all the fish and Catville had nothing to exchange for the fish it wanted. That allocation would be Pareto efficient: neither community could be made better off unless Bassville gave some fish to Catville, which would make Bassville worse off. But if there are other interests—such as fairness or the maximization of society's overall benefits—alternative criteria are needed to guide the decision of how to allocate bass and catfish.

There are differing schools of thought about the objective when dividing benefits. For example, Chapter 16 explains the **utilitarian** goal of maximizing the sum of everyone's utility, the **egalitarian** goal of dividing benefits evenly among members of society, and the **Rawlsian** goal of maximizing the utility of the least-well-off person. Efficiency alone does not ensure any such outcome. The types of efficiency discussed in this chapter are branches of Pareto efficiency in that they focus on opportunities for net gain from an activity. In theory, that gain can be divided to satisfy other objectives as well. Next, we will see that what happens in practice is another matter.

Suppose there are two possible sites for a new landfill, one in a low-income area and one in a wealthy area. The property needed to build the landfill in the low-income area would cost $300,000; the property needed to build the landfill in the wealthy area would cost $2 million. There are presently 30 families living on each of the sites, all of whom would need to move elsewhere if their site were chosen for the landfill.

The NIMBY (not in my backyard) effect makes it difficult to find locations for new landfills. Economists seek efficient solutions that maximize the net gain for society.

Near each site are an additional 100 families that would not be relocated but would face risks of groundwater contamination, property devaluation, odor, flying debris, and ugliness. If the landfill provides benefits that exceed the associated costs, and if the price of the land accurately reflects the cost to society of using that land, then the net gain is maximized by placing the landfill where the land has the lowest price.

The appropriate selection of a landfill site might not be as clear-cut as it seems. A careful assessment will include at least two other considerations. First, there is nothing about the *creation* of net benefits that causes the benefits to be allocated to serve society's goals. For example, the 30 relocated families might not receive full compensation for their emotional and financial costs of moving. And those who are not relocated might not be compensated for their unsavory neighbor and for related losses in property value. Most of the benefits from the landfill might go to the people in the wealthy area, including the owners of the sanitation companies that send their trash to the landfill. If the goal is to provide equitable outcomes for the broader well-being of society, a second step of *distributing* the benefits must follow. Economists generally focus on efficiency and leave decisions about equity to others. A good policymaker understands that both equity and efficiency are important to social welfare.

A second consideration is that the property valuations are driven by the incomes and wealth of the people interested in living in these areas. Although some people who desire to live in the wealthy area might pay far more for property in that area than people would pay for property in the low-income area, it is not necessarily the case that the loss of happiness or "utility" as the result of relocation would be larger for the wealthy people than for the low-income people. It is possible that people in the low-income area receive much of their happiness from the history, community, and unique geographical features associated with their location and not from anything money can buy. If the low-income people derive greater utility from their property than the wealthy people gain from the wealthy area, and if the amount of money the wealthy people are willing to pay is inadequate to compensate the low-income people for their relatively large losses, placement of the landfill in the low-income area will not maximize the well-being of society.

Another caveat regarding the search for efficiency as commonly conceived is that, although it is not all about money, it does tend to be about people. In other words, efficiency targets are anthropocentric. Natural resources and the environment are considered in calculations of efficiency, but generally in terms of their value to humans. For example, when the Meli Bees Network discusses the protection of native bees in Argentina, they emphasize human interests in bees, pollination, and agricultural jobs. The bees' interests in bee preservation, pollination, and jobs may hold less sway. If the interests of bees carried more weight, the current numbers of bees, agricultural jobs, and humans for that matter would probably be different. Given conflicting interests, humans tend to do what is best for humans. This isn't surprising, but it is important to understand when seeking context for outcomes described as "efficient," "optimal," or "the best possible."

Despite these limitations, efficiency is a common goal and a defensible starting point. While efficiency guidelines do not resolve the equity issue, they can accomplish the important first step of maximizing the size of the net gains to be divided. The landfill example showed that practical applications of efficiency may have more to do with maximizing net monetary benefits than utility. However, the virtual impossibility of comparing utility across people complicates more direct attempts to maximize utility. For instance, if I say I'm really happy and you say you're really happy, do we really know that we are equally happy? And likewise, in the landfill story, it may be impossible to determine which group would lose more utility if forced to move.

As discussed in later chapters, applications of efficiency are also limited by the difficulty of measuring all the costs and benefits associated with an action. Nonetheless, informative cost-benefit analysis can be conducted in many situations, even if the data may not be perfect. And while efficiency criteria tend to be anthropocentric, that may be fitting because our goals tend to be equally self-serving.

Cost-Benefit Analysis

Media reports lamenting drops in the gross domestic product or celebrating increases in the number of housing starts can leave the impression that boundless increases in production and development would necessarily be a good thing. Addressing environmental issues with the goal of economic efficiency improves upon more simplistic views that production should be maximized or pollution should be minimized. The best outcome for society is neither the highest possible level of production nor zero pollution. An all-out effort to make as much output as possible would decimate social welfare because it would result in smoke-blackened skies, failing health, the end of leisure time, and the eradication of wildlife. But some pollution is better than none. Chapter 1 explained that to live is to pollute. No food, shelter, or medicine could be produced without creating some pollution.

To determine whether contemplated pollution, production, or anything else is worthwhile, the benefits can be weighed against the costs. The U.S. Environmental Protection Agency (EPA) estimates that the benefits of Clean Air Act regulations between 1990 and 2020 were worth $2 trillion and the costs were $65 billion. If legislation like this were an all-or-nothing proposition, we would want to take it all because the benefits outweigh the costs.

Often decisions are about *how much* of an activity to engage in. For example, how many trees should be planted? How much carbon dioxide should be released? And how many regulations should be added to the Clean Air Act? For decisions about how much, the net gain is maximized by comparing the *additional* benefits and costs of each incremental unit. The additional benefit of one more unit is called the **marginal benefit** (*MB*); the additional cost of one more unit is called the **marginal cost** (*MC*). If the marginal benefit of a unit exceeds the marginal cost, that unit provides a net gain.

You may recall from your introductory economics course that the marginal benefit of most things diminishes as more units are consumed. For example, the first tree in your yard brings new greenery and natural beauty to the landscape. A second tree is nice, but it adds nothing new and is likely to contribute less additional benefit than the first tree. A third tree is also beneficial, but less so because you need it less and the growing number of trees may start to impede desired open space for recreation.

While the marginal benefit generally falls with an increase in quantity, the marginal cost generally rises. Suppose you're a tree

These orcas promote tourism. The global whale watching industry generates over $2 billion in revenues and employs about 13,000 workers.

farmer, looking to increase from growing 1,000 trees to growing 2,000 and then 3,000 trees. You would start on the available land with the lowest cost and the least need for expensive clearing and fertilizers. As you grew more trees, you would resort to using more expensive, less productive land, so the second 1,000 trees would cost more than the first 1,000 trees, and the third 1,000 trees would cost more still. That is, the marginal cost of each batch of trees would exceed the marginal cost of the previous batch, so you would experience increasing marginal cost.

The comparisons of marginal benefits and marginal costs that constitute **marginal analysis** can determine the efficient stopping point for all sorts of activities. Let's take a deeper dive into the importance of marginal analysis by considering the decision of how many whales to "harvest" from the sea each year. The International Whaling Commission governs whaling around the world, and many countries, including Norway, the United States, and Japan, permit limited whaling. The marginal benefit from the first few whales harvested is high, as they allow for scientific research and the subsistence of native cultures that rely on whaling. The marginal benefit from harvesting whales diminishes as labs get enough research subjects, native cultures get plenty of whale oil to last through the winter, and whale meat lovers are satiated. The blue line in the top graph of Figure 2.1 illustrates the decreasing marginal benefit from "harvested" whales.

As more whales are harvested, the marginal benefit of another whale decreases and the marginal cost of harvesting another whale increases. Efficiency is achieved if whales are harvested until the marginal benefit equals the marginal cost. An efficient whale harvest maximizes the net benefit, which is the difference between the total benefit and the total cost.

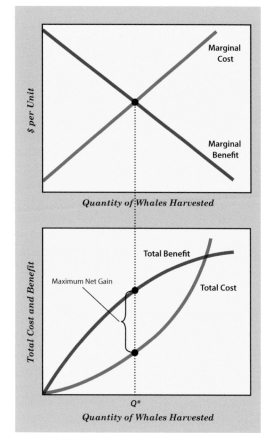

Figure 2.1
An Efficient Whale Harvest

As the whale population decreases, the added costs of additional whaling include the loss of genetic variation and herd viability and the associated risk of extinction. There are also increasing costs to future generations of whale harvesters and eaters, present and future generations of whale watchers, and workers in the multibillion-dollar whale watching industry. And excessive whale harvests cause imbalance in the ecosystem. With fewer whales, an overabundance of the types of fish whales eat could lead to the decimation of the smaller forms of marine life which those fish eat. For all these reasons, the marginal cost of harvested whales increases while the marginal benefit decreases, as illustrated in the top graph of Figure 2.1.

The bottom graph in Figure 2.1 allows us to see precisely why net gains are maximized when marginal benefit equals marginal cost. The definitions of those terms in this section remind us that marginal values are changes in total values when the quantity increases by one. Conveniently, that makes the marginal benefit simply the slope of the total benefit curve, a measure of the rate at which total benefit is increasing. The marginal cost is likewise the slope of the total cost curve, indicating the rate at which total cost is increasing. So when the marginal benefit exceeds the marginal cost, the total benefit is rising faster than the total cost, meaning the difference between them—the net gain—must be rising. To maximize the net gain in that case, the

23

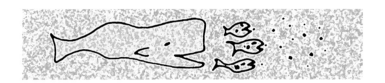

quantity should increase. When the marginal cost exceeds the marginal benefit, the total cost is rising faster than the total benefit, and the difference between them is falling. That means fewer is better: The net gain would increase if the quantity decreased. Combining those two findings, the lesson is that the net gain is maximized by increasing the quantity up until the marginal benefit equals the marginal cost. Look at Figure 2.1 to see that where marginal benefit equals marginal cost, the gap between the total benefit and the total cost that represents the net gain is maximized.

Types of Efficiency

The equation of marginal cost and marginal benefit forms the basis for several types of efficiency. The efficiency story is made even more interesting by the fact that the relevant costs and benefits depend on the parties involved. The marginal benefit firms care about is called *marginal revenue*—the additional revenue received by selling one more unit of a good. Efficiency for firms typically corresponds with maximizing profit. Firms equate their marginal revenue with their marginal cost of production to find the profit-maximizing level of output.

The marginal benefit individuals care about is **marginal utility**, the additional utility received from one more unit. As with the marginal benefit of planting trees and harvesting whales discussed earlier, the marginal utility received from most goods diminishes as consumption increases because additional units serve decreasing needs and wants. Individuals should continue each activity until the value of the marginal utility received equals the marginal cost.

An economy's achievement of three key types of efficiency depends on the answers to three fundamental questions:

1. What will be produced?

2. What resources will be employed?

3. Who will receive the final goods and services?

This section describes the associated types of efficiency.

Under ideal conditions, markets can achieve efficiency on their own. Among these conditions is *perfect competition*. Recall from your introduction to economics that in a perfectly competitive market (1) many firms sell identical products; (2) it is easy for firms to enter or leave the market; and (3) competition prevents any firm from charging a higher price than other firms. We'll touch on how market forces promote efficiency in this chapter. The appendix to this chapter provides a detailed explanation of the more graphical and mathematical side of this story. Chapter 3 discusses the market failure that occurs when market conditions are not ideal.

What Goods and Services Should Be Produced?

Let's first explore the decision of what goods and services an economy should produce. Each economy has a set of *inputs* to work with, also referred to as *resources* or *factors of production*. Economists categorize inputs broadly as *land* (including all natural resources), *labor* (the efforts of workers), *capital* (human-made inputs such as buildings and equipment), and *entrepreneurship* (people's willingness to take risks and organize inputs for productive purposes).

Allocative efficiency is achieved when an economy allocates its available inputs to produce the goods and services that best serve society's needs and preferences. To examine this goal, we will model the simplified economy of a national park that produces only two goods: bears and multiuse trails for hiking, Jeep riding, cross-country skiing, and snowmobiling.

Figure 2.2 illustrates a **production possibilities frontier** (PPF), which represents the various combinations of output an economy can produce if all inputs are used to their full potential. For example, point *A* indicates that the park could have 15 miles of trails if all its inputs were devoted to trail making. Point *B* indicates that the park could

Figure 2.2

A Production Possibilities Frontier

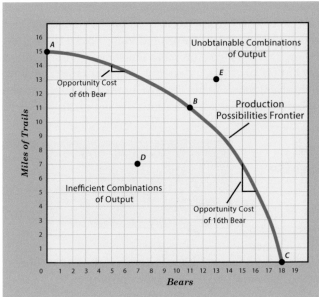

A production possibilities frontier represents all the combinations of two goods that can be produced using every available input to their full potential. The opportunity cost of each good increases as more of it is made due to the increased use of inputs specialized for making the other good.

have a combination of 12 miles of trails and 9 bears. And point C indicates that the park could have 18 bears if all inputs were devoted to bear habitat. Points such as *D* that lie beneath the PPF represent combinations of the two goods that are attainable but inefficient because more of one or both goods could be made without giving up any of either good. Points such as *E* that lie outside the PPF represent combinations of output that cannot be made with the available resources. Allocative efficiency amounts to selecting society's most favored combination of the two goods among all the possibilities shown along the PPF.

To understand the shape of the PPF, we must consider the set of inputs that could be allocated to either trail creation or bear habitat. The creation of trails can require the clearing and grooming of land and the production of signage and parking areas at trailheads. Successful bear protection may necessitate not only land use, but also equipment for reforestation and people for research, planning, and implementation.

Unlike animals such as hedgehogs and ground squirrels, bears sleep lightly when they hibernate, which makes them vulnerable to disturbance. Noise from nearby trail activity can cause a bear to burn critical stores of energy, lose its cubs, or abandon its den. This means the same area cannot serve well for both multi-use trails and bear habitat. Difficult decisions must be made about the amount of land, labor, and capital to devote to trails and to bears.

The *opportunity cost* of a decision is the value of the best alternative not chosen. The opportunity cost of an acre of land devoted to trails is the number of bears that land could have supported instead. To pursue allocative efficiency, we begin with attention to the opportunity cost of making more of either good, which we can learn from the PPF. The first trails should go where the opportunity cost is the lowest in terms of lost bear habitat—on land far from rugged bear country that requires minimal labor and capital for trail clearing. The first land preserved for bears should be backcountry where rocks and streams make life wonderful for bears but difficult for trail makers, thus minimizing the opportunity cost in terms of forgone trails. The specialization of other inputs for producing trails or bear habitat is also a consideration. For example, the first labor and capital assigned to trail making should be the workers and equipment most useful for making trails and least useful for creating bear habitat.

The opportunity cost of one more of either good will increase as growing quantities necessitate the use of inputs less suited for making that good and more valuable for making the other good. Each additional mile of trail encroaches into more pristine wilderness areas, displacing more bears and requiring more labor and capital to clear trails and install infrastructure. And the habitat for each additional bear must take up increasingly trial-friendly land and other inputs that could otherwise support more trails than the inputs used for the previous bear habitat.

The increasing opportunity cost of each activity is reflected in the shape of the production possibilities frontier. Looking again at Figure 2.2, notice what happens as you move from left to right along the PFF, increasing the number of bears: The vertical drop that represents the opportunity cost (lost miles of trail) per bear steadily increases. For example, the opportunity cost of the sixth bear is 0.5 miles of trail, whereas the opportunity cost of the sixteenth bear is 2 miles of trail. Likewise, the increasing opportunity cost of trails is observed by moving from right to left along the PPF: The number of bears lost per mile of trail grows steadily as more trails are made.

We have seen that the slope of the PPF indicates the opportunity cost of bears in terms of trails, which is the rate at which the available inputs make it possible to gain bears in exchange for trails. Suppose we start at point A. How far should we move along the PPF, gaining bears at the expense of trails, to achieve allocative efficiency? Until the rate at which society is willing to substitute one good for another equals the rate at which production possibilities allow one good to be made instead of the other. If these rates are not equal, society can achieve a higher level of satisfaction by reallocating some inputs from one good to the other.

As a numerical example, suppose that at a particular point on the PPF, one more bear would require the inputs needed to make 1 mile of

The Efficiency of Drugs, Sex, and Partying Until You Puke

Although efficiency is a common goal, mistakes are made. Consider criminal activity, which sometimes satisfies efficiency criteria and oftentimes does not. If you are late for an important interview, the benefit of parking illegally might exceed the cost. But many lawbreakers overestimate the efficient level of crime due to ignorance of the repercussions of their behavior, heat-of-the-moment decisions, perceptions of invincibility, or drug-induced irrationality (see Anderson, 2002). Sexual activity sometimes exceeds the efficient level for similar reasons, contributing to the widespread cases of sexually transmitted diseases and unwanted pregnancies. Addictions to drugs, alcohol, food, or shopping can also result from flawed or absent cost-benefit analysis. Yet addictions may not always be irrational. Economists including Gary Becker and Kevin Murphy (1988) famously discussed the possibility of rational addiction.

Sometimes marginal costs are experienced firsthand but quickly forgotten. You may have heard people say "never again" in the midst of experiencing the marginal cost of running a marathon, throwing up due to alcohol consumption, or giving birth to a child. For better or worse, the same people often pay the marginal cost again, despite their earlier assertions that it exceeded the marginal benefit.

At times, the absence of cost-benefit analysis is deliberate. The U.S. Supreme Court has upheld federal air quality standards that, in accordance with the Clean Air Act of 1970, do not take compliance costs into account. Although the quest for efficiency is sometimes imperfect or absent, economists generally see it as the best target among the available alternatives.

trail. Suppose also that at that point, society would be willing to substitute one bear for 2 miles of trail, meaning one bear makes society as happy as 2 miles of trail. By adding one bear, society gains the happiness provided by 2 miles of trail at the cost of only 1 mile of trail, so society is better off with the additional bear. At another point further to the right along the PPF, due to increasing opportunity costs, one more bear requires the inputs needed for 3 miles of trail. But having a relative abundance of bears at that point, suppose society would only trade

0.5 mile of trail for one bear. In that case, by giving up a bear, society would gain 3 miles of trail while losing only the happiness provided by 0.5 mile of trail. The opportunities for net gains end, and allocative efficiency is achieved, when *the rate at which consumers would willingly trade one good for the other equals the rate at which available inputs allow one good to be produced instead of the other*. The appendix to this chapter provides additional examples and a more rigorous explanation of this efficiency condition.

With What Resources Should Goods and Services Be Produced?

After the decision of what to make comes the question of how to make it. Suppose an economy decides to make paper products such as books, posters, and stationery. Paper can be made by various machines or by hand. It can be made from cotton, wood, hemp, bamboo, or even elephant dung. And each of those natural resources can be produced with a variety of inputs, as we will soon see in the case of cotton. To produce anywhere on its production possibilities frontier, the economy must allocate inputs appropriately and use them to their full potential, which takes a second type of efficiency. With **productive efficiency**, it is not possible to produce more of any good without producing less of something else. Without productive efficiency, an economy finds itself at a point below its production possibilities frontier, missing opportunities to make more of either or both goods without making less of anything.

We will again look at trade-offs to find efficiency: Productive efficiency requires that *the rate at which one input can be substituted for another is the same for the production of each good*. Consider the cultivation of hydroponic cotton and tomatoes, which are grown without soil. The plants are given the water and nutrients they require, while the roots rest in a growing medium such as coconut fiber or sand. Hydroponic farming does require the use of green-houses and workers. Substitutions can be made between the inputs. For example, the number of plants grown could remain the same with more greenhouses and fewer workers because extra greenhouses ease space constraints and can make up for lost labor. Suppose that with the current allocation of inputs for cotton production, cotton output would be unchanged if three workers were removed and one additional greenhouse were added. In tomato production the current trade-offs are

Productive efficiency is about making the right choices among inputs. In this greenhouse, AppHarvest grows hydroponic tomatoes without soil.

Photo courtesy of AppHarvest.

different: Suppose tomato output would be unchanged if one worker were added and one greenhouse were taken away.

Given these trade-offs, assuming the same workers and greenhouses could serve either cotton or tomato cultivation, the exchange of one cotton worker for one tomato greenhouse would allow more cotton to be grown with no loss of tomatoes. The added tomato worker would just make up for the loss of a greenhouse, and the new cotton greenhouse would make up for the loss of *three* workers, whereas only *one* was lost. The cotton producers would then have two more workers than are needed to maintain the current production level. Those two extra workers could continue to produce cotton and thereby increase production. Alternatively, the two workers could switch to tomato production and increase the number of tomatoes. A third option would be to split the workers between cotton and tomatoes and produce more of *both* goods. These opportunities for gains without losses demonstrate that the initial allocation of resources brought the economy to a point below the production possibilities frontier and did not achieve productive efficiency.

As with paper, cotton, and tomatoes, most goods and services can be produced with a variety of input combinations. Clothing can be hand sewn or machine sewn. A road through the forest can be forged by many workers wielding machetes or a few workers operating bulldozers. And windows can be cleaned by workers or machines. In every case, productive efficiency is achieved by adjusting the allocation of inputs until the rate at which one input can be substituted for another is consistent across outputs. See the appendix to this chapter for a more rigorous discussion of productive efficiency.

Who Will Receive the Final Products?

If the optimal mix of goods and services is produced with the optimal combination of inputs, the final efficiency hurdle is to get the output to the right consumers. **Distributive efficiency** is achieved when no allocation of goods or services could make anyone better off without making someone worse off. You might recognize this as an application of Pareto efficiency with a direct focus on the distribution of consumer goods.

As with the other types of efficiency, distributive efficiency comes down to a comparison of trade-offs. Consider the distribution of medicines to treat malaria and tuberculosis (TB)—diseases that kill over a million people each year. Suppose the United Nations has treatments to divide between Kenya and Tanzania and begins by distributing half of each type of treatment to each country. Because the incidence of these diseases differs between the countries, the rate at which consumers would trade one type of treatment for the other also differs between the countries. If consumers in Kenya are willing to provide one

TB treatment in exchange for 50 malaria treatments, and consumers in Tanzania are willing to provide 100 malaria treatments in exchange for one TB treatment, both countries can be made better off by an exchange of one TB treatment from Kenya for a quantity of malaria treatments from Tanzania that exceeds 50 but falls short of 100. For example, if 75 malaria treatments from Tanzania were exchanged for one TB treatment from Kenya, the Tanzanians would be better off because they gave up 25 fewer malaria treatments than what the TB treatment is worth to them. At the same time, the Kenyans would be better off because they received 25 more malaria treatments than what the TB treatment is worth to them.

As Tanzanians receive more TB treatments and trade away more malaria treatments, their willingness to pay for TB treatments will decrease. For Kenyans, with a growing number of malaria treatments and a dwindling number of TB treatments, their willingness to pay for malaria treatments will decrease. *Distributive efficiency is achieved when the rate at which each consumer would trade one good for the other is the same.* The appendix to this chapter explains distributive efficiency in greater detail.

Supply and Demand

If you've taken an economics class in the past, you've surely seen supply curves and demand curves. Why are supply and demand central to economic analysis? For starters, we'll see that under the right conditions, efficiency is achieved at the market equilibrium where the quantity supplied equals the quantity demanded. This section provides a brief overview of supply and demand curves and highlights their role

A demand curve indicates the quantity that would be demanded at each price. The demand curve's height above each unit is the most a consumer would be willing to pay for that unit, which is a measure of the marginal benefit of that unit.

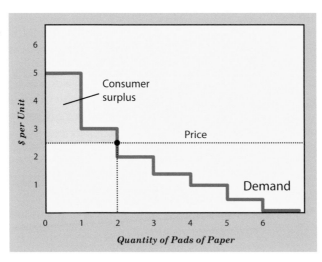

Figure 2.3
Erick's Demand Curve

in the story of efficiency. If you have never taken an economics class, you may want to consult a principles of economics textbook for a broader introduction to these concepts.

A **demand curve** represents the relationship between the price of a good or service and the quantity demanded within a given period. Figure 2.3 illustrates Erick's demand curve for pads of recycled notebook paper. The height of the demand curve above each unit represents the most Erick would be willing to pay for that unit. Because a first pad of paper would give Erick $5 worth of utility, his demand curve has a height of $5 at the quantity of one. Erick would receive $3 worth of additional utility from a second pad, so $3 is the height of his demand curve at the quantity of two, and so on.

If the price of pads is $6, Erick will buy zero pads because even the first one is only worth $5 to him. If the price is $4.50, he will purchase one and receive $0.50 worth of **consumer surplus**, which is the difference between the most a consumer would pay for a purchase and the actual amount paid. If the price is $2.50, Erick will purchase two units because each of the first two units is worth more to him than the price. His consumer surplus with a price of $2.50 will be $5.00 – $2.50 = $2.50 for the first pad, plus $3.00 – $2.50 = $0.50 for the second pad, for a total of $3.00. In general, to find the quantity Erick would purchase at any given price, first draw a horizontal price line with the height of the price, as shown in Figure 2.3 for a price of $2.50. Then draw a vertical line down from the intersection of the price line and the demand curve to the quantity axis to find the quantity. Consumer surplus is the area below the demand curve and above the price line.

The market demand curve for pads of recycled notebook paper holds the same information for all the consumers in the market that the individual demand curve does for Erick. It indicates the total number of pads that would be sold at each price. For each quantity, the demand curve's height is the number of dollars' worth of marginal utility that one more pad would provide to some consumer. The market demand curve is found by adding all of the individual demand curves horizontally, meaning that at each price, the quantities demanded by each of the individuals in the market are added up to find the total quantity demanded in the market. For example, suppose Erick and Margaret are the only consumers in a small market. If, at a price of $2.50, Erick demands two pads and Margaret demands three, then the market demand at that price is 2 + 3 = 5 pads.

The supply curve for a competitive firm shows the quantity that would be supplied at each price. The supply curve's height at each unit is the marginal cost of that unit.

The **supply curve** indicates the relationship between the price of a good or service and the quantity supplied within a given period. Suppose Recycled Paper International (RPI) makes recycled notebook paper in runs of 100 pads at a time. RPI's marginal costs for the first six runs

The supply curve for a competitive firm is the marginal cost curve above the minimum average variable cost (AVC). When the price falls below the minimum AVC, the firm should shut down immediately to cut its losses.

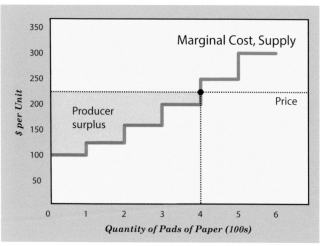

Figure 2.4
RPI's
Supply
Curve

are $100, $125, $160, $200, $250, and $300, respectively, as shown in Figure 2.4.

If the price is high enough to justify RPI's participation in this competitive market, it should keep producing more until the marginal cost rises to equal or exceed the market price. If the market price is $2.25 per pad ($225 per hundred), RPI should produce the first four runs, all of which have a marginal cost below the price. For example, the marginal cost of the first 100 pads is $100 and the price is $225, so RPI receives $225 − $100 = $125 more than the marginal cost for those pads. When the price exceeds the marginal cost, the difference between the price and the marginal cost is called **producer surplus**. The area shaded purple in Figure 2.4 shows the producer surplus for each of the first four runs when the price is $225. RPI should not produce the fifth run, because the marginal cost of $250 exceeds the price received of $225.

To find the quantity supplied at any given price, draw a horizontal price line at the height of the price as shown in Figure 2.4. Then draw a vertical line from the intersection of the price line and the marginal cost curve down to the quantity axis to find the quantity. So if a firm operates in a perfectly competitive market, the supply curve that indicates the quantity supplied at any given price is the marginal cost curve. The market supply curve is found by adding all of the individual firm supply curves horizontally, meaning that at each price, the quantities supplied by each of the firms are added up to find the total market supply.

Monopolies do not have a supply curve because rather than taking the market price as given, they charge as much as market demand will allow for the quantity that equates the marginal cost and the marginal revenue. Since a supply curve indicates one particular quantity that would be produced at each price, and the quantity a monopoly would

Figure 2.5
*Market
Equilibrium*

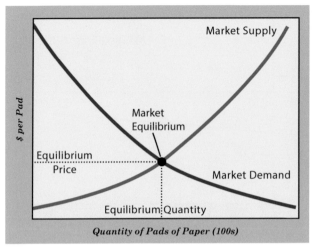

When the market is in equilibrium, the marginal benefit equals the marginal cost. The market equilibrium is thus efficient *if* all marginal costs and benefits are included and measured accurately.

produce for a particular price changes when the demand curve changes, one cannot specify a supply curve for a monopoly. (See https://youtu.be/XKdoe5SL_Bc for a detailed explanation of this point.)

Market equilibrium occurs where market supply equals market demand, as shown in Figure 2.5. Keep in mind that the demand curve represents the dollars' worth of marginal utility gained from each additional unit, and the supply curve represents the marginal cost of each unit. The market equilibrium thus equates marginal benefit and marginal cost to yield the quantity that maximizes net benefits as shown in Figure 2.1. The appendix explains that collectively, markets operating at equilibrium yield allocative efficiency in an economy (baring market failure, as discussed in Chapters 1 and 3).

Expected-Value Calculations

In mid-June of 2023, the weather forecast for Southern Brazil was for heavy rain, possible flooding, and landslides. Forecasts can be wrong, and some storm warnings precede a sunny day. This time, residents who took full precautions were glad they did—a major cyclone slammed the area, wreaking havoc on power supplies, transportation systems, and lives. Like storms that may or may not strike, climate change may occur faster or slower than predicted, endangered plant species may or may not hold remedies for long-fought diseases, and indeed, few things in life are certain. Consequently, decisions must be made about approaches to problems of unknown severity.

Suppose meteorologists estimate a 65 percent chance of a cyclone hitting your area tomorrow. Would you take the precaution of stocking up on bottled water in case the storm shuts down or contaminates water supplies? That probably depends on your expected losses with no water and your attitude toward risk taking. Chapter 5 takes up the topic of attitudes toward risk. This section explains how to find the *expected value* of a scenario involving uncertainty. An **expected value** is an average of possible outcomes, each weighted by the probably of that outcome occurring.

Imagine a game in which a coin is flipped to determine whether you win or lose. You will win $100 if the coin lands on heads, and you will lose $100 if the coin lands on tails. The probably of each outcome is 50 percent. Although you will either win or lose $100, neither outcome is certain, so you should not expect to win $100 or to lose $100. If you played this game many times, you would expect to win half the time and lose half the time, and you would expect to walk away with nothing. That expected gain of $0 is the weighted average outcome, with each outcome weighted by its probability: $0.50 \times \$100 + 0.50 \times -\$100 = \$0$. So, $0 is the expected value of this game. If there were any charge to play this game, you wouldn't want to play unless you particularly enjoy taking risks because any positive price would exceed the expected value of your winnings.

Now let's explore a citizen's expected value of bottled water given the possibility another cyclone will disrupt the water supply in Brazil. As a cyclone nears, Murilo must decide whether to purchase bottled water based on the projections of meteorologists, past experience, and the unfolding story in the skies. Murilo estimates a 5 percent chance the storm will shut down access to running water for a week, a 15 percent chance of no running water for a day, and an 80 percent chance the storm will not disrupt the water supply. Murilo and his family need water to drink, cook, clean, and flush toilets. It would be worth $1,000 to Murilo to avoid the loss of water for a whole week. He would be willing to pay $100 to avoid the relatively minor loss of water for a day.

We can formalize the expected-value formula as the sum, for each possible outcome, of the estimated probability of that outcome multiplied by the value of that outcome:

$$\text{Expected Value} = \left(\begin{array}{c}\text{Probability}\\\text{of Outcome 1}\end{array}\right)\left(\begin{array}{c}\text{Value of}\\\text{Outcome 1}\end{array}\right)$$
$$+ \left(\begin{array}{c}\text{Probability}\\\text{of Outcome 2}\end{array}\right)\left(\begin{array}{c}\text{Value of}\\\text{Outcome 2}\end{array}\right) + \ldots$$

For Murilo, the expected value of bottled water is the 5 percent chance of a 1-week water disruption times the $1,000 value of having water in that event, plus the 15 percent chance of a 1-day disruption times the $100 value of having water for that day, plus the 80 percent chance of no damage times the zero value of bottled water when running water is steadily available:

$$\text{Expected value of bottled water} = (.05)(\$1{,}000) + (.15)(\$100) + (.80)(\$0)$$
$$= \$65.$$

Murilo should compare the expected value of bottled water with the cost of bottled water when deciding whether to make the purchase. If a stock of water sufficient for these contingencies would cost less than $65, Murilo should make the purchase, unless he has a particular desire to take risks.

Similar calculations are useful to inform decisions about a wide range of uncertain scenarios. Governments make expected-value calculations for environmental policies to accommodate uncertainty about potential outcomes, including the outcomes of species extinctions and climate change. Corporations have an eye on expected values when considering environmental and safety precautions. In a famous example from the 1970s, the Ford Motor Company looked at expected values when deciding whether to invest in safety equipment to avoid gas tank explosions on its Pinto model. Sadly, Ford underestimated the value of harm and under-invested in safety. And individuals are responding to expected values if their decision to obey rules for auto emissions testing or driving speed hinges on the probability of being caught and the fine for noncompliance.

Summary

If you are wise, you have decided to study today until the marginal benefit of studying equals the marginal cost. That is the path to efficiency and the maximization of net benefits. Efficiency for firms means maximizing profit by equating marginal revenue and marginal cost. Efficiency for individuals means maximizing utility by equating the value of marginal utility and marginal cost.

The fundamental economic questions of what, how, and for whom to produce are answered by satisfying the criteria of allocative efficiency, productive efficiency, and distributive efficiency, respectively. In a perfectly competitive market in the absence of market failure, the equation of supply and demand is synonymous with the equation of marginal cost and marginal benefit and yields all three types of efficiency. However, few markets are perfectly competitive and market failure is common. Flawed assessments of costs and benefits can cause environmental missteps.

We have addressed the efficiency of hiking, chocolate, kissing, landfill sites, clean air, whaling, T-shirts, fish, tomatoes, cotton, drugs, sex, drinking, recycled paper, and bottled water. The central objective of equating marginal cost and marginal benefit is critical to decisions of all sorts. Of course, real-world solutions aren't as simple as these models suggest. The next chapter builds additional realism into these models.

When allocating resources for trails, or time for hiking them, efficiency comes from attention to the trade-offs involved. Would a larger allocation create an additional benefit that exceeds the additional cost? If so, more is better.

Problems for Review

1. Suppose that while walking down a beach you find a seashell you like very much. A moment later you find another seashell. The second seashell adds to your happiness but not as much as the first seashell did.

   a) *When you found the second seashell, did your marginal benefit from seashells increase, decrease, or stay the same? Explain.*

   b) *When you found the second seashell, did your total benefit from seashells increase, decrease, or stay the same? Explain.*

2. Explain how you could use marginal analysis to determine the optimal amount of time to spend studying environmental economics today. Illustrate your answer with a graph.

3. The marginal utility Hugo receives from planting flowers diminishes as he plants more flowers. On a graph, draw a hypothetical flower demand curve for Hugo. Explain why Hugo should not purchase the quantity of flowers that maximizes his marginal utility.

4. Draw a graph to show the general shape of the production possibilities frontier for carrots and apartments. Explain why the opportunity cost of apartments increases as more are built.

5. Ally would pay up to $40 for her first compost bin and nothing for any additional compost bins. Every pair of socks Ally purchases gives her $12 worth of utility, and she could never have too many. Draw Ally's demand curves for compost bins and pairs of socks.

6. This chapter explains that efficiency does not imply equity. What is one example, from the chapter or from your own observations, of an efficient solution to an environmental problem that is not an equitable solution?

7. A tank behind Núria's service station has leaked a large quantity of motor oil into the soil below. If left uncontained, the oil could contaminate local drinking water and possibly even reach a nearby river. Experts estimate a 6 percent chance that the oil will cause $1 million worth of damage, a 2 percent chance that the oil will instead cause $3 million worth of damage, and a 92 percent chance that the oil will cause no damage. What is the expected value of damage from Núria's tank?

8. The state of Colorado requires campers to purchase a pass before using a campsite. The fine for not purchasing a pass is five times the cost of the pass. Suppose a camper decides whether to buy a pass by comparing the cost of the pass and the expected value of the fine if caught without a pass. That

camper will certainly purchase a pass if the probability of being caught without a pass exceeds what percentage?

9. Allocative efficiency is achieved when the rate at which consumers would willingly trade one good for another equals what other rate?

10. Describe a situation you've learned about in your area or in the national news in which a maximization of net financial gain might differ from a maximization of social welfare.

The last question draws from information in the appendix.

11. At her current consumption levels, Ava receives 15 utils from an additional cup of tea and 300 utils from an additional pair of sandals. The price of tea is $1, and the price of sandals is $30 per pair. To maximize her utility, should Ava purchase this combination of the two goods, more tea and fewer pairs of sandals, or less tea and more pairs of sandals? Explain your answer.

websurfer's challenge

1. Find a description of the "Water-Diamond Paradox" and explain the paradox in your own words.

2. Find an example of economic research that considers *non-monetary* costs and benefits.

3. Find an online newspaper article that uses the word "efficiency" and compare the meaning of the word in that article to the meaning of the word in this chapter.

Key Terms

Allocative efficiency
Consumer surplus
Demand curve
Distributive efficiency
Expected value
Marginal analysis
Marginal benefit
Marginal cost

Marginal utility
Market equilibrium
Pareto efficiency
Producer surplus
Production possibilities frontier
Social welfare
Supply curve
Utility

Internet Resources

Advice on cost-benefit analysis used for environmental policymaking: *www.brookings.edu/articles/three-steps-to-improving-cost-benefit-analysis-of-environmental-regulatory-rulemaking/*

EPA guidelines on cost-benefit analysis: *www.epa.gov/environmental-economics/guidelines-preparing-economic-analyses*

The International Whaling Commission: *http://iwc.int*

Further Reading

Anderson, David A. "The Deterrence Hypothesis and Picking Pockets at the Pickpocket's Hanging." *American Law and Economics Review 4*, no. *2* (2002): 295–313. A study of the capacity of criminals to perform cost-benefit analysis.

Becker, Gary S., and Kevin M. Murphy. "A Theory of Rational Addiction." *Journal of Political Economy 96*, no. *4* (August 1988): 675–700. A theoretical discussion of the possibility that even addiction is rational.

Dowie, Mark. "Pinto Madness." *Mother Jones* (September/October 1977): 18–32. An overview of Ford Motor Company's ill-fated cost-benefit analysis of the advisability of additional safety features.

Environmental Protection Agency. "Benefits and Costs of the Clean Air Act 1990–2020, the Second Prospective Study." Overviews and Factsheets. Last updated August 10, 2022. www.epa.gov/clean-air-act-overview/benefits-and-costs-clean-air-act-1990-2020-second-prospective-study. A brief summary of cost-benefit findings from an EPA study of the Clean Air Act.

Appendix

Efficiency Criteria in Greater Detail

Allocative Efficiency

Further discussion of the criteria for allocative efficiency provides an opportunity to introduce a few new terms and renew acquaintances with some old ones.[1]

The chapter explained that *marginal utility (MU)* is the additional utility gained by consuming one more unit of a good or service. Suppose consumers purchase food (*F*) and clothing (*C*). The ratio of the marginal utilities of food and clothing is called the *marginal rate of substitution (MRS)*, and it represents consumers' willingness to pay for one item (food) by giving up another item (clothing):

$$MRS_{FC} = \frac{MU_F}{MU_C}.$$

The *FC* subscript indicates that food is being substituted for clothing. Suppose the food units are pizzas and the clothing units are T-shirts. If Jen's marginal utility from pizzas is 4 and her marginal utility from T-shirts is 2, her $MRS_{FC} = 4/2 = 2$, meaning that she would trade at most two T-shirts for another pizza. By giving up two T-shirts for one pizza, she obtains 4 "utils" (*utils* are simply an arbitrary measure of utility) worth of pizza in exchange for 4 utils worth of T-shirts, and her total utility is unchanged. If she paid three T-shirts for a pizza, she would be giving up $3 \times 2 = 6$ utils worth of clothing for 4 utils worth of food, resulting in a net loss.[2] The last section of this appendix provides

1 If you have never seen any of these concepts before, you may want to review an introductory economics book in addition to reading this section.

2 These calculations treat the marginal utility from these items as constant, which is reasonable for very small changes in the quantities. In reality, the marginal utility is expected to decrease as more items are acquired.

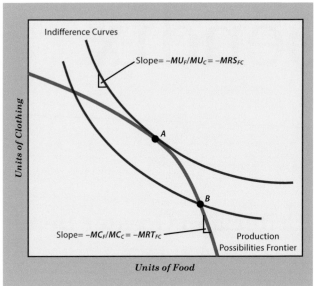

Indifference Curves

Slope= $-MU_F/MU_C = -MRS_{FC}$

Units of Clothing

A

B

Slope= $-MC_F/MC_C = -MRT_{FC}$

Production
Possibilities Frontier

Units of Food

The slope of an indifference curve indicates the rate at which an individual is willing to trade one good for another. The slope of a production possibilities frontier indicates the rate at which one good can be substituted for another in production. When these rates are the same, as at the tangency point *A*, resources are allocated efficiently. When the *MRS* differs from the *MRT*, as at point *B*, there is opportunity for a Pareto-optimal move to point *A*.

further discussion of the marginal rate of substitution for anyone interested.

Figure 2.6 illustrates an *indifference curve* depicting all of the combinations of food and clothing that make the representative consumer equally well-off. Since the *MRS* represents the consumer's willingness to exchange clothing for food at a constant level of utility, the slope of the line (the change in clothing divided by the change in food) is equivalent to the negative of the *MRS*:

$$\text{Slope of indifference curve} = -MRS_{FC} = -\frac{MU_F}{MU_C}.$$

The indifference curve in Figure 2.6 is drawn assuming the consumer can freely dispose of any undesired units of either good. In that case, the consumer never has too much of anything. In the absence of free disposal, a consumer with more of a good than desired would need to be compensated for receiving more of that good by receiving more of something that is still desired or less of something that is undesired. For example, garbage pickup is a good thing up to a point, and Rondi might be indifferent about trading a large bag of rice for the second garbage pickup in a month. If the garbage truck visited Rondi's house 40 times per month, it might become more of a noisy nuisance than a beneficial service. Rather than being willing to give up some rice in exchange for additional pickups, she would have to receive more rice to compensate her for a forty-first garbage truck visit, assuming she doesn't have too much rice, too. Under the more common assumption of free disposal, Rondi could just say "stop" after the last beneficial pickup, and the truck would stop coming. Indifference curves between a good, like the number of smog-free days in a year, and a bad, like the number of miles from one's home to the nearest hospital, have a positive slope to reflect the need to receive more of the good in exchange for enduring more of the bad.

Figure 2.6 also illustrates a production possibilities frontier, which depicts the combinations of food and clothing that could be produced using all available resources efficiently. The slope of this line (again, the change in clothing divided by the change in food) is determined by the amount of clothing that must be forgone due to constraints on resources and production technology in order to produce one more unit of food. In effect, the slope of the PPF measures the marginal cost of producing food relative to the marginal cost of producing clothing. That slope— the ratio of the marginal costs of producing the two goods—is called the *marginal rate of transformation (MRT)*:

$$\text{Slope of production possibility frontier} = -MRT_{FC} = -\frac{MC_F}{MC_C}.$$

If you have never encountered the *MRT* before, you may want to read more about it in the last section of this appendix.

Allocative efficiency is achieved when, for a representative consumer, $MRS = MRT$. This condition is met when the *PPF* is tangent to the indifference curve, as shown by point A in Figure 2.6. If *MRS* doesn't equal *MRT*, say, because $MRS_{FC} = 0.5$ and $MRT_{FC} = 3$ as shown by point B, then the typical consumer would be willing to exchange up to two pizzas for one T-shirt, whereas one more T-shirt could be produced at the cost of just one-third of a pizza. A net gain in utility could be realized if resources were reallocated to produce more T-shirts and fewer pizzas because a new T-shirt would generate six times as many utils as would be lost due to a one-third unit decrease in pizza production. In this case, more T-shirts and fewer pizzas should be made until $MRS = MRT$. If, for example, $MRS = MRT = 2$, a new pizza would generate twice as much utility as one T-shirt, but the production of a new pizza would require the loss of two T-shirts, thus providing no net gain in utility.

An economy achieves allocative efficiency when $MRS = MRT$, but individuals and firms have their own private objectives apart from the efficiency goals of society. Let us now consider how the self-centered workings of the market might bring about the best result for society.

The benefit from a first meal or article of clothing is tremendous, providing life or decency. The second meal or article of clothing is surely nice to have but less vital. The fiftieth of either item provides far less marginal utility than earlier units because the most urgent needs and wants are already satisfied. This reasoning supports the *law of diminishing marginal utility*, which states that the satisfaction gained from additional units of a good consumed within a given period of time decreases as more units are consumed.

If Alexandra gains 12 utils per additional dollar spent on food and 9 utils per additional dollar spent on clothing, taking a dollar away from clothing and spending it on food would create a net increase in total utility of 12–9 = 3. More money should be spent on food and less should be spent on clothing until, as the marginal utility from food decreases with increased consumption (according to the law of diminishing marginal utility) and the marginal utility from clothing increases with decreased consumption, the marginal utility per dollar is the same for food and clothing. *Consumers get the most utility from any given amount of money by equating the marginal utility per dollar spent on each good purchased.* For a consumer who purchases food and clothing for the prices of P_F and P_C per unit, respectively, utility is maximized when

$$\frac{MU_F}{P_F} = \frac{MU_C}{P_C}.$$

Multiplying both sides by P_C and dividing both sides by MU_F, this equation becomes

$$\frac{P_C}{P_F} = \frac{MU_C}{MU_F}.$$

Recall that, in a perfectly competitive market, price equals marginal cost. Firms have no incentive to produce and sell a product for a price below marginal cost. Firms maximize their profit by producing more until their marginal cost rises to equal the price, and that price is their marginal benefit from selling a unit. If $P_C = MC_C$ and $P_F = MC_F$, then the ratios of these equalities will also be equal:

$$\frac{P_C}{P_F} = \frac{MC_C}{MC_F}.$$

From these two equations, it is clear that the price ratio equals both the marginal cost ratio (*MRT*) and the marginal utility ratio (*MRS*):

$$\frac{MU_C}{MU_F} = \frac{P_C}{P_F} = \frac{MC_C}{MC_F}$$

thus satisfying the criterion for allocative efficiency.

The left side of this equation is easily satisfied because consumers always have the incentive to maximize utility by equating the marginal utility per dollar spent on each good. However, when markets are not perfectly competitive, firms maximize profit by setting price *above* marginal cost. For that reason, the $P = MC$ condition that leads to allocative efficiency is harder to achieve. Potential pitfalls that can cause price to differ from marginal cost, or cause perceived costs and benefits to differ from reality, are detailed in Chapter 3.

Productive Efficiency

The tomatoes and cotton story earlier in this chapter illustrates how it is sometimes possible to produce more of one good without producing less of any other good. This is the case if the rate at which one input can be substituted for another—the *rate of technical substitution (RTS)*—differs among producers. The rate of technical substitution between two inputs is equal to the ratio of the *marginal products* of the inputs. For example, the *marginal product of labor (MP$_L$)* is the additional output produced by one more unit of labor and the *marginal product of capital (MP$_K$)* is the additional output produced by one more unit of capital. If labor (*L*) and capital (*K*) are the two inputs, then

$$RTS_{LK} = \frac{MP_L}{MP_K}.$$

If this is your first introduction to the *RTS*, you may want to read the more in-depth explanation of this concept a bit later in this appendix. Productive efficiency requires the *RTS* to be equal across all products. Our task now is to determine whether profit-maximizing behavior by firms will lead to this equality.

According to the *law of diminishing marginal returns*, as the quantity of one input increases, holding other input levels constant, the marginal product will eventually decrease. Consider the classic example of growing flowers in a flowerpot. Holding the size of the flowerpot and the amount of sunlight, water, soil, and seeds constant, the marginal product of fertilizer will eventually decrease. The first application of fertilizer in a week is likely to increase growth by more than the second and the third, and the twentieth application of fertilizer may do more harm than good, meaning that the marginal product of fertilizer becomes negative. Conversely, given diminishing marginal returns, as less fertilizer is used, its marginal product will increase. The same story of diminishing returns could be told in regard to increases in a different input, such as seeds or water.

Figure 2.7 illustrates diminishing marginal returns for workers and bulldozers. To minimize the cost of producing a given quantity of output, a firm adjusts the quantity of each input to equate the marginal product per dollar spent on each input. If more roadway is produced per additional dollar spent on workers than per additional dollar spent on bulldozers, workers should be substituted for bulldozers until another dollar spent on workers would no longer increase production more than another dollar spent on bulldozers. If labor and capital (such as bulldozers) are the only two inputs, the cost-minimizing production condition is that the "bang per buck," meaning the marginal product per dollar spent, is the same for both inputs. Letting w represent the price of labor and r represent the price of capital, the cost-minimizing condition is

$$\frac{MP_K}{r} = \frac{MP_L}{w}.$$

Multiplying both sides by w and dividing both sides by the MP_K yields

$$\frac{w}{r} = \frac{MP_L}{MP_K}.$$

If each firm minimizes costs by equating its ratio of marginal products with the ratio of input costs, under the assumption that wage and rental rates are constant across firms, the ratio of marginal products will be the same for all firms regardless of what they produce. That means that the marginal product of labor divided by the marginal product of capital will be the same for the production of clothing and roadways and tomatoes and cotton and so on. Hence, this condition for productive efficiency is met as the result of profit-maximizing behavior by firms.

The law of diminishing marginal returns holds that as the amount of one input increases, holding other input levels constant, the marginal product of the increasing input will eventually decrease. For example, as the quantity of labor increases, the marginal product of labor decreases. Going in the other direction, if the marginal product curve is downward sloping and the quantity of an input decreases, the marginal product increases, as shown here for capital.

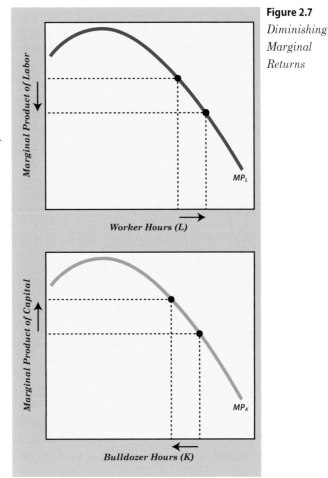

Figure 2.7
Diminishing Marginal Returns

Distributive Efficiency

Distributive efficiency follows from consumers making purchases that maximize their utility subject to budgetary constraints. The formal condition for distributive efficiency is that every consumer has the same marginal rate of substitution. If MRSs are not equal, there are opportunities to exchange goods that will yield improvements in at least one consumer's utility without decreasing anyone's utility.

Suppose Sunyi's marginal utility from T-shirts is 1 and his marginal utility from pizza slices is 2, while Forrest's marginal utility from both T-shirts and pizza slices is 1. Sunyi and Forrest have differing marginal rates of substitution of 1/2 and 1, respectively. An exchange of one T-shirt from Sunyi for one pizza slice from Forrest would have no effect on Forrest's total utility, while making Sunyi better off. Forrest loses 1 util by giving up a slice and gains 1 util from the new T-shirt; Sunyi loses 1 util by giving up the T-shirt but gains 2 utils from the new pizza slice. Similar Pareto improvements—trades that make at least one party better off without harming anyone—will be possible as long as Sunyi's and Forrest's marginal rates of substitution differ.

47

EFFICIENCY AND CHOICE

How can we expect millions of consumers to converge on similar marginal rates of substitution as required for distributive efficiency? Remember that consumers maximize their utility by equating their marginal rate of substitution to the price ratio: $MRS = P_C/P_F$. To the extent that consumers face the same prices, as is the case in a perfectly competitive market, P_C/P_F will be the same for every consumer. Thus, profit-maximizing consumers facing the same prices will have the same MRS, and distributive efficiency will be achieved.

In summary, with firms minimizing their costs and consumers maximizing their profits, economic theory holds that allocative, productive, and distributive efficiency are all achieved simultaneously in a perfectly competitive market in long-run equilibrium.

More on MRS, MRT, and RTS

Marginal Rate of Substitution

The marginal rate of substitution (MRS) is the rate at which an individual can substitute one good for another without altering the individual's level of happiness. Economists refer to happiness as "utility," and units of happiness as "utils." Suppose you are stranded on a deserted island with a lot of matches but only a few pieces of firewood. You are good at catching fish, but you do not like sashimi (raw seafood). Thus, the ability to build fires is a good thing. A man named Friday, your cohabitant of the island, also has logs you can use as firewood and offers you the opportunity to exchange matches for logs. To find your MRS between logs and matches, consider the largest number of matches you would trade for a log.

Suppose that one more log would give you 20 utils, and that the first match you give up is worth 5 utils to you. This is another way of saying that for you, the marginal utility of firewood (MU_F) is 20 and the marginal utility of matches (MU_M) is 5. What is the largest number of matches you would trade for a log, assuming these MU levels hold? Four. How is this number determined? By dividing the utils gained from a log by the utils lost per match. In exchange for the 20 utils from a log, you would exchange at most $20/5 = 4$ matches worth 5 utils each. If you traded fewer than four matches, say, three, you would gain 20 utils from the log and only give up $3 \times 5 = 15$ utils, for a net gain of $20-15 = 5$ utils. If you traded more than four matches, say, five, you would gain 20 utils from the log and give up $5 \times 5 = 25$ utils from the matches, for a net loss of 5 utils. Of course, you would like to give up as few matches as possible, but the most you would give up for a log is four matches, so that is your MRS_{FM}. This explains why the marginal rate of substitution is found as

$$MRS_{FM} = \frac{MU_F}{MU_M}.$$

As more firewood is gained and more matches are traded away, the law of diminishing marginal utility tells us that the marginal utility of firewood declines and the marginal utility of matches increases. The more firewood obtained, the less valuable is one more log, and the fewer matches held, the more valuable is the next match to be parted with. As MU_F decreases and MU_M increases for these reasons, the value of their ratio, MRS_{FM}, will decrease. Since $-MRS$ is the slope of an indifference curve, this explains why the slopes of indifference curves decrease (the curves become flatter) as one obtains more of the x-axis good (firewood) and less of the y-axis good (matches).

Marginal Rate of Transformation

The marginal rate of transformation (MRT) is the rate at which one good can be produced in place of another. Because simple models are easier to work with, we will again examine the relevant trade-offs between just two goods. Suppose an economy makes only jeans and scarves, and at the current production levels, it takes $30 worth of inputs to make another pair of jeans and $10 worth of inputs to make another scarf. In other words, the marginal cost of jeans (MC_J) is $30, and the marginal cost of scarves (MC_S) is $10. We will examine the trade-off between producing these two goods within a range of the production possibilities frontier so small that the marginal costs can be considered constant.

If one fewer pair of jeans is produced, that frees up $30 worth of inputs, with which $30/$10 = 3 scarves can be made at a cost of $10 worth of inputs per scarf. Thus, at this point on the production possibilities frontier, scarves can be produced instead of jeans at a rate of three additional scarves for every one pair of jeans forgone:

$$\text{Trade-off} = \frac{\text{Change in scarves}}{\text{Change in jeans}} = -\frac{MC_J}{MC_S} = -\frac{\$30}{\$10} = -MRT_{JS} = -3.$$

On a production possibilities frontier graph with the quantity of scarves measured on the vertical axis and the quantity of jeans measured on the horizontal axis, $-MRT_{JS}$ would be the slope of the frontier. The negative sign reflects the fact that either the change in scarves or the change in jeans is always negative and the other change is positive. This makes sense because when the quantity of scarves increases, the number of pairs of jeans decreases, and with more jeans must come fewer scarves.

If there is some specialization of inputs, as is generally the case, the marginal cost of producing each product will increase as more of it is made. For example, if some workers are better at knitting scarves and others are better at sewing jeans, the workers most skilled at sewing will make the first pairs of jeans. Slower sewers will be asked to make jeans as more jeans are made, so it will take relatively more labor hours and correspondingly higher payments to workers to make each additional pair of jeans. As the marginal costs change, the ratio of the marginal costs—the marginal rate of transformation—will also change along a production possibilities frontier, making the slope steeper. If the inputs are exactly the same for the two goods, the *MRT* will be constant and the PPF will be a straight line.

Marginal Rate of Technical Substitution

The marginal rate of technical substitution (*RTS*) is the rate at which one input can be substituted for another without changing the output level of the good being produced. Consider the planting of tree seedlings. Seedlings can be planted with various combinations of capital and labor. For example, 100 seedlings could be planted in an hour by a team of ten workers each using a small shovel or by a single worker operating a properly equipped tractor.

Suppose that with a given combination of capital and labor, 1 additional unit of capital (perhaps a shovel or a tractor component) would increase the number of seedlings planted per hour by 2, assuming that the amount of labor did not change. Alternatively, an additional worker would increase the number of seedlings planted per hour by 8 if the amount of capital did not change. This means that the marginal product of capital (MP_K) is 2, and the marginal product of labor (MP_L) is 8.

To find the rate at which capital could replace labor without changing the quantity of output, notice that the loss of a worker's contribution of 8 seedlings would require the addition of $8/2 = MP_L/MP_K = 4$ units of capital. Four units of capital adding 2 seedlings each would contribute a total of $4 \times 2 = 8$ seedlings, just equal to the loss from the departing worker. This confirms that the marginal rate of technical substitution (*RTS*) is

$$RTS_{LK} = \frac{MP_L}{MP_K} = \frac{8}{2} = 4 \cdot$$

A curve that includes all of the combinations of two inputs that create the same quantity of output is called an *isoquant*. If capital is on the vertical axis and labor is on the horizontal axis, as in Figure 2.8, the slope of the isoquant is the change in capital necessary to compensate

An isoquant curve shows all of the combinations of two inputs (capital and labor in this case) that can be used to produce the same level of output, such as planting 100 tree seedlings. The slope of any line is "rise over run." For this isoquant, a run of −1 represents the loss of one worker, and a rise of RTS represents the number of units of capital required to replace the loss of one worker and reach another point on the isoquant with output unchanged. Thus, the slope of the isoquant is equal to rise/run = RTS/−1 = −RTS.

Figure 2.8
An Isoquant Curve

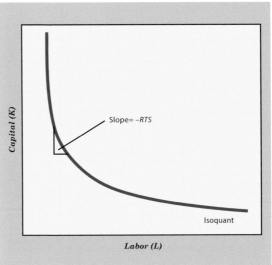

for the loss of one worker, which is exactly what the RTS tells us (with a negative sign in front to indicate the loss of the marginal productivity of a worker). The RTS at a particular point on the isoquant curve is calculated for infinitely small changes in labor and capital, for which the marginal products of the inputs can be treated as constants. As shown in Figure 2.7, the marginal product of an input eventually decreases as more of the input is employed. Thus, the RTS changes along the isoquant curve, decreasing as more labor and less capital is employed, and increasing as less labor and more capital is employed. For example, when seedlings are planted using relatively more labor and less capital than at the point analyzed here, the MP_L might equal 6 and the MP_K might equal 3, making the $RTS = 6/3 = 2$.

A classic source of market failure, for reasons explained in this chapter.

Market Failure

*I*f Charles Dickens had written about the free market, he might have put it this way: It is the best of forces, it is the worst of forces, it is the embodiment of wisdom, it is the embodiment of foolishness, it is the source of wealth, it is the source of poverty, it is the root of progress, it is the root of destruction, it is the answer to resource allocation, it is the bane of all resources, in short—it is everything we need and less.

In theory, the market is miraculous: Individual consumers and firms, all acting in their own self-interest in the market, can bring about an efficient allocation of goods and services with no more than the worthwhile levels of stress on the environment and natural resources. The logic behind market efficiency follows from the underpinnings of supply and demand. As explained in Chapter 2, the supply curve reflects the marginal cost of production, and the demand curve reflects the marginal benefit of consumption. When market forces establish the price and quantity of a good at the intersection of the supply and demand curves, the coveted efficiency criterion is achieved: Marginal cost equals marginal benefit. This is the result economist Adam Smith reveled about in his famous 1776 discussion of the "unseen hand" that seems to guide the economy to efficiency. Unfortunately, the invisible hand can be misguided by unseen environmental costs and benefits, unfettered market power, unrealized information, and unwillingness to pay for goods whose benefits cannot be withheld.

DOI: 10.4324/9781003428732-3

Why Markets Fail

Market failure is the failure of the free market to allocate resources efficiently. Market failure can result from

- *Imperfect competition*

- *Imperfect information*

- *Externalities*

- *Public goods.*

This section explains the culprits of inefficiency and their influence on the market's unseen hand.

Imperfect Competition

There is **imperfect competition** when sellers have sufficient market power to charge prices above marginal cost, limit the quantity produced, or produce goods of inferior quality. These practices can lead to an inefficient allocation of resources that does not equate marginal cost and marginal benefit—one of the essential criteria for efficiency. Imperfect competition is particularly relevant to environmental economics because major polluters such as electricity and fuel producers tend to have considerable market power.

Chapter 2 discussed supply and demand in a perfectly competitive market with many firms selling an identical product. The opposite extreme is a *monopoly* market in which a single firm, known as a *monopolist*, is the only seller. Like perfectly competitive firms, monopolists maximize profit or minimize loss by producing the quantity that equates marginal cost and marginal revenue. Facing the entire downward-sloping market demand curve, a monopolist must lower its price on all units in order to sell another unit of output. As a result of this price decrease, the marginal revenue gained from selling the additional unit is not simply the price received for that unit. Instead, it is the price minus the revenue lost from lowering the price on all the other units. For example, suppose a monopolist could sell ten birdhouses for $100 each, and a price cut to $95 would allow it to sell an eleventh. The marginal revenue from the eleventh birdhouse would not be the $95 price a customer pays for it. Instead, the marginal revenue would be the $95 price minus the $5 lost on each of the ten birdhouses that would otherwise have sold for $100 and now sell for $95 each. So the marginal revenue is $95 – ($5 × 10) = $45.

Figure 3.1 shows the demand, marginal revenue, and marginal cost for a typical monopoly. The profit-maximizing quantity, Q_m, is found on the quantity axis directly below the intersection of the marginal cost curve

Relative to a perfectly competitive market with the same cost structure, a monopoly will have a higher price and produce less. Because $MB > MC$ for a monopoly, the output level is inefficient and there is deadweight loss, represented by the shaded triangle.

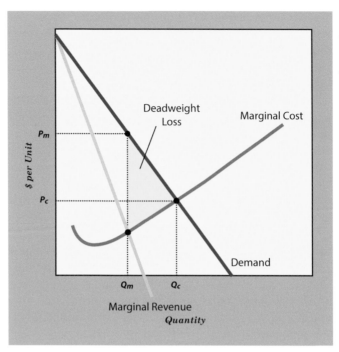

Figure 3.1
*Monopoly
Versus
Perfect
Competition*

and the marginal revenue curve. The price, P_m, is found as the height of the demand curve above the profit-maximizing quantity of Q_m. Although the monopolist stops producing at Q_m, additional units up to Q_c would provide a marginal benefit to consumers that exceeds the marginal cost of producing them. We know this because the demand curve—an indication of marginal benefit—is above the marginal cost curve for the first Q_c units. The result of not producing the units between Q_m and Q_c is **deadweight loss**, a loss of producer and/or consumer surplus caused by an inefficient allocation of resources. The yellow deadweight loss area on the graph represents the net loss to society due to the monopolist's restriction of output to only Q_m units.

Chapter 2 explained that, in a perfectly competitive market, marginal cost determines supply, and the equilibrium of supply and demand determines price and quantity. If the market depicted in Figure 3.1 were perfectly competitive and the production costs remained the same, market equilibrium would occur at the intersection of the demand curve and the marginal cost curve. Production would *increase* from Q_m to Q_c, and the price would *fall* from P_m to P_c. Competition also encourages quality enhancements, whereas firmly entrenched monopolists have little reason to improve existing products.

The U.S. Congress responded to the problems of imperfect competition in 1890 with the enactment of the first U.S. federal antitrust legislation, the Sherman Act. This was followed in 1914 with the Clayton Act and the Federal Trade Commission Act. With the Federal

Trade Commission (FTC) as chief enforcer, these laws and subsequent enhancements punish unfair methods of competition and unfair or deceptive business practices. Although the FTC is effective in many circumstances, legal and financial constraints prevent it from deterring all abuses of market power.

Market power has its virtues as well. The ability of a monopoly to sustain profits in the long run is a clear incentive for innovation. To promote this, the European Patent Convention and the U.S. Patent Act permit inventors to apply for *patent protection* as (typically) a 20-year barrier to competition. In order to be patented, an invention must be novel, useful, and not of an obvious nature. Patents are issued for machines such as electric engines, human-made products such as biodegradable food containers, compositions of matter such as sunscreen, processes such as Amazon's One-Click ordering system, designs such as the computer mouse, and new plant species such as the CrimsonCrisp apple.

Sometimes the output restrictions of monopolies are beneficial. Consider a good whose production generates *negative externalities* such as air pollution, noise, or other detrimental side effects. A competitive market would produce inefficiently large quantities of such a good, and the lower quantities produced by a monopoly might be closer to the optimal quantity for society. For example, a monopoly would produce less petroleum at a higher price than a perfectly competitive market, resulting in fewer of the negative externalities from petroleum production and combustion. Goods with beneficial *positive externalities*, such as flooring made of bamboo (a thick grass that substitutes for wood and helps prevent deforestation), are underproduced in a perfectly

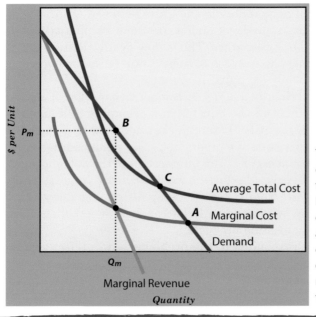

Figure 3.2

Natural Monopoly

The average total cost of a natural monopoly falls as output rises, because large fixed costs are spread across a growing quantity of output. This means one firm can serve the market at a lower cost than multiple firms.

competitive market. In these cases, the restricted quantity produced by a monopoly is *even further* below the social optimum.

Antitrust legislation also leaves room for **natural monopolies**, which are monopolies in industries with such high start-up costs that it would be impractical for several firms to share the same market. The markets for nuclear power and water treatment are examples of natural monopolies. As a natural monopoly increases its production, the high start-up or *fixed* costs are spread across more and more units of output. This spreading of costs causes the **average total cost**—the total cost divided by the quantity—to fall throughout the relevant range of production as shown in Figure 3.2. If only one firm serves such a market, its average and marginal costs will be lower than if there were multiple firms. If that monopolist acted like a competitive firm and charged the price at which marginal cost equals demand, as shown by point A in Figure 3.2, the average total cost would exceed the price and the firm would experience losses. As a monopolist, the firm would like to operate at point B, charging P_m and producing Q_m. To prevent such high prices and low quantities, regulatory boards typically control prices and output levels for natural monopolies so that they operate at a point like C, where the price equals the average total cost and output is as large as possible without causing losses.

Imperfect Information

Did you know that computer components like the ones held by the children in the photo contain heavy metals including arsenic, lead, and mercury? Or that cotton farming is one of the most chemically dependent forms of agriculture? Did you know that common coffee cups, opaque plastic cutlery, and to-go containers made of polystyrene can leach a colorless, sweet-smelling chemical called styrene into food and beverages? According to the Environmental Protection Agency, styrene exposure may increase the risk of leukemia and lymphoma, not to mention hearing loss, headache, depression, and weakness.[1]

Unaware of the dangers involved, children in India play with computer parts containing heavy metals.

---

1 See www.epa.gov/sites/default/files/2020-05/documents/styrene_update_2a.pdf.

Imperfect information exists when consumers or firms have inadequate knowledge to appropriately weigh all of their buying and selling options. In the past, incomplete information about the environmental and health effects of products such as tobacco, asbestos, lead, solvents, dioxins, and furans may have resulted in inefficient consumption levels. Consumers who do not grasp a product's dangers will purchase too much of it. Inefficiencies also stem from consumers paying too much for products because they don't know about lower-priced alternatives. And producers can make too much of one product and not enough of another because they hold inadequate information about consumer demand. Solutions to these types of imperfect information include requirements for product testing, truth-in-advertising regulations, consumer information services, and market surveys by firms.

As another important example, consider the immeasurable volumes of water, land, labor, fuel, and fertilizer used to produce food that is never eaten. The United Nations reports that 14 percent of food is lost before reaching stores or restaurants. The reasons include problems with weather, pests, storage, transportation, and processing. Another 17 percent of food is wasted by households, restaurants, or stores. That includes food that is discarded because it is out-of-date, unattractive, or over-purchased. Food loss and food waste account for an estimated 38 percent of all energy devoted to food production globally.[2]

Inefficient levels of food loss and waste can result from imperfect information. Uncertainty about market demand leads farmers to overproduce and stores and restaurants to overstock food. Uncertainty about market supply, as occurred during the COVID pandemic, leads households to over-purchase food. And misunderstandings about food expiration labels causes people to discard useful food.

Solutions for food waste include improved technology and education. In India, fishers' access to cell phones has allowed them to customize their harvests to the needs of local markets and reduce fish waste.[3] In Egypt, increased awareness of opportunities to dry tomatoes and grapes has avoided the loss of rotten produce.[4] And many countries seek to revise freshness dating policies to avoid the loss of edible food.[5]

Information problems also plague decision-makers trying to coordinate their strategies. The classic case is a **prisoner's dilemma**, in which two parties acting in their own best interest end up worse

2 See www.un.org/en/observances/end-food-waste-day.

3 See www.academic.oup.com/qje/article-abstract/122/3/879/1879540.

4 See www.un.org/en/observances/end-food-waste-day/stories.

5 See www.weforum.org/agenda/2022/08/waitrose-scrap-best-before-dates-cut-food-waste/.

		Crook B	
		Confess	*Deny*
Crook A	**Confess**	4,4	1,5
	Deny	5,1	2,2

Table 3.1

Classic
Prisoner's
Dilemma

off than if they could have cooperated. Suppose two crooks are apprehended with limited evidence linking them to a robbery they committed. Placed in separate rooms for interrogation, they must decide whether to confess to the crime or deny involvement. Table 3.1 illustrates the possible strategies and outcomes for Crook A and Crook B in a *payoff matrix*. The numbers indicate the resulting years spent in prison. Within each box, the number on the left is for Crook A and the number on the right is for Crook B. If each crook denies involvement in the crime, the sentence is only 2 years in prison due to a lack of compelling evidence. If one crook confesses and the other denies, the confessor will receive an even lighter 1-year sentence in exchange for the confession, and the liar will get a heavy 5-year sentence for robbing someone and then lying about it. If they both confess, each will go to prison for 4 years.

What should the two crooks do? Each sees that if the other denies, it is better to confess in order to receive one year in prison rather than two. If the other confesses, it is better to confess in order to avoid the heavy sentence for being a known liar (four years is better than five). Confession is a **dominant strategy** because it is better than the alternative of denial regardless of the other crook's strategy. Given their inability to cooperate, we can expect the two crooks to confess and spend 4 years in prison. Yet if each knew the other would cooperate by denying, they could each spend only two years in prison.

The prisoner's dilemma causes inefficiency in other instances as well. Warring nations may be safer if neither side has atomic weaponry than if both do, but the downside risk of unilateral disarmament leads each side to maintain arms. From the standpoint of stockholders in a firm, it can be similarly undesirable for the firm to unilaterally reduce pollution when the cleanup costs are substantial. Consider a dilemma facing competing tire manufacturers about whether to invest in a more environmentally friendly manufacturing process. The investment could go toward cleaner fuels, recycled materials, or advanced filtration systems for emissions. Some, but not all, of the cleanup cost could be passed on to consumers in the form of higher prices. If consumers worry mostly about price when they pick a tire supplier, the payoffs will resemble those in Table 3.2. The numbers represent the firms' profits in thousands of dollars. In each box, the number on the left is Firm 1's profit and the number on the right is Firm 2's profit.

Table 3.2

Consumers Care Primarily About Price

			Firm 2	
			Clean	**Dirty**
Firm 1	**Clean**		10,10	8,14
	Dirty		14,8	12,12

If both firms run clean, the added costs borne by the firms and the decrease in the quantity demanded resulting from higher prices result in a profit of $10,000 for each firm. If both firms run dirty, lower costs and higher sales give each firm $12,000 in profit. To be the only clean firm when customers care mostly about price results in the lowest possible profit. The clean firm would experience higher costs and earn a profit of only $8,000 from loyal customers, while the dirty firm with lower costs and lower prices would earn a profit of $14,000. The dominant strategy in this case is to run dirty.

Pollution is not always a dominant strategy. If customers exhibit a preference to buy from clean firms despite sharing some of the added expense, the payoff matrix might resemble that in Table 3.3. If both firms are clean or both firms are dirty, and there are no alternatives to purchasing from these two firms, then the outcomes are no different than before. Customers have no opportunity to choose between clean and dirty firms, and profits are higher in the dirty firms because costs are lower. But if one firm stays dirty and the other runs clean, the clean firm enjoys greater popularity and earns a profit of $14,000 while the dirty firm earns $8,000. This presents a prisoner's dilemma that works in society's favor. If the firms could cooperate and both run dirty, they would each receive $12,000 in profit, but the risk that the other firm will clean up and lure customers away causes each firm to clean up. Cleaning up in this case is a dominant strategy because it yields the highest profit regardless of what the other firm does. With customers seeking clean production and firms unable to trust each other to cooperate, firms are motivated to clean up even when they would earn more profit with everyone running dirty. Have you noticed that fast-food restaurants use very little non-biodegradable Styrofoam packaging these days? If so, you've seen the motivating effect of consumer preferences.

Table 3.3

Consumers Prefer Clean Firms

			Firm 2	
			Clean	**Dirty**
Firm 1	**Clean**		10,10	14,8
	Dirty		8,14	12,12

reality check

Cooperative Games at Sea

The bottlenose dolphin is second only to humans in the ratio of brain size to body size, and dolphins apparently outdo humans in some cooperative games. The prisoner's dilemma reveals the value and diffi-culty of cooperation among players when there is an incentive to cheat. Likewise, members of oil cartels try to cooperate to restrict output and raise prices, but selfish cartel members often undermine cooperative strategies by selling more than they're supposed to. Firms that can't cooperate on production or environmental strategies frequently take actions that ultimately have inferior outcomes for everyone involved. Dolphins face similar dilemmas. When eating from a school of fish, dolphins encircle the fish and take turns eating, one dolphin at a time. There is an incentive for the circling dolphins to cheat by eating while on duty. However, if a significant number of dolphins followed that incentive, the fish would disperse and the benefits from coordination would be lost. In reality, the trustworthiness of on-duty dolphins pre-vails to benefit all of dolphin society.

Marine biologist Pieter A. Folkens shares this anecdote about dolphin economics: Trash is dangerous to dolphins if they ingest it, so dolphins at Marine World were trained to remove trash from their tank and bring it to their trainer for a reward of fish. One dolphin kept appearing with trash even when the tank appeared clean. An underwater view revealed the dolphin's strategy. The dolphin had established an underwater sav-ings account. He collected all of the available trash and deposited it in a bag at the bottom of the tank. When he made a withdrawal, he did not bring up a whole piece of trash; rather, he tore off a small bit of what remained to increase and prolong his return. With this behavior, the dolphin exhibited a capacity for delayed gratification and the ability to plan for the future.

Perhaps dolphins could teach humans a few tricks.

Externalities

A third source of market failure, **externalities**, consist of the costs or benefits felt beyond or "external to" those causing the effects. When such side effects are detrimental, they are called **negative externalities**; beneficial side effects are called **positive externalities**. Inefficiencies arise from externalities when decision-makers do not consider all the repercussions of their behavior. Do you weigh the health and safety costs imposed on others when you decide how many miles to drive in your car? Likewise, developers may not consider the detriments of habitat loss when deciding where to locate homes. Manufacturers may overlook the costs of pollution when deciding production levels. And homeowners may ignore the value to others of beauty and clean air when they decide how many trees to plant on their property. If these external costs and benefits are neglected during the decision-making process, from a societal standpoint, too many developments crop up in sensitive wilderness areas, production levels are too high, and too few trees are planted.

As another example, honeybees create honey for apiarists (beekeepers), but they also pollinate fruit trees for nearby farmers. When they decide how many hives to purchase, the apiarists may not consider the positive externalities their hives provide. The external benefit of one more unit of a good is called the **marginal external benefit** (*MEB*). In this case, the *MEB* is the benefit to farmers of one more hive. The **private marginal benefit** ($MB_{private}$) is the benefit of one more unit to the consumer—the apiarist in this case. The **social marginal**

Figure 3.3

A Positive Externality

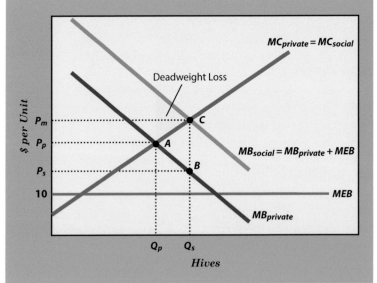

Marginal external benefit drives a wedge between private marginal benefit and social marginal benefit. Disregarding the marginal external benefit, the apiarist will purchase hives until her private marginal benefit equals her private marginal cost at the quantity of Q_p. If the apiarist internalizes the externalities due to subsidies or side payments, she will produce the socially optimal quantity, Q_s, that equates the social marginal benefit and the social marginal cost.

Bees provide positive externalities by pollinating farmers' crops.

benefit (MB_{social}) is the benefit of one more unit to everyone affected, so it includes the $MB_{private}$ and the MEB. Figure 3.3 illustrates the divergence of MB_{social} and $MB_{private}$ caused by a positive externality. The private marginal benefit is decreasing due to diminishing marginal returns. The MEB is constant at \$10 per hive. The social marginal benefit in this case is the private marginal benefit plus \$10, because $MB_{social} = MB_{private} + MEB$.

When negative externalities are present, there is again a separation of private and social marginal costs. The external cost of one more unit is called the **marginal external cost** (MEC). The **private marginal cost** ($MC_{private}$) is the cost of making one more unit, not including the MEC. The **social marginal cost** (MC_{social}) is the cost of another unit to everyone affected, so it includes the $MC_{private}$ and the MEC. When there is no external cost, $MEC = 0$ and $MC_{private} = MC_{social}$.

Private incentives guide the apiarist to point A at the intersection of $MB_{private}$ and $MC_{private}$, and an output of Q_p. For quantities higher than Q_p, the private marginal cost of another hive exceeds the private marginal benefit. The efficient quantity of hives from a societal stand-point is Q_s, the quantity at which $MB_{social} = MC_{social}$. As discussed later, the apiarist would purchase the socially optimal quantity if a subsidy or similar payment reduced the private marginal cost to P_s, as shown at point B.

In another example, flights into Quebec City, Canada, serve travelers and provide profits for airlines but cause the negative externalities of noise and air pollution. During each brief passage of a plane, the St. Lawrence River loses a bit of its serenity, and the Château Frontenac in Old Quebec loses some of its eighteenth-century charm. When airlines decide how many flights to schedule into Quebec City, they may not consider these or other pollution costs. A negative externality such as pollution creates a

Figure 3.4
*A Negative
Externality*

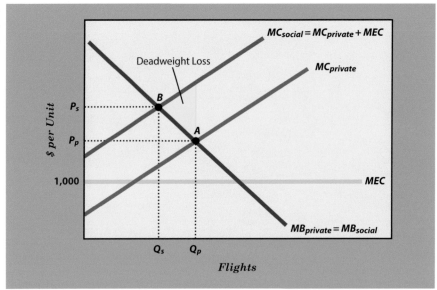

Marginal external cost separates private marginal cost and social marginal cost. Airlines will add flights until the private marginal benefit equals the private marginal cost at quantity Q_p. The socially optimal quantity of flights, Q_s, equates the social marginal cost and the social marginal benefit. Airlines would choose the socially optimal quantity if they internalized the marginal external cost of their decisions, as would occur if they paid a tax equal to the marginal external cost.

marginal external cost that brings the social marginal cost up above the private marginal cost because $MC_{social} = MC_{private} + MEC$.

If each flight causes $1,000 worth of noise and air pollution, the marginal external cost curve is horizontal with a height of $1,000 and the social marginal cost curve is above the private marginal cost curve by $1,000 as shown in Figure 3.4. Airlines will choose to operate Q_p flights—the number that brings the private marginal cost and the private marginal benefit into equilibrium at point *A*. (The private marginal benefit for the airline is its marginal revenue.)

For flights and other goods and services that impose negative externalities, the privately optimal quantity, Q_p, exceeds the socially optimal quantity, Q_s. One solution is for the government to limit the quantity to Q_s. Many city governments do impose flight restrictions. An alternative is to bring the decision-makers to bear the social marginal cost of their behavior. A private marginal cost of P_s would lead the airlines to point *B* at the socially optimal quantity of flights. Next, we'll see how the assignment of property rights, voting, Pigou taxes, or private negotiations can lead to efficiency in some cases.

Property Rights Inefficiency resulting from externalities is often avoidable. The socially optimal resource allocation can be achieved if decision-makers internalize, or feel themselves, the costs and benefits they bring to society. Biologist Garrett Hardin (1968) prescribed the privatization of assets as one route toward internalization. He referred to externalities produced on publicly held property, such as open grazing lands, the seas, the air, and national parks, as the **tragedy of the commons**, and held that abuses of open-access areas could be curtailed if the areas were privately held.

Consider a household that dumps sewage into a public lake rather than purchasing a septic system to process and store the waste. This "straight pipe" method of disposal damages the lake's value as a source of drinking water and as a venue for water sports. Although the social cost of dumping sewage is larger than the cost of a septic system, the household's private cost of dumping is not, because the household bears only a fraction of the overall damage of dumping. If the lake area belonged to the household dumping the sewage, that household would internalize the full social cost of dumping and invest in a septic system. If the lake area belonged to someone outside the household, that person would have an incentive to prohibit and carefully monitor dumping to avoid damage to the owner's property value. Given these incentive effects of ownership, Hardin felt that by assigning property rights to land, water, and air, society could avoid externalities caused by everything from factories to obnoxious music. As evidence that Hardin's point applies in some cases, poaching is a far greater problem in countries where property rights are weak than in countries where they are well-defined and strictly enforced.

The Efficient Voter Rule   If the externalities affect an identifiable group of people similarly, voting on a policy change can lead to an efficient solution. According to the **efficient voter rule**, *when individuals who receive the same harm from a problem vote on whether to eliminate the problem at a uniform cost per individual, the outcome will be efficient, regardless of each individual's contribution to the problem.* This could apply, for example, to policy decisions about noise pollution, overfishing, fuel economy standards, smoking bans, or zoning, but let's apply this to the straight pipe problem discussed earlier.

Suppose that three lakeside residents pipe their toilet waste directly into a shared lake. As an alternative to dumping waste into the lake, for $4,000 each, the residents could purchase septic systems for their households that would receive and process the waste. Over its lifetime, each septic system would prevent damage to the lake with a value today of $6,000 (Chapter 5 explains how to calculate the present value of future costs and benefits). The damage from dumping waste is distributed evenly, with each resident incurring $1/3 \times \$6,000 = \$2,000$ worth of damage from each straight pipe. Each resident decides between a $4,000 septic system and the $2,000 worth of damage the resident incurs from his or her own straight pipe, so it is rational from each resident's standpoint to choose the straight pipe. However, it is best for society if residents use septic systems because the $4,000 price is below the $6,000 worth of total damage each straight pipe imposes—the $2,000 worth of damage that is internalized by each household plus the $4,000 worth of damage that is imposed on the other households as an externality.

The neighborhood association that sets rules for the lakeside residents could hold a vote on a policy to require septic systems. If enacted, the policy would cost each resident $4,000. The policy would also mean septic systems for all, so it would eliminate the $2,000 in damage each resident incurs from each of the three straight pipes, for a total benefit of $6,000 to each resident. So when deciding which way to vote, each household will weigh a $4,000 cost against a $6,000 gain, effectively internalizing the full $6,000 benefit of purchasing a septic system. The privately rational choice now becomes the socially optimal choice: vote in favor of the policy.

Suppose instead that the total damage from a straight pipe is $3,000 rather than $6,000. In that case, it would be socially optimal for each resident *not* to purchase a septic system for $4,000 because the $4,000 cost exceeds the $3,000 benefit. Again, a vote would yield the socially optimal solution: By voting in favor of the policy, a resident would avoid $1/3 \times \$3,000 = \$1,000$ in damage from each of the three straight pipes, but this $3,000 benefit is less than the $4,000 cost of a septic system. So each resident would vote against the requirement. Anderson (2020) explains more applications of voting in the context of environmental economics.

Pigou Taxes Cambridge economist Arthur C. Pigou (1932) suggested taxes and subsidies to bring private and social costs into line. A **Pigou tax** is intended to equal the marginal external cost of the behavior being taxed. If developers paid a tax equal to the marginal external cost of repercussions such as habitat loss and pollution, they would internalize this cost and expand their development only if the social marginal benefit exceeded the social marginal cost. In the airline example from earlier, a tax of $1,000 per flight would cause the airlines to internalize the full cost of their behavior and operate at point B in Figure 3.4. Likewise, activities that provide positive externalities, such as education, the raising of bees, and the planting of trees, can be subsidized by the value of the marginal external benefit so that the private marginal benefits to the relevant decision-makers equal the marginal benefits to society. In the earlier beehive example, a subsidy of $10 per hive would bring the private marginal benefit up to the value of the social marginal benefit. The subsidized apiarist would purchase hives until the now-internalized social marginal benefit equaled the social marginal cost. This outcome is illustrated by point C in Figure 3.3, with an efficient quantity of Q_s and a market price of P_m. Likewise, Pigou has inspired

Some lighting companies really do emit lead into the air. Do you think citizens really bargain with them about production levels as Coase suggested would happen?

efficiency-seeking governments around the world to tax gasoline and other sources of negative externalities and subsidize solar panels and other sources of positive externalities. Chapter 4 covers government solutions in greater detail.

Coasian Solutions   English economist Ronald Coase (1960) argued that market failure will not result from externalities if the affected parties can bargain to an efficient solution. Coase's argument hinged on *transaction costs*, which are the costs of identifying and contacting the relevant parties, assessing costs and benefits, and creating and enforcing contracts. An issue involving two parties requires only one line of communication, perhaps a visit, an email, or a text, for the sharing of ideas. With ten parties involved, 45 lines of communication are needed in order for each party to share ideas with each of the other parties. With 100 parties, 4,950 lines of communication are required. Clearly the transaction costs associated with negotiations increase exponentially as the number of interested parties increases. This means efficient private solutions are less likely when the underlying problems involve many parties or complex issues.

According to the **Coase theorem**, *if property rights are clearly defined and transaction costs are negligible, bargaining between the parties involved can yield efficient solutions to externality problems.* Suppose the Good Idea Lightbulb Company releases lead (a carcinogenic pollutant) into the air during its production process. Let's consider how Coasian bargaining could lead to an efficient outcome.

For simplicity, assume the social marginal benefit from another truckload of lightbulbs is the profit received by the lighting company, and the

Table 3.4 Costs and Benefits of Lightbulb Production	Production Level (Truckloads)	Social Marginal Benefit (Additional Profit) ($)	Social Marginal Cost (Additional Damage) ($)	Total Profit ($)	Total Damage ($)	Total Profit Minus Total Damage ($)
	0	—	—	0	0	0
	1	4,000	500	4,000	500	3,500
	2	3,000	1,000	7,000	1,500	5,500
	3	2,000	3,000	9,000	4,500	4,500
	4	1,000	5,000	10,000	9,500	500

social marginal cost is the external health and environmental damage caused by the production of those lightbulbs. Table 3.4 indicates the marginal and total values of profit and damage at each production level. The company can produce up to four truckloads of lightbulbs. The net benefit to society is maximized with the production of two truckloads of lightbulbs, as shown in the last column, which lists the total profit minus the total damage. You can also find this socially optimal production level by comparing the social marginal benefit and the social marginal cost of each unit and noting that the second truckload is that last unit for which the social marginal benefit exceeds the social marginal cost.

In this model, the social marginal benefit of another truckload of lightbulbs is the additional profit for the firm. The social marginal cost is the pollution damage another truckload imposes on residents. The efficient quantity is two truckloads because the social marginal cost exceeds the social marginal benefit for each truckload beyond two.

Figure 3.5 provides a model of the lightbulb story. The blue line shows the diminishing social marginal benefit received as profit from each truckload of lightbulbs. The red line shows the increasing social marginal cost of each truckload, incurred by the residents as damage from pollution. For the first two truckloads, the social marginal benefit exceeds the social marginal cost, creating a net gain for society represented by the area between the red and blue lines. Each of the next two truckloads creates a net loss for society because the social marginal cost of those truckloads is higher than the social marginal benefit. Social welfare is maximized with the production of two truckloads of lightbulbs, at which point the social marginal cost curve and the social marginal benefit curve intersect.

Now the question is, in the absence of transaction costs, can private bargaining lead to the socially optimal production level? The answer, at least in theory, is yes. The Coase theorem applies under a variety of legal rules about who decides the production level and what compensation must be paid for losses or damages. The outcomes under two

Figure 3.5

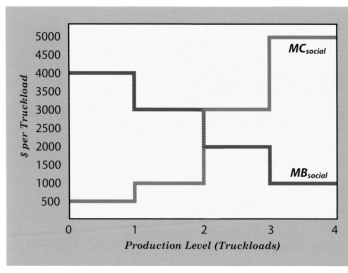

This graph shows the diminishing marginal benefit of light bulb production in terms of the additional profit for the firm per truckload of bulbs. The increasing marginal cost here is the additional pollution damage caused to residents per truckload. In this model, the social marginal benefit of another truckload of lightbulbs is the additional profit for the firm The social marginal cost is the pollution damage another truckload imposes on residents. The efficient quantity is two truckloads because the social marginal cost exceeds the social marginal benefit.

scenarios are described here; two other scenarios are included for your enjoyment among the problems for review.

> *The trick to Coasian bargaining is to start with the ideal output level for the party that receives compensation (no production if the victims receive payments, the profit-maximizing quantity if the injurers receive payments) and consider changes from that position, one unit at a time.*

Suppose the citizens can decide the production level and the light-bulb company must negotiate payments to the citizens in exchange for any permission to pollute. We start with the citizens' ideal of no production and consider how that might change. The first truckload provides $4,000 in profit to the company and only causes $500 in damage to the citizens. So any payment from the company to the citizens between $500 and $4,000 for permission to produce one truckload will make both sides better off than with no production. For example, if the company pays the citizens $1,500 for permission to produce the first truckload, the citizens receive $1,500 for $500 worth of damage, making them better off by $1,000. At the same time, the company pays $1,500 to earn $4,000 in profit, making the company better off by $2,500 relative to their no-production profit of zero. Whether the company's payment to the citizens is closer to $501 or $3,999 depends on the relative negotiation skills and bargaining positions of the two parties.

Having negotiated to produce the first truckload of lightbulbs, Good Idea Lightbulb Company will naturally seek more profit from increased production. Table 3.4 indicates that the production of a second truckload of bulbs would provide $3,000 in additional profit and cause $1,000 in additional damage. Any payment from the company to the citizens between $1,000 and $3,000 for permission to produce a second truckload would make both sides better off than they were with one truckload, so Coase would anticipate an agreement somewhere in that range. After the second unit, there are no further opportunities for mutually beneficial bargaining because the third and fourth units create more damage than profit. The citizens would require at least $3,000 to compensate them for the damage from the third truckload, which the company would be unwilling to pay for $2,000 in profit. Likewise, the company would not pay $5,000 to earn $1,000 from the fourth truckload.

Notice that this discussion revolves around the additional (marginal) profit and damage from incremental units. Looking just at the total profit and damage, it might appear that even the fourth truckload is negotiable because the total profit still exceeds the total damage at that production level. However, since the citizens would accept up to $5,000 less if the fourth truckload were not produced, and it is worth only $1,000 to the company, both sides would prefer a lower production level. The same argument can be made against the third truckload.

As another example, suppose citizens again have the right to restrict production, but the law requires that for every unit of production they forbid, they must compensate the lightbulb company for lost profit. We start with the company's ideal of four truckloads because now the company receives compensation. The citizens will gladly pay the $1,000 in lost profit to prevent the fourth truckload and its $5,000 worth of damage. The citizens are also willing to pay the $2,000 to prevent the third truckload and its $3,000 worth of damage. They will not, however, pay $3,000 to avoid $1,000 in damage from the second truckload or $4,000 to avoid $500 in damage from the first truckload. Thus, the citizens will restrict production to two truckloads—*the efficient level once again*—and compensate the lightbulb company for its total of $3,000 in lost profit from not producing the third and fourth truckloads.

Several complications could foil the Coase theorem. For instance, *strategic behavior* could rear its ugly head. Consider the example in which the citizens receive payments from the lightbulb company. Although any payment between $500 and $4,000 for the first truckload would make both sides better off than no production, the citizens might demand no less than $3,500 and the company might stubbornly refuse to pay more than $1,000. In such situations it can be difficult to settle on a payment that is agreeable to all parties.

Despite the deadly nature of externalities such as secondhand cigarette smoke, social norms alone can prevent fruitful negotiations and payoffs. That is, it would be abnormal and probably awkward to offer someone a bribe to stop smoking. Also, if payoffs to deter smokers and other creators of externalities became commonplace, people might create more externalities simply to collect the payments for stopping. Director Penny Marshall told a story of a fellow to whom, in true Coasian form, she paid $100 to stop using his chainsaw while Marshall was filming her movie *A League of Their Own* nearby. Unfortunately, this fellow's strategy was to create his noise pollution repeatedly over several days to receive additional payments.

Modern society has developed few efficient approaches to financial negotiation over the behavior of others. It is uncomfortable enough to ask a roommate to stop playing music during study periods, much less offering them money not to do so. Have you ever offered someone a bribe to stop talking or smacking their gum in the library? Neighbors complain about the unsightly house across the street, but do the complainers ever knock on their neighbor's door with an offer of $1,000 in exchange for a less neon color of house paint? When development or industry threatens natural habitat, the transaction costs of organizing everyone involved and negotiating an agreement make a private solution even less likely.

The following is a summary of common factors that can derail Coasian bargaining as a private solution to externalities:

- **Multiple sources of an externality**, *making it hard to know with whom to bargain*

- **Multiple victims**, *making it difficult to organize the affected parties*

- **Incomplete information** *about the costs or benefits of the externality*

- **Strategic behavior** *involving aggressive attempts to gain larger portions of the benefits from negotiation*

- **Time lags** *between the cause and the effect of the externality, making it difficult to identify the externality and its source*

- **Asymmetric information**, *which can mean that those creating a negative externality are aware of its danger, but those affected by it are not*

- **Transaction costs**, *due to the location, availability, and opportunity costs of those involved, making it difficult to assemble the parties for negotiations*

- **Social mores**, *making it uncomfortable for the victims to confront those causing the externalities.*

Chapter 4 discusses the alternative of government solutions.

Public Goods

Two people cannot wear the same watch, receive the same medical checkup, or drink the same cup of coffee. Yet in some cases, many people can share the benefits of the same good or service. And sometimes, people who did not purchase a good or service can use it regardless of the owner's wishes. Both of these descriptions apply to *public goods*.

Public goods are characterized as being *nonrival* in consumption and *nonexcludable*. A good is nonrival if one person's consumption of the good does not detract from other peoples' consumption of it. This is common for animal species that are appreciated for their mere existence. Your pleasure in knowing that polar bears still frolic in the arctic does not affect your neighbor's appreciation for the same bears. Multiple users, on the other hand, cannot consume the same *rival* good, such as a fried chicken leg or a housing site.

A good is nonexcludable if it is not possible to prevent people from benefiting from it. If environmental protection efforts stabilize or reverse global climate change, it will be impossible to prevent particular individuals from benefiting from that accomplishment. On the other hand, one can, with few exceptions, exclude others from consuming one's lunch or entering one's home.

Examples of public goods include streetlights, military protection, pollution abatement, airborne radio and television signals, fireworks displays, and disease control. Rivalry and excludability do not always go together. Online music subscriptions and scenic views from private lands are examples of **club goods**, which are nonrival but excludable. Club goods are made artificially scarce when the owners prevent people who don't pay for the goods from using them. **Open-access goods** (also called common resources) such as timber on public lands and fish in the sea are rival but nonexcludable. **Private goods** such as sandwiches and dental care are rival and excludable. Figure 3.6 summarizes these four types of goods.

Because multiple users can benefit from a public good without affecting the consumption of others, the value to society of each additional unit of the public good is found by summing the values of that unit to each person who benefits. If 30 million people each place a $10 value on the existence of the last humpback whale herd, the existence value of those whales to society is 30 million times $10, or $300 million. That is, society would be willing to pay at least $300 million to preserve

		Excludable	Nonexcludable
Figure 3.6 *Four Types of Goods*	Rival	Private Goods	Open-Access Goods
	Nonrival	Club Goods	Public Goods

the humpback whale. This is different from calculating the value of a private good. If there are 30 million people who each place a $10 value on the last bucket of chicken at Friendly Fried Food, the total value of that bucket to society is still just $10 because only one of those people will be able to enjoy the chicken.

The market demand curve for a public good is *not* found by adding the quantities demanded by each consumer at each price, as with private goods. Consider *green space*, which is land that serves as a public good because it is left undeveloped to provide beauty, wildlife habitat, and recreational space. Suppose that Andrew and Shakti are the only people in a very small community, and that nearby land suitable to become green space sells for $14,000 per acre. Figure 3.7 shows Andrew's and Shakti's demand curves for green space. Since the benefits of green space are nonrival and nonexcludable, both Andrew and Shakti can enjoy each acre. The value to society of each acre of green space is the sum of the most Andrew would pay for that acre and the most Shakti would pay for that acre, as indicated by the heights of their demand curves. More generally, the market demand curve for a public good is the vertical sum of the demand curves of the individuals in the market.

In Figure 3.7, the blue market demand curve is found by adding the height of Andrew's demand curve and the height of Shakti's demand curve at each quantity. For example, for the 250th acre, Andrew would pay up to $9,000 and Shakti would pay up to $5,000, so the height of the market demand curve is $9,000 + $5,000 = $14,000. The socially optimal quantity of acres is 250 because every acre up to the 250th provides a benefit to society that exceeds the $14,000 price. That is, the market demand curve is above the price line up to a quantity of 250.

The problem with public goods is that individuals know they can benefit from the existence of these goods whether or not they pay

Unlike the market demand curve for a private good, the market demand curve for a public good is the vertical sum of the demand curves of the individuals in the market. To find the height of the market demand curve at any quantity, simply add up the heights of the individual demand curves.

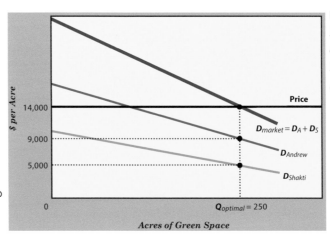

Figure 3.7
Individual and Market Demand for a Public Good

for them. A door-to-door collection to pay for green space or whale preservation would likely come up short due to the number of **free riders** who attempt to benefit from a public good without paying for it. Given the nonrival and nonexcludable nature of many natural resources, individuals have little incentive to reveal their true preferences. Instead, they may feign disinterest, only to benefit from the preservation efforts and expenditures of others. Most people enjoy green space, but few people have voluntarily purchased it, so communities have a hard time reaching the socially optimal quantity with private purchases. A solution to the free rider problem is to have the government provide public goods and pay for them with taxes collected from everyone who stands to benefit. This is how many green spaces and wildlife preservation efforts are funded today.

Summary

Market incentives aren't always what they should be. The efficiency of a free market can be marred by imperfect information, imperfect competition, externalities, and public goods. If consumers have incomplete information about the health or environmental benefits of a good, they may purchase too few. If the firm that produces the good is a monopolist, it will charge a relatively high price that leads to an inefficiently low level of production and consumption. If the firm does not fully internalize the burden of its emissions, is may not purchase filters for its smokestacks even if their benefit to society exceeds their cost. If the good provides benefits that everyone can enjoy regardless of whether they buy the good, consumers will prefer to free ride on the expenditures of others and purchase an inefficiently small quantity of the good.

There are several solutions to market failure that can work under the right circumstances. Arthur C. Pigou suggested taxes and subsidies that would cause decision-makers to internalize the full costs and benefits of their behavior. Ronald Coase saw private bargaining between decision-makers and victims as a path to efficiency. Garrett Hardin advocated the assignment of private property rights as a solution to the tragedy of the commons. The efficient voter rule applies to many situations in which voters could make socially optimal decisions on policy changes. Governments can fend off market failure by providing public goods with tax dollars, enacting and enforcing antitrust legislation, and requiring transparency on product information. Market failure persists, as evidenced by uninformed consumers, existing monopolies, mispriced sources of externalities, and underfunded public goods. In light of these problems, the benefit of studying environmental economics far exceeds the cost.

Markets fail when we don't pay the full price of our behavior. Would you ride a bicycle more often if you had to pay the full price of the externalities caused by driving a car?

• • • • • • • • • • • • • • • • • •

Problems for Review

1. As discussed in this chapter, under what circumstances could society be better off to have a good produced in a monopoly rather than in a perfectly competitive market?

2. Food waste is a global problem with repercussions for the environment.

    a) *Which type of market failure does the chapter associate with food waste?*

    b) *As discussed in the chapter, what is one specific solution to food waste caused by the type*
    of market failure identified in part (a)?

    c) *Suppose food waste left in a restaurant's garbage bin attracts rodents to the surrounding neighborhood. Which type of market failure does this exemplify?*

    d) *The chapter identified four categories of solutions to the type of market failure discussed in part (c). Which solution would you choose to apply if you were in the local government for the neighborhood in part (c)? Explain your reasoning.*

3. Driving causes the negative externalities of congestion and pollution.

   a) *Draw a graph with "miles driven" on the horizontal axis and "$ per mile" on the vertical axis. Draw curves to represent the private marginal cost, the social marginal cost, and the marginal benefit of driving. Label the socially optimal quantity Q_s and label the quantity drivers will actually choose Q_p.*

   b) *Show how the graph would change as the result of a new, cleaner-burning fuel. Label the new socially optimal quantity Q_{s2}.*

   c) *Of the solutions suggested by Pigou, Hardin, and Coase, which is the most likely to achieve efficiency in this situation? Explain your answer.*

4. Consider the situation described in Table 3.4 under the legal rule that the Good Idea Lightbulb Company can choose the production level and the citizens must negotiate payments to the company to lower production. Describe how the Coasian bargaining process would proceed and discuss the efficiency of the expected final outcome.

5. Consider once more the situation described in Table 3.4 under the legal rule that the Good Idea Lightbulb Company can choose the production level, but the company must compensate the citizens for the pollution damages they incur. Describe how the company will choose its production level and discuss the efficiency of the expected final outcome.

6. The moon is a public good. Advertising agencies have looked into projecting logos on the moon. Suppose Oksana and Ted are the only two people who care about the appearance of the moon, and they prefer no logos. Oksana's demand curve for logo-free days each year is a straight line starting at $365 with a slope of -1. Ted's demand curve for logo-free days is a straight line starting at $182.50 with a slope of $-1/2$. Suppose the advertisers are entitled to project logos, but citizens can purchase logo-free days for a price of $200 each, which equals the advertisers' benefit per day of showing logos.

   a) *Draw Oksana's and Ted's demand curves and the market demand curve on a graph with the quantity of logo-free days measured on the horizontal axis.*

   b) *Add the price line to your graph and label the socially optimal quantity of logo-free days Q_s.*

   c) *Suppose Ted is a free rider and Oksana is not. Label the quantity of logo-free days that Oksana would buy Q_O.*

7. The only two grocery stores in FoodVille, Chowarama and

FoodieFood, are remodeling their retail spaces. Each store must decide whether to allocate space for an organic foods section without knowing whether the other store will do the same. If there are no organic options, each store will earn $6 million in profit next year. Many customers prefer organic food, so if one store has organic food and the other does not, the store with organic food will draw many more customers and earn $9 million in profit next year while the one without organic food will earn $4 million in profit. If both stores have organic food, neither will gain a competitive advantage despite expenditures on added space, so each store will earn $5 million in profit next year.

a) Draw a payoff matrix for this situation (like the ones in the chapter).

b) Identify the dominant strategy for each store.

8. For each of the following externalities, explain the underlying source of inefficiency (that is, explain what effects are felt beyond the decision-maker) and suggest a solution that could bring the activity to its efficient level:

a) Billboards along the highway detract from natural beauty.

b) Vaccinations received by domestic animals help wild animals avoid disease because they won't catch the disease from those vaccinated.

c) The use of email communications as a substitute for letters reduces the number of trees harvested for paper production.

9. Identify two barriers to successful Coasian bargaining in the real world.

10. Suppose the 25 households in an economy could each rent their own windmill for $3,200 that provides $3,000 worth of energy to the household, making the net cost to each household $200 per year. Each windmill would prevent $300 worth of pollution damage from fossil fuel energy production each year. The pollution is distributed evenly in the economy, so each of the 25 households feels $1/25 \times \$300 = \$12$ worth of the harm that could be avoided by each windmill annually.

a) Use a comparison of specific numbers to explain why households would not independently make the socially optimal decision regarding renting a windmill.

b) Suppose the households vote on whether every household should be required to rent a windmill. Specify the dollar amount of the yearly cost and benefit to each household of enacting the windmill requirement.

c) How should each household vote if they only care about themselves?

websurfer's challenge

1. Find a website that describes legislation designed to curtail one of the sources of market failure.

2. Find a website that speaks in favor of free-market policies.

3. Find a website that speaks against free-market policies.

Key Terms

Average total cost
Club goods
Coase Theorem
Deadweight loss
Dominant strategy
Efficient voter rule
Externalities
Free riders
Imperfect competition
Imperfect information
Marginal external benefit
Marginal external cost
Market failure

Natural monopolies
Negative externalities
Open-access goods
Pigou tax
Positive externalities
Prisoner's dilemma
Private goods
Private marginal benefit
Private marginal cost
Public goods
Social marginal benefit
Social marginal cost
Tragedy of the commons

Internet Resources

Video: Solutions to Externalities:
https://youtu.be/hKjhOxEuG5g

Video: Externalities and Deadweight Loss:
https://youtu.be/wcEIHs4daYo

Further Reading

Anderson, David A. "Environmental Exigencies and the Efficient Voter Rule." *Economies 8*, no. *4* (2020): 100. https://doi.org/10.3390/economies8040100. An explanation of environmental applications of the efficient voting rule.

Cherry, Todd L., Steffen Kallbekken, and Stephan Kroll. "Accepting Market Failure: Cultural Worldviews and the Opposition to Corrective Environmental Policies." *Journal of Environmental Economics and Management 85* (September 2017): 193–204. An examination of why people sometimes oppose environmental policies that would improve their material outcomes.

Coase, Ronald H. "The Problem of Social Cost." *Journal of Law and Economics 3* (October 1960): 1–44. An influential defense of private remedies.

Collart, Alba J., and Matthew G. Interis. "Consumer Imperfect Information in the Market for Expired and Nearly Expired Foods and Implications for Reducing Food Waste." *Sustainability 10*, no. *11* (2018): 3835. https://doi.org/10.3390/su10113835. A discussion of market failure in the context of food waste caused by imperfect labelling information.

Hardin, Garrett. "The Tragedy of the Commons." *Science 162* (1968): 1243–1248. A brief article outlining Hardin's famous concept.

Pigou, Arthur C. *The Economics of Welfare*. London: Macmillan, 1932. Pigou's insightful argument for the use of taxes and subsidies to equate private and social costs and benefits.

A "free rider" problem? Without government, who would provide the bike lanes and roads?

The Role of Government

The movies typically portray government, like romance, as being either idyllic or disastrous. In life, both tend to be a blend of the essential and the absurd. Government's role in the efficient allocation of environmental resources depends on the effectiveness of the "free market" alternative. When successful, free-market mechanisms can allocate resources to those who value them the most, invite efficient production methods, and price goods to reflect a combination of value and scarcity. As a particular resource becomes relatively scarce, its price will rise, inducing greater efforts to supply the product or find viable replacements. This pricing mechanism goes on display every time rising gas prices motivate research into alternative energy, energy conservation, and new sources of oil.

Chapter 3 explained why free markets struggle to allocate resources efficiently in the presence of imperfect information, imperfect competition, externalities, or free riders. Some goods, like air and river water, have no formal markets to address their allocation. When markets are inefficient or non-existent, government can provide remedies. If mismanaged, government can also come with excessive bureaucracy, ineptitude, misguidance, and corruption.

Being the product of human will, government is potent and pliable. Citizens with courage and insight can shape government. In a democracy, if voters and leaders understand the potential virtues and blunders of government, they are better prepared to exercise collective authority and promote efficient environmental policy. For these reasons, this chapter explains the role of government. The chapters that follow address related issues of policy and procedure.

DOI: 10.4324/9781003428732-4

The U.S. Capital Building in Washington, D.C.

The Meaning and Purpose of Government

What Is Government?

Saint Augustine quotes a captured pirate as saying to Alexander the Great, "Because I [keep hostile possession of the sea] with a petty ship, I am called a robber, whilst thou who dost it with a great fleet art styled emperor." In perilous times, the primary distinction between a government and a gang can be size. Strength and acceptance also convey authority, and a **government** is a body with the authority to govern. Any type of government can go astray of environmental and social goals, although democratic elections provide incentives for even the most selfish leaders to address their constituents' needs. In modern times, environmentalism has become a political ideology in itself, embraced by political parties such as the Green Parties of Britain, Senegal, Australia, Tiawan, and the United States.

Political-economic systems parade in myriad forms. What follows is a thumbnail sketch of several of the most prominent models.

An **autocracy** is governed by one individual with unlimited power. North Korea under Kim Jong Un and Saudi Arabia under Mohammed bin Salman resemble autocracies, although modern dictators often lack the absolute power of a true autocrat like King Louis XIV of France. A **theocracy** is a government run by priests or clergy, such as Iran under Ayatollah Ali Khamenei. Theocracies have become more common with the growth of Islamic fundamentalism and more relevant to the West since the terrorist attacks of September 11, 2001.

A **monarchy** is ruled by a king, queen, emperor, or empress. In this century most European monarchies, such as the British monarchy, have limited power. Some monarchies in Africa, the Middle East, and Asia continue to hold absolute, usually inherited, power. A **plutocracy** is governed by the wealthy or by those primarily influenced by the wealthy. Efforts to avoid plutocratic tendencies in the United States include frequent proposals for campaign finance reform that include limits on large political donations and provide matching funds or tax credits for small donations.

Communism is a system designed to eliminate material inequities via collective ownership of property. Legislators from a single political party—the Communist Party—make production decisions and divide wealth for equal advantage among citizens. The pitfalls of communism include a lack of incentives for added effort, risk taking, and innovation. The critical role of the central government in allocating wealth and setting production quotas makes this scheme particularly vulnerable to corruption.

Socialism shares with communism the goal of fair distribution and the stumbling block of limited incentives. Major industries are collectively owned, with production decisions made by the government or by *worker cooperatives* whose priority is to serve workers and the community. Wages are determined by negotiations between trade unions and management, rather than being controlled by the government as under a communist system. Another difference is that a single political party does not rule the economy under socialism. Systems with strong elements of socialism exist in Canada, India, Tanzania, Sweden, and elsewhere.

Under **pluralism**, governmental decisions are based on negotiations among leaders from business, government, labor groups, and other constituencies. Power is in a state of flux, with no one group retaining a dominant position for long.

Populism describes a political movement purported to serve the interests of ordinary citizens rather than the established elite. The impetus for populism can be concerns over government systems, economic insecurity, and immorality among those in power. Populism encompasses a broad range of political ideologies, with leaders that include Italy's conservative prime minister Giorgia Meloni and Brazil's relatively liberal president Luiz Inácio Lula da Silva. Populist leaders are often accused of misrepresenting their commitment to ordinary citizens, playing on emotions or downplaying problems such as climate change. These shortcomings of populism influence environmental policy. For example, research by Tobias Böhmelt (2021) on over 200 leaders in 66 countries found that populist leadership has a strong negative influence on environmental quality.

Imperialism is a policy of expansion and domination by a nation's authority. The European imperialism of the fifteenth to nineteenth centuries involved notable territorial acquisition, which was largely dismantled in the twentieth century.

Classical liberalism is a political philosophy that embraces freedom from church and state authority, free-enterprise economics, and individual freedoms. It was the foundation of parliamentary democracy in Britain and the philosophy of choice among many founding fathers of the United States. Unlike classical liberals, **modern liberals** are sometimes characterized as favoring a more substantial role for government and some limited elements of socialism.

While communism places individuals second to society, **fascism** (also known as national socialism or Nazism) places individuals second to the state or race. In a rejection of democracy, fascists emphasize loyalty to a strong leader (Mussolini or Hitler, for example), national pride, and a collective view of society. The absence of checks and balances in fascist states and the incentives for leaders to abuse their power for national and personal gain are among the reasons such forms of government have infamous legacies. The relevant catch phrase may be, "Power corrupts, and absolute power corrupts absolutely."

A **democracy** is governed by the people or their elected representatives. Decisions are made by majority rule, which does not prevent social injustices but may make them less likely. Chile, the United States, and Japan are among the many successful democracies.

The economies in democracies tend to have strong elements of **capitalism**, under which private individuals own land and businesses and operate them in the pursuit of profit. In a capitalist system, markets determine the prices and quantities of goods and services. Wages are influenced by market forces and, in some cases, by negotiations between employees or their unions and management. The government may regulate businesses and provide tax-supported social benefits. Most modern economic systems are **mixed economies**, which operate under the combined influences of socialism, capitalism, and adherence to tradition.

Is Government Necessary?

Philosophers Thomas Hobbes and John Locke argued that government is essential. Hobbes (1946) wrote that

> *during the time men live without a common power to keep them all in awe, they are in that condition which is called war. . . .. In such a condition there is no place for industry; . . . no arts; no letters; no society; and which is worst of all, continual fear, and danger of violent death; and the life of man, solitary, poor, nasty, brutish, short.*

Karl Marx argued that government is simply an instrument of class domination and would "wither away" with the abolition of distinct classes under communism. *Authoritarians* question the practicality of self-government and look to centralized government for the advancement of society.

Conservatives express concern about government's efficiency in the role of environmental steward. They stress the possibility of *government failure*, meaning that the cost of intervention exceeds the benefit. Conservatives sometimes advocate privatized or decentralized decision-making, arguing that government officials are out of touch or that the process of government is unwieldy. *Libertarians* call for self-government, believing that government's role should be limited to protecting citizens from coercion and violence. *Anarchists* feel that all forms of government are oppressive and should be abolished. Anarchists have played a significant role in recent anti-globalization protests, which also focus on environmental issues.[1]

In the context of environmental economics, the call for government involvement ranges from a whisper to a shout, depending on the state of the nation in question. Factors that influence the need for government involvement in environmental protection include

- *Population density*

- *Religious and social culture*

- *Education*

- *Wealth*

- *Degree of industrialization*

- *Sensitivity of the existing ecosystem.*

In the early eighteenth century, 3 million people lived on the North American continent. With no industrialization, the collective environmental impact was negligible, and there was little need for government to resolve externalities. The same is true today in a decreasing number of isolated regions in the Arctic, Central America, and Africa.

French philosopher Jean-Jacques Rousseau argued that humans are essentially good but corrupted by society. A society's religious and cultural stance toward the environment can obviate or escalate the need for government intervention. For example, the teachings of Buddhism prohibit harm to all sentient (conscious) beings whether human or animal. As a result, there is little need for the likes of the Endangered Species Act among devout Buddhists. In other situations, government can play a critical role in shaping society. Through education and exemplary leaders, government itself can help foster the type of moral climate that makes its role as forceful defender less necessary.

---

1   For an anarchist's view of the link between environmentalism and anarchy, see www.spunk.org/library/intro/sp001695.html.

The environmental concerns associated with sensitive ecosystems and industrialization are clear. The influences of education and wealth are less certain. Education systems can disseminate information from anywhere along the spectrum of attitudes toward the environment. Wealth is a double-edged sword. Having more money facilitates overindulgence and waste. Yet wealth also makes expenditures on environmental protection relatively painless. For a typical family in Bangladesh, the price of an emissions-reducing catalytic fireplace represents a full year's income. For a wealthy family, that same price may seem negligible. Clean air and water are classified as *normal goods*, meaning that people with higher incomes are willing to spend more on them. A study that sorted out the effects of income and education on the generation of waste in the form of trash found that waste decreased with education but increased with income.[2] So support of education, and environmental curricula in particular, may be among the solutions for governments in countries facing rapid industrialization, population growth, or soaring incomes.

The Role of Government: Bit Part? Supporting Actor? Lead?

Historical Ideologies

Advocates of a *laissez-faire* or *free-market* approach have long trumpeted the virtues of a market unfettered by government intervention. In the words of the third U.S. president, Thomas Jefferson, "That government is best that governs least." A contemporary of Jefferson's and a fellow classical liberal, Scottish economist Adam Smith, thought that a free market driven by the actions of self-interested individuals would regulate and correct itself. Classical conservatives including Irish economist Edmund Burke felt that classical liberals placed too much faith in human rationality. They argued that people are prone to bouts of unreasonable behavior, irrational passions, and immorality. The classical conservatives favored adherence to the institutions of government and the church as well as to societal traditions and standards to avoid the chaotic results of freewheeling human impulses.

The nineteenth century brought with it monopolies, deception of consumers, gross economic inequities, and abuse of the environment. Modern liberals, like classical conservatives, became leery of the

---

2 See Anderson (2005). Although waste levels appear to increase with income levels, as discussed in Chapter 9, the environmental Kuznets curve suggests that pollution might decrease with GDP when per capita income exceeds about $13,000.

Thomas Jefferson, the third president of the United States, stated "that government is best that governs least." But abuses of market power in the nineteenth century led to calls for greater government involvement in the marketplace.

laissez-faire approach. They felt that without the influence of regulations and liability, firms did not perform adequate testing to determine the health and environmental safety of their products. Under these conditions, pertinent information held by firms was less likely to be passed on to customers. Led by British philosopher Thomas Hill Green in the late-nineteenth century, and the likes of U.S. presidents Woodrow Wilson and Franklin D. Roosevelt in the early twentieth century, modern liberals championed the causes of labor, education, the environment, and the freedoms of speech and press. With these efforts, the U.S. government received significant new roles in tempering market power, promoting transparency, and combating externalities and free rider problems. The role of government progressed similarly in other industrialized nations during the nineteenth and twentieth centuries. For example, Britain passed the Smoke Nuisance Abatement Act in 1853, the Rivers Pollution Prevention Act in 1876, the Protection of Birds Act in 1954, and the Clean Air Act in 1956.

Chapter 3 explained problems with private solutions to externality problems. Citizens in countries around the world now look to various forms of government to solve environmental problems. Students bothered by noise pollution in their residence hall might look to their resident adviser or student government representative—the local government authority—to set out new hall policies about quiet times. National governments tax, fine, or regulate sources of negative externalities ranging from unleashed dogs to hazardous waste while subsidizing sources of positive externalities such as geothermal heating systems and reforestation. The next section explains these and other public approaches to market failure.

Governmental Solutions to Market Failure

Enforcement of Property Rights

In the selfishness that is human nature, and with notable exceptions, humans tend to take better care of what they own than what they do not own. People are unlikely to litter in their own yards, dump toxins in their own ponds, or overhunt their own forests. Of course, all of the above occur with open-access resources such as publicly held lands. American ecologist Garrett Hardin used the phrase **tragedy of the commons** to describe the overexploitation of open-access resources. His inspiration was herdsmen he observed allowing their cattle to overgraze common pastures because the burden was shared by many.

To understand the tragedy of the commons more clearly, consider harvesters of wild blackberry leaves for tea. For simplicity, suppose each blackberry plant has two leaves. If the plant retains at least one leaf, it will survive and grow a second leaf in the following season. Figure 4.1 illustrates a publicly owned tract with four herbal tea collectors and four blackberry plants. If the pickers want to maximize their harvests over the foreseeable future, how many leaves should the tea collectors pick? If they are subsequently moving on to another region and will never pick in this blackberry patch again, they should pick every leaf because the loss of plants is felt only by others. Similarly, if it is likely that other pickers will pick any remaining second leaves off the plants, it is in each picker's best interest to pick as many leaves as possible. In that case, the loss of the plants is inevitable, and leaves that are left by one picker merely end up in the harvest of another picker. Only if all the leaf pickers want to return to that location to receive benefits from future harvests, *and* if each party can trust each of the other parties to pursue a plan of leaving one leaf per plant, is it selfishly rational to leave some leaves unpicked.

When open access prevents users from *internalizing*, or feeling for themselves, the full costs of their behavior, one solution is the extension and enforcement of private property rights. Figure 4.2 presents the blackberry leaf scenario with the modification that property lines

With no property rights, tea-leaf harvesters might not internalize the effects of their behavior. If a picker will not be returning to the same patch again, or if other pickers are likely to pick leaves that are left behind, the picker has no incentive to conserve the resource. The resulting excessive leaf collection is one example of what Hardin called the tragedy of the commons.

Figure 4.1
Harvesting on Open-Access Property

Figure 4.2
Harvesting on Private Property

If each tea-leaf picker owns a particular patch of land, the effects of harvest behavior are internalized. To over-harvest is to decrease the value of the picker's own resource. Pickers thus have an incentive to harvest only the efficient quantity of leaves. Similarly, property rights convey incentives for the efficient treatment of fisheries, forests, oil pools, and dumpsites, among other resources.

have been established to divide the land among the tea harvesters. With each picker owning a blackberry plant, the entire loss from picking both leaves is felt by the owner. Even if the picker will be moving on before the next harvest, the value of the resources left behind will be reflected in the selling price the picker can obtain for the property, and thus the losses will be internalized. Like this blackberry patch, ocean fisheries, stocks of hunted game, and subsurface pools of oil owned by multiple parties are vulnerable to exploitation when property rights are not well-defined and enforced.

The transferability of property rights is critical to efficiency because it allows property to be owned by those who can make the best use of it. In the absence of market failure and its miscues, the market will allocate property to those who value it the most, be they farmers, developers, or conservation groups such as the Nature Conservancy. Efficient transfers are impeded by incomplete information about property ownership, boundaries, and value, and by all sources of transaction costs as discussed in Chapter 3. Governments can facilitate the transfer of property rights by keeping accurate records of exactly what is owned by whom and how much was paid for it. Governments can also provide courts for the civilized resolution of boundary disputes and other conflicts that arise.

Provision of Public Goods

Beyond defining and protecting property rights, a government can allocate goods and services that would be over- or underproduced by private

Where the Buffalo Roamed

Property rights change behavior, sometimes in favor of wildlife living on the property. Few people would treat their own property the way they treat open-access areas. Trees accessible to loggers in national parks are more likely to become lumber than trees  in our own backyards. Animals, too, can benefit from property rights that apply to them and the soil they tread upon.

Before property rights were clearly defined in the western United States, buffalo were open-access resources. No one had a personal stake in their survival, and buffalo were slaughtered by the thousands—sometimes purely for sport by riflemen riding passenger trains. Then the Homestead Act of 1862 privatized much of the land west of the Mississippi River. This act divided the land into 160-acre tracts that any citizen could claim. If the homesteaders "improved" the property with a dwelling, grew crops, and remained for at least 5 years, the tract became their property free and clear. The government then proceeded to support private property ownership with the provision of courts to help resolve land disputes, laws to protect property values, and police to deter trespassers and enforce the findings of the courts. With the enforcement of property rights came private (and public) pollution cleanups, reforestation, and a resurgence of buffalo in the West.

Today it would be impossible to hunt buffalo without violating property rights. Fences prevent buffalo from migrating the way they used to, but the fences also assure protection by the private ranchers who have a personal stake in the natural assets on their property. The ranchers may still turn their stakes into steaks, but they do so at a sustainable rate.

For a detailed examination of the relationship between buffalo (bison) and property rights, see Lueck (2018).

markets. Chapter 3 explained that public goods such as national defense and streetlights would be underprovided without government assistance because "free riders" would misrepresent their interests and seek benefits from the purchases of others. There are many public goods in the realm of the environment and natural resources. Society as a whole benefits when hazardous waste sites are cleaned up, government-sponsored research addresses climate change, or animal species are protected. The benefits from these activities are nonrival and nonexcludable, in that one's enjoyment of a healthy and biologically diverse environment does not preclude other people's enjoyment, and no one can exclude others from this enjoyment. We've already seen that investments in public goods would be inadequate without intervention. Government's role includes the provision of public goods, the collection of taxes to pay for them, and the enactment and enforcement of regulations to sustain them.

Taxes and Subsidies

Chapters 1 and 3 discussed externalities as a source of market failure. Gasoline is a source of negative externalities because its combustion creates marginal external costs to health and the environment apart from the private marginal costs of producing gasoline. Figure 4.3 depicts a competitive market for gasoline. For simplicity, assume the marginal external cost is constant at $1 per gallon (actual estimates exceed $2 per gallon[3]). There are no marginal external benefits, so the social marginal benefit curve (MB_{social}) is also the private marginal benefit curve ($MB_{private}$), which represents market demand as explained in Chapter 2.

The $1 per gallon marginal external cost separates the social marginal cost curve (MC_{social}) and the private marginal cost curve ($MC_{private}$), which also represents market supply. Without government intervention, equilibrium occurs at point A, with 1,000 gallons of gasoline produced and consumed at a price of $3.75 per gallon. Note that the social marginal cost exceeds the social marginal benefit for every gallon in excess of the socially optimal quantity of 850, resulting in deadweight loss as shown by the yellow triangle. Although the social marginal cost would continue to exceed the social marginal benefit for units beyond 1,000, the deadweight loss triangle ends at a quantity of 1,000 because costs and benefits do not exist for units that are not produced or consumed.

One remedy for this case of overindulgence would be a tax equal to the marginal external cost of gasoline. This would constitute a Pigou tax as discussed in Chapter 3. A $1 per gallon tax collected from the sellers of gasoline would cause sellers to pay the external cost of gasoline and thereby equate the social marginal cost with the private marginal cost. In Figure 4.3, this means that the red private marginal cost curve would shift up by $1 to coincide with the purple social marginal cost curve. The equilibrium between the new private (and social) marginal cost and the private (and social) marginal benefit would occur at point B, and output would be at the efficient level of 850 gallons.

---

3 For example, see Resources for the Future's 2007 report, "Automobile Externalities and Policies," downloadable at https://media.rff.org/documents/RFF-DP-06-26-REV.pdf.

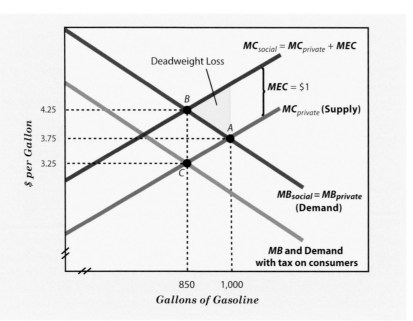

Figure 4.3

Negative Externalities and Pigouvian Taxes

The pre-tax equilibrium in this market occurs at point *A*, with 1,000 gallons of gasoline selling for $3.75 per gallon. Due to a negative externality of $1 per gallon, the social marginal cost of the last 250 gallons sold exceeds the marginal benefit. To correct for this inefficiency, a Pigovian tax equal to the $1 marginal external cost could be imposed on gasoline sales. This would lead consumers to internalize the full cost of their behavior and choose the efficient level of consumption—850 gallons per period. The post-tax price to consumers would be $4.25; sellers would receive $3.25 per gallon.

The same result would occur if a $1 per gallon Pigouvian tax were imposed on the consumers of gasoline. In this event, consumers would internalize the external cost. The market demand curve would shift *down* by $1, as illustrated by the downward-sloping light blue line. Why? Because the new private marginal benefit would be the original private marginal benefit minus the $1 tax payment necessary for each gallon. The new equilibrium between the private marginal cost and the post-tax private marginal benefit would occur at point *C*.

This example illustrates why, regardless of whether the tax is imposed on the buyers or the sellers, the equilibrium quantity is 850, the buyers' total payment is $4.25 per gallon, and the sellers' post-tax revenue is $3.25 per gallon. The division of the tax burden depends on the shapes of the supply and demand curves and not on which party actually pays the tax. This is valuable information for policymakers, who can collect the tax from whichever party is easiest to collect from, knowing that the outcomes will be the same either way.

The pre-tax equilibrium in this market occurs at point *A*, with 1,000 gallons of gasoline selling for $3.75 per gallon. Due to a negative externality of $1 per gallon, the social marginal cost of the last 250 gallons sold exceeds the social marginal benefit. To avoid the resulting inefficiency and deadweight loss, the government could impose a Pigouvian tax equal to the $1 marginal external cost per gallon. Depending on who pays the tax, this would cause either buyers or sellers to internalize the external cost of gasoline and lead to the efficient level of consumption −850 gallons per period.

Government can also address positive externalities. Consider the market for solar panels. In addition to the private benefits from solar panels in terms of energy provision, they provide positive environmental externalities by decreasing the need for fossil fuel extraction and combustion. These benefits go largely to society and are not fully internalized by those deciding how many solar panels to purchase. Figure 4.4 illustrates a competitive market for solar panels. A $100 marginal external benefit per panel separates the social marginal benefit and the private marginal benefit. Supposing no marginal external costs are involved for simplicity, the social marginal cost curve coincides with the private marginal cost curve, which is also the market supply curve.

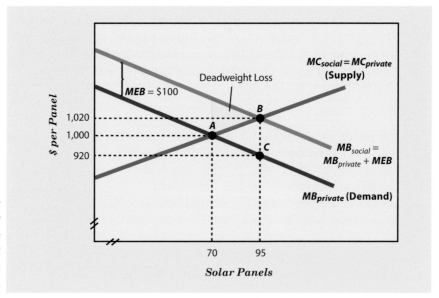

Figure 4.4
Positive
Externalities
and
Government
Subsidies

Without intervention, 70 solar panels would be purchased in this market for $1,000 each. The positive externalities associated with solar panels cause the social marginal benefit to exceed the private marginal benefit by $100 per panel. Neglecting the externalities, consumers purchase fewer than the socially efficient quantity of 95 panels. The government could eliminate the inefficiency and deadweight loss with a subsidy equal to the marginal external benefit per panel. The subsidy would cause its recipients, whether consumers or producers, to internalize the externalities and bring about the efficient quantity of panels.

With no intervention, equilibrium occurs at point A, with 70 panels purchased for $1,000 each. At this level of consumption, the red marginal social cost curve is below the light blue marginal social benefit curve, meaning the installation of more panels would provide benefits that exceed their costs. The yellow deadweight loss triangle shows the forgone net gains that could be achieved by increasing output to the socially optimal quantity of 95 panels.

If consumers received a subsidy for each panel equal to the $100 value of the positive externality, they would internalize the benefits panels provide to society. The private marginal benefit curve would shift up by $100 to coincide with the social marginal benefit curve, and equilibrium would occur at point B. With that subsidy in place, consumers would purchase the socially optimal quantity of 95 panels. The same outcome would result if producers received a $100 subsidy for each panel: The subsidy would shift the marginal cost curve down by $100, causing the marginal cost curve to intersect the demand curve at point C, again at the efficient quantity of 95 panels.

The governments of many nations including Germany, China, and Serbia subsidize solar panels as well as research on alternative fuels, public transportation, and related sources of positive environmental externalities. Upcoming chapters discuss various policy mechanisms

Solar panels and other products that provide positive externalities are under-consumed in a free market. Properly conceived government subsidies can help remedy the resulting misallocation of resources.

by which governments can implement taxes and subsidies, including emission charges for pollution, deposit-refund systems for bottles, tax credits for electric vehicles, and equipment subsidies for emission-abatement equipment.

Liability

Governments provide the systems that create, interpret, and enforce the law. With roughly 18 million new civil cases filed each year in the United States, litigation adds to the risk and expense of harming people or the environment. Lawsuits over how cleanup costs will be shared have added more than $10 billion to expenditures on hazardous waste sites administered by the federal government's *Superfund* program.[4] At issue is the optimal level of litigation. The open-access nature of the court system may invite overuse. Does the current threat of litigation cause more harm than good? Critics claim excessive litigation destroys international competitiveness, hinders innovation, and creeps into the prices we pay for most goods and services. Advocates argue that the threat and cost of litigation are the prices we must pay to promote responsible behavior.

Private wrongs called *torts*, such as someone polluting their neighbor's pond, are a primary source of civil litigation. *Mass toxic torts* involve large-scale exposure to dangerous substances and can generate a tremendous volume of litigation. For example, health problems associated with asbestos exposure generated legal claims from an estimated 340,000 parties. BP paid $20.8 billion to settle lawsuits over the Deepwater Horizon oil spill in the Gulf of Mexico. And water contamination at Marine Corps Base Camp Lejeune in Jacksonville, North Carolina, led to over 70,000 legal claims.

Litigation can inhibit behavior that unduly jeopardizes humans or the environment. Suppose a mining company is deciding whether to drill into a mountain to look for coal at a cost of $2 million. The company estimates a 1-in-3 chance of finding $6.9 million worth of coal. There is a 2-in-3 chance that no coal will be found, and there is a 1-in-10,000 chance that the drilling will cause $6 billion worth of environmental damage that lawsuits would force the company to pay for. As explained in Chapter 2, we find the company's expected cost and expected benefit by multiplying the probability of each possible outcome by the value of that outcome:

$$\text{Expected benefit} = \frac{1}{3}(\$6,900,000) + \frac{2}{3}(0) = \$2.3 \text{ million}$$

$$\text{Expected cost} = \$2,000,000 + \frac{1}{10,000}(\$6,000,000,000) = \$2.6 \text{ million.}$$

---

4 See www.epa.gov/superfund/.

Although the $2.3 million expected benefit exceeds the $2 million drilling cost, the $600,000 expected litigation cost forces the company to internalize the risk of environmental damage and disincentivizes inefficient drilling unless the company has a preference for risk taking.

Attitudes toward risk influence the size of litigation's threat. A firm is *risk neutral* in this context if it cares only about the expected value of its litigation cost and not about the range of possible costs, which in this case is from zero to $6 billion. A firm is *risk averse* if the mere uncertainty involved with litigation imposes a burden on the firm beyond the expected litigation cost. A risk-averse firm is sufficiently troubled by the possibility of a large expense that it would pay extra to know for sure that the actual cost would equal the expected cost. That certainty about the cost would eliminate stress and the need to prepare for the worst possible litigation outcome, such as paying $6 billion in damages.

The harm from uncertainty about an outcome is called the **risk burden**. The value of the risk burden for a firm facing litigation is the largest amount the firm would be willing to pay *in addition to the expected cost* to avoid uncertainty about the outcome. Consider a firm with a 90 percent chance of avoiding costly litigation and a 10 percent chance of having to pay a $1 million damage award at trial. This firm faces an expected cost of $(0.9 \times \$0) + (0.1 \times \$1 \text{ million}) = \$100,000$. If the firm would be willing to pay $110,000, $10,000 more than the expected cost, to avoid the 10 percent chance of having to pay $1 million, the risk burden for the firm is $10,000. The firm would be willing to pay up to $110,000 for insurance that covered the damage award, if there is one, thereby removing all uncertainty for the firm. Companies selling insurance for fire, theft, illness, and other potential threats are able to earn profits only because risk-averse customers are willing to pay more than the expected cost to avoid uncertainty about the relevant outcomes.

Risk burdens and excessive damage awards can impede the efficiency of decisions to innovate. Developers of new products and processes not known to be completely safe, including medicinal cures and alternate energies, may shun beneficial investments in new products to avoid the risk of costly lawsuits. Suppose a firm is pondering a $10 million expenditure on next-generation lithium-ion batteries for electric cars. The firm estimates a 50 percent chance that the new battery will provide $24 million in revenue for the firm and a 50 chance that the battery will fail and provide no revenue. There is also a 1-in-100 chance that the project will cause $100 million in environmental damage that the firm will end up paying due to litigation. The expected benefit is $(0.5 \times \$24 \text{ million}) + (0.5 \times 0) = \$12 \text{ million}$. Not including the risk burden, the expected cost is $\$10 \text{ million} + (0.001 \times \$100 \text{ million}) = \$11 \text{ million}$, so the project passes a cost-benefit test. However, if the risk burden exceeds

$1 million, the expected cost exceeds the $12 million expected benefit, and the firm will not develop the battery. The project will be scrapped even without a risk burden if the penalty for causing $100 million in damage would exceed $200 million, because that, too, would bring the expected cost above $12 million.

Like markets and governments, litigation is vulnerable to many types of inefficiency. The determination of appropriate damage awards requires estimates of the likelihood of harm, the likelihood of identifying the offending firms and successfully litigating against them, the dollar value of the damages, the firms' attitudes toward risk taking, and their associated risk burdens. The possibilities for frivolous litigation and claims against the wrong parties further cloud the quest for efficient incentives from litigation. For those firms whose levels of precaution would be appropriate *without* the added incentives that litigation provides, the threat of litigation causes excessive precaution, as addressed in Chapter 12.

Regulations

Some experts advocate regulations as an alternative to liability. For example, economist W. Kip Viscusi suggests that the vast uncertainties of litigation weaken its deterrent effect and that the best solution to mass toxic torts and related problems is the regulation of risks. Regulations could set standards for environmental safety that are *exculpatory*, meaning that those who comply with designated safety rules would be spared from liability for the associated outcomes. There is no limit to the standard of precaution that could be required under such a policy, and by meeting that standard, firms could avoid the risk of paying highly uncertain damage awards.

Looking again at the example of fuel consumption, instead of taxing gasoline, there are several regulatory options that could bring gasoline consumption closer to an efficient level. Iceland has the lowest highway speed limit in Europe, 90 kilometers per hour, which reduces fuel consumption. The U.S. Congress has debated energy bills that would reduce externalities by requiring cleaner-burning fuels, alternative energy sources, and lower-emission engines. Regulations could also limit gasoline consumption or emission levels per firm or household. Another regulatory option is to promote alternative forms of transportation. For example, requirements for bike lanes or sidewalks along commuter routes would encourage biking and walking and thereby reduce gasoline consumption. Currently, while cities such as London have bicycle superhighways with safe, direct routes between homes and jobs, workers in many other places don't even have a paved shoulder of a road to bike on.

Education and Moral Leadership

Education and moral leadership can improve citizens' decision-making even without taxes, laws, or regulations. Government authorities set examples with their actions. It sends a signal when a leader

- *makes a major city's most congested streets bicycle friendly, as Paris Mayor Anne Hidalgo did*

- *earns the United Nations Environmental Program's Champion of the Earth award, as Barbados's Prime Minister Mia Mottley did*

- *makes his city the first in the world to comply with the Global Covenant of Mayors for Climate & Energy, as Rio De Janeiro mayor Eduardo Paes did*

- *implores every citizen to actively promote sustainable development, as Singapore's Prime Minister Lee Hsien Loong did.*

When officials call for their own thermostats to be set at a moderate level, their cars to be electric, or their paper to be recycled, this sets a tone for the country and invigorates efforts on the environmental front.

Targeted educational efforts can encourage people to think creatively about solutions to environmental problems. These college students are visiting the Natural Energy Lab of Hawai'i, which implements innovative strategies for energy generation and conservation.

Governments also provide schooling. Education is a means of affecting the skills and attitudes of society and promoting the efficient use of environmental resources. Out of concern for India's troubled environment, the country introduced environmental education into all levels of education in 2004. Since the energy crises of the 1970s, public school curricula in the United States have included units on how to conserve fossil fuels. More generally, education can help more people be aware of the costs and benefits of their actions and think creatively about solutions to environmental threats and resource scarcity.

Dispute Resolution

A government can work to limit the expense and uncertainty of dispute resolution when there is disagreement over the use of a forest, the interpretation of environmental regulations, or the liability for hazardous waste cleanup. Civilized dispute resolution is less likely in the absence of government enforcement mechanisms. Even if courts or other authorities can make rulings in such a situation, decisions have little influence without the backing of government institutions for monitoring and policing.

International conflicts reveal the dangers of limited authority. With no global support system, the decisions of international courts involving, for example, whale hunting, are often ignored with few repercussions. As discussed in Chapter 11, the relative impotence of international law is unfortunate given the global scale of many environmental problems. Chapter 15 explains more of the sources and repercussions of environmental dispute resolution, as well as a variety of solutions.

Environmental Legislation

Consider the United States as a case study for environmental legislation. The Environmental Protection Agency (EPA) is the backbone of the U.S. government's efforts to protect the environment. The EPA was established in 1970, the same year as the first Earth Day and the beginning of an unprecedented decade of environment legislation summarized in the Reality Check. The environmental initiatives of that period included restrictions on lead-based paint, a ban on the pesticide DDT, phaseouts of leaded gasoline and the persistent organic pollutant PCB, and a phase-in of fuel economy standards for cars.

The momentum for these breakthroughs began mounting in 1962 after the publication of Rachel Carson's best-selling book *Silent Spring*. The book warned of the demise of songbirds due to organic phosphate insecticides, which also contaminate the human food supply. Calls for new environmental policy were rekindled by uproar over the use of an

Major U.S. Environmental Laws

National Environmental Policy Act *42 USC s/s 4321 et seq. (1969)*
This Act requires all branches of government to file environmental assessments and environmental impact statements prior to taking actions that could significantly affect the environment. For example, these reports precede the construction of highways, airports, and military bases.

Occupational Safety and Health Act *29 USC 651 et seq. (1970)*
This legislation is intended to secure workplace safety for employees. The act's standards limit exposure to toxic materials, excessive noise levels, extreme temperatures, unsanitary conditions, and mechanical dangers. OSHA also created the National Institute for Occupational Safety and Health to conduct research on employment hazards.

Clean Air Act *42 USC s/s 7401 et seq. (1970)*
This Act authorizes the EPA to establish national ambient air quality standards (NAAQS) for every state and direct the states to develop state implementation plans applicable to industrial pollution sources within each state. The act was amended in 1990 to better address associated problems, including acid rain, airborne toxins, ground-level ozone, and stratospheric ozone depletion.

Federal Insecticide, Fungicide, and Rodenticide Act *7 USC s/s 136 et seq. (1972)*
FIFRA gives the EPA authority to license pesticides, register major users (farmers, exterminators, and others), and study the effects of pesticides on health and the environment. Major users must take certification exams, and products must carry sufficient hazard warning labels. If used properly, the products cannot cause unreasonable harm to the environment.

Endangered Species Act *7 USC 136; 16 USC 460 et seq. (1973)*
This Act protects threatened and endangered plants and animals by prohibiting their elimination, import, export, or interstate sale. The law prohibits any action that results in the loss of a listed species or its habitat.

(continued)

Safe Drinking Water Act *42 USC s/s 300f et seq. (1974)*
This legislation authorizes the EPA to establish purity standards for water that is potentially designed for human consumption, regardless of the source.

Resource Conservation and Recovery Act
42 USC s/s 6901 et seq. (1976)
RCRA ("rick-rah") authorizes the EPA to oversee hazardous waste generation, transportation, treatment, storage, and disposal. This Act does not cover abandoned hazardous waste, which is addressed by the Superfund under CERCLA (see below).

Toxic Substances Control Act *15 USC s/s 2601 et seq. (1976)*
The TSCA enables the EPA to track and require testing of some 75,000 industrial chemicals currently in use. The EPA can prohibit the manufacture and import of chemicals that pose an unreasonable risk to health or the environment.

Clean Water Act *33 USC s/s 1251 et seq. (1977)*
The CWA is a 1977 amendment to the Federal Water Pollution Control Act, originated in 1948. Under the CWA, the EPA can set effluent standards for industries on the basis of available technology and set surface water quality standards for all contaminants. The act makes it illegal to discharge any pollutant into navigable waters without a permit.

Comprehensive Environmental Response, Compensation, and Liability Act *42 USC s/s 9601 et seq. (1980)*, reauthorized by the **Superfund Amendments and Reauthorization Act** *42 USC 9601 et seq. (1986)*. CERCLA ("sir-cla") created the federal Superfund, intended for the cleanup of abandoned, uncontrolled, and emergency releases of hazardous materials into the environment. The acts also empower the EPA to seek those responsible for the releases and elicit their cooperation in the cleanup.

Emergency Planning and Community Right-to-Know Act *42 USC 11001 et seq. (1986)*
With Title III of the Superfund Amendments and Reauthorization Act, Congress enacted this legislation to help protect communities from chemical hazards. The act requires each state to name a state emergency response commission that divides the state into emergency

(continued)

planning districts and names a local emergency planning committee for each district.

Marine Protection, Research, and Sanctuaries Act (Ocean Dumping Act)
16 USC § 1431 et seq. and 33 USC §1401 et seq. (1988)
This Act prohibits ocean dumping (1) in the U.S. territorial sea by anyone; (2) of material from the United States anywhere; and (3) of material from anywhere by U.S. agencies or vessels, without a permit. The decision standard for permit requests is whether the dumping will "unreasonably degrade or endanger" human health, welfare, or the marine environment.

Oil Pollution Act of 1990 *33 USC 2702 to 2761*
As the first line of defense against oil spills, this Act requires operators of oil storage facilities and vessels to submit oil-spill response plans to the federal government. Oil-spill contingency plans are required for vulnerable areas. The EPA publishes guidelines for above-ground oil storage facilities, and the Coast Guard does the same for tankers. The Act also provides that a tax on oil will finance a fund available to clean up spills when the responsible party will not or cannot do so.

Pollution Prevention Act *42 USC 13101 and 13102, s/s et seq. (1990)*
This Act addresses concern that industries lack the information, technology, or focus to reduce pollution at the source, where it may be more easily addressed than at the stages of treatment and disposal. In an effort to promote cost-effective changes in production, operation, and raw materials use, the EPA is authorized to establish a source-reduction program that collects and disseminates information and provides financial assistance to the states.

Energy Policy Act *42 USC §13201 et seq. (2005)*
This Act addresses energy efficiency, renewable energy, oil and gas, coal, tribal energy, nuclear issues, vehicles and motor fuels, hydrogen, electricity, energy tax incentives, hydropower, geothermal energy, and climate change technology. The Act provides loan guarantees for entities that develop or adopt new technologies that reduce greenhouse gases emissions. The Act also increases the amount of biofuel that must be mixed with gasoline sold in the United States.

(continued)

Energy Independence and Security Act *42 USC 17001 (2007)*
The purpose of this legislation was to promote energy independence and security in the United States by increasing the production of clean renewable fuels, protecting consumers, and improving the efficiency of products, buildings, and vehicles. The Act also advanced the capture of greenhouse gases and improved the energy efficiency of the Federal Government.

Inflation Reduction Act *42 PUB. L. 117–169 (2022)*
This Act provides the largest ever U.S. investment to fight climate change. The law includes tax credits for electric vehicles, solar panels, geothermal heating, and energy-efficient appliances. Collectively, the spending on green energy and conservation may reduce greenhouse gas emissions by 40 percent below 2005 levels by 2030.

herbicidal chemical called Agent Orange to destroy forest coverage and food crops during the Vietnam War. Intense public interest in the environment took President Richard Nixon's attention away from the war long enough for him to sign the National Environmental Policy Act on the first day of 1970. Later that year, he called for a strong, independent agency to establish and enforce environmental protection standards, conduct environmental research, and provide assistance to those working to improve the environment. On December 1, 1970, William "The Enforcer" Ruckelshaus received Senate confirmation as the first EPA administrator.

The U.S. government pooled most of the responsibility for protecting the environment under the auspices of the EPA. This included programs previously controlled by the Department of Health, Education and Welfare, the Department of the Interior, and the Food and Drug Administration. The EPA works to support the approaches to environmental inefficiencies explained in the previous section. Among its other activities, the EPA establishes and enforces environmental regulations, participates in lawsuits to establish liability for hazardous waste cleanups, and sponsors research and consumer education to provide information critical in a market system in which consumers make most of their own decisions about product use.

Similar agencies enjoy varying degrees of success in most developed nations. These include the aptly named Environment Canada, Environmental Agency of England and Wales, Abu Dhabi Environmental Agency, and Environment Australia. The emergence of these agencies over the past five decades is symbolic of government's growing role in the protection and allocation of natural resources.

In the United States, as is typical elsewhere, several departments that report to the President share responsibility for policies and research on the environment:

- The **Department of Energy** operates the Office of Energy Efficiency and Renewable Energy, which funds research into alternative fuels.[5]

- The **Department of the Interior** oversees the Office of Surface Mining, the Bureau of Land Management, the U.S. Fish and Wildlife Service, and the National Park Service among other agencies concerned with the environment.[6]

- The **Department of Agriculture** operates the Forest Service and the Natural Resource Conservation Service, which in turn oversees the Wetland Reserve programs, the National Resources Inventory, and Backyard Conservation programs.[7]

- The **Department of Health and Human Services** oversees the Agency for Toxic Substances and Disease Registry and the National Institute of Environmental Health Sciences, which studies the effects of pollution.[8]

- The **Department of Commerce** houses the Office of Environmental Technologies Industries and the National Oceanic and Atmospheric Administration, a sponsor of research on oceans, fisheries, and climate change.[9]

- The **Department of Housing and Urban Development** handles brownfield cleanup and recovery.[10]

This broad oversight of the environmental agenda creates broad opportunities for the government to apply economic tools to environmental problems.

Summary

Modern conservatives would like a smaller government. Modern liberals would like it refocused. Communists think it will wither away. Anarchists want to do away with it. But government seems to be here to stay. While laissez-faire approaches have their time and place, market failure is a persistent nemesis in modern times, particularly in

5 See www.energy.gov/.

6 See www.doi.gov/.

7 See www.usda.gov/.

8 See www.dhhs.gov/.

9 See www.noaa.gov/.

10 See www.hudexchange.info/programs/bedi/.

developed places. Government brings tools to the table with which to restrain market failure, including opportunities for

- *Pigouvian taxes and subsidies*
- *Legal remedies for torts*
- *Regulations*
- *Education*
- *Moral leadership*
- *Dispute-resolution mechanisms*
- *The protection of property rights*
- *The provision of public goods*

Government is a malleable human construct. By popular vote, brute force, or shift in public sentiment, the governments of the world are molded—for better or worse—by human desires. We the people can influence governments, and governments can influence the environment. So to contemplate the appropriate role of government is to prepare for some of our greatest opportunities to effect environmental change.

• • • • • • • • • • • • • • • • • • •

Problems for Review

1. Under capitalism, socialism, and communism, most productive resources are owned, respectively, by private individuals, the government, and the Communist Party. Which of these political-economic systems is the least vulnerable to the tragedy of the commons? Explain.

2. Identify four different types of government solutions to market failure caused by negative externalities.

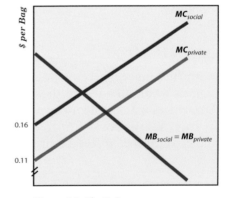

Figure 4.5 *Plastic Bags*

3. Suppose that 5 percent of crude oil pipelines leak, and that litigation over a leaky pipeline results in the oil company paying an average of $10 million in damage awards. If a pipeline company would be willing to pay $600,000 to eliminate all risks of litigation related to a pipeline, what is that firm's risk burden?

4. Suppose that as an alternative to transportation powered by fossil fuels, each bicycle provides $50 worth of pollution reduction. In the perfectly competitive bicycle market, the private marginal cost is upward sloping, and the private marginal benefit is downward sloping. For simplicity, assume there are no negative externalities from bicycles.

    a) Draw a graph showing the social marginal cost, the private marginal cost, the social marginal benefit, and the private marginal benefit of bicycles.

    b) Label the supply curve for bicycles S, the demand curve D, the quantity that will be bought and sold in the market Q_M, and the socially optimal quantity Q_S. Shade-in and label the deadweight loss in this market.

    c) Suppose there is a bicycle subsidy for consumers equal to the marginal external benefit of each bicycle. Label the new demand curve for bicycles D_2 and the quantity of bicycles that will be purchased Q_2.

5. Use the graph to answer the following questions:

    a) Is there an external benefit from plastic bags? How can you tell by looking at the graph?

    b) What is the marginal external cost of each plastic bag?

    c) What specific government policy would cause plastic bag producers to internalize the external cost of plastic bags?

6. Identify one pro and one con of both litigation and regulation as deterrents to environmental wrongdoing.

7. With the 1978 Tellico Dam decision (TVA v. Hill, 437 U.S. 187) the U.S. Supreme Court determined that the Endangered Species Act (ESA) defines the value of endangered species as "incalculable" and that endangered species are to be saved "whatever the cost."

    a) This ruling affects development projects when the clearing of land threatens animal species. If endangered species have a finite value, does this decision lead to the efficient amount of development? Explain.

    b) In what way does this decision simplify or eliminate litigation over development that affects endangered species?

8. There is current debate in many nations over the private use of public lands for activities such as logging in national parks. Based on their general philosophies, how would classical liberals most likely differ from modern liberals on this issue?

9. In the country where you live, what aspects of the government resemble capitalism? What aspects resemble socialism? Are there aspects of other systems of government?

10. A lumber company would cause $10 million worth of health costs by secretly substituting arsenic (a carcinogen) for safer, more expensive preservatives in its pressure-treated lumber. The estimated probability that victims will discover the use of arsenic and successfully litigate against the firm is 20 percent. If arsenic is used, the litigation risk burden for the firm is $1 million. Liability is the only threat to the firm because no enforceable regulation is in place. What specific damage award, if known in advance, would lead the firm to make the socially efficient choice?

websurfer's challenge

1. Find one argument for and one against a larger role for government in protecting the environment. Identify the strengths and weaknesses of these arguments.

2. Find a discussion on pending environmental legislation and analyze the legislation from an efficiency standpoint.

Key Terms

Autocracy
Capitalism
Classical liberalism
Communism
Democracy
Fascism
Government
Imperialism
Mixed economies

Modern liberals
Monarchy
Pluralism
Plutocracy
Populism
Risk burden
Socialism
Theocracy
Tragedy of the commons

Internet Resources

United Nations:
www.un.org

World Bank:
www.worldbank.org

Economic Policy Institute:
www.epinet.org

League of Conservation Voters:
www.lcv.org/

South African Department of
Environmental Affairs and Tourism:
www.environment.gov.za/

European Environment Agency:
www.eea.europa.eu

Indian Ministry of
Environment and Forests:
www.envfor.nic.in

Environment Canada:
www.ec.gc.ca

Singapore Ministry of the
Environment and Water Resources:
www.mse.gov.sg/

New Zealand Ministry for the
Environment:
www.mfe.govt.nz

U.S. Code, including current
environmental legislation:
www.law.cornell.edu/uscode/

White House:
www.whitehouse.gov

U.S. House of Representatives:
www.house.gov

U.S. Senate:
www.senate.gov

Australian Department of the Environment, Water, Heritage and the Arts:
www.environment.gov.au

*Government involvement in environmental initiatives, such as the climate change
conference organized by Fiji, can provide both progress and pride.*

Further Reading

Amico, Laura. "Do Democracy and Capitalism Really Need Each Other?" *Harvard Business Review* (March 11, 2020). https://hbr.org/2020/03/do-democracy-and-capitalism-really-need-each-other. An article spotlighting a range of perspectives on the interdependency of democracy and capitalism.

Anderson, David A. "Government's Role in Property Ownership." In *The Fundamental Interrelationships Between Government and Property*, edited by N. Mercuro and W. J. Samuels. Stamford, CT: JAI Press, 1999. A discussion of the benefits and costs of collective authority.

Anderson, David A. "The Determinants of Municipal Solid Waste." *Journal of Applied Economics and Policy 24*, no. 2 (2005): 23–29. An empirical study of the determinants of municipal solid waste levels.

Augustine, St. *The City of God.* Translated by M. Dods. New York: Charles Scribner's Sons, 1924. A self-proclaimed "giant of a book," this is a work of philosophy, history, and religion that contrasts Roman and Christian cultures.

Bistline, John, Neil R. Mehrotra, and Catherine Wolfram. 2023. "Economic Implications of the Climate Provisions of the Inflation Reduction Act." BPEA Conference Draft, Spring. www.brookings.edu/articles/economic-implications-of-the-climate-provisions-of-the-inflation-reduction-act/. A detailed overview of the anticipated effects of new U.S. policies to reduce greenhouse gas emissions.

Böhmelt, Tobias. "Populism and Environmental Performance." *Global Environmental Politics 21*, no. 2 (2021): 97–123. An empirical study of the influence of populist leadership on environmental quality.

Carson, Rachel. *Silent Spring.* New York: Houghton-Mifflin, 1962. An influential book about the trade-offs between agricultural chemicals and the environment.

Green, T. H. *The Political Theory of T. H. Green*, edited by J. R. Rodman. New York: Meredith, 1964. Readings about the influential nineteenth-century "modern liberal" Thomas Hill Green.

Hobbes, Thomas. *Leviathan*, edited by M. Oakeshott. Oxford: Blackwell, 1946. A classic work of political philosophy.

International Center for Technology Assessment. *The Real Price of Gasoline.* Washington, DC: Center for Technology Assessment, 1998. www.icta.org/projects/trans/index.htm. A comprehensive study of the external cost of a gallon of gasoline.

Locke, John. *Two Treatises of Government.* New York: Hafner, 1947. Locke's foundation for classical liberalism, including his conception of natural laws and natural rights.

Lueck, Dean. "The Comparative Institutions Approach to Wildlife Governance." *Texas A&M Law Revue 6*, no. *147* (2018): 148–178. https://doi.org/10.37419/LR.V6.I1.6. This article examines how property rights affect wildlife populations. Examples include the dramatic near extinction of the American bison during a period of open access to them in the nineteenth century.

Lueck, Dean, and Jeffrey Michael. "Preemptive Habitat Destruction under the Endangered Species Act." *Journal of Law & Economics 46*, no. *27* (2003): 27–60. Provides evidence that landowners destroy wildlife out of a fear that governmental restrictions may be imposed if endangered species are found on their property.

Marx, Karl, and Friedrich Engels. *The Communist Manifesto.* New York: International, 1948. A critique of capitalism and a call for a new political and economic system in response to the inequities of the industrial revolution.

Menger, Carl. *Problems of Economics and Sociology*, edited by F. J. Nock. Urbana, IL: University of Illinois Press, 1963. The thoughts of an influential economist and founder of the Austrian school of thought.

Ravindranath, Shailaja. "Environmental Education in India—The Shifting Paradigm." In *Reorienting Educational Efforts for Sustainable Development*, edited by R. Gorana and P. Kanaujia. Dordrecht: Springer, 2016. A review of India's advancements in environmental education.

Ricardo, David. *The Principles of Political Economy and Taxation.* New York: E. P. Dutton, 1911. The chief work of a principal founder of classical economics who influenced Karl Marx, John Stuart Mill, and Alfred Marshall, among others.

Rousseau, Jean-Jacques. "The Social Contract." In *Famous Utopias.* New York: Tudor, 1901. A prominent discussion of various forms of government, social engineering, and the general will of mankind.

Smith, Adam. *The Wealth of Nations*. New York: Penguin, 1970. First published in 1776, this is arguably the most important account of the principles and practice of modern capitalism.

Viscusi, W. Kip. *Reforming Products Liability*. Cambridge, MA: Harvard University Press, 1991. Proposes solutions to the overlapping influence of liability risks and regulation.

Weber, Max. *Economy and Society*, edited by G. Roth and C. Wittich. Berkeley, CA: University California Press, 1978. A study of the institutional foundations of the modern economy.

Trade-Offs and the Economy

There is no such thing as a free lunch. Every decision is complicated by trade-offs between benefits and costs that can include direct monetary expenditures, externalities, and opportunity costs. Many of the greatest dilemmas in environmental economics involve choices between long-run and short-run benefits and between financial and environmental gains. To harvest a tree today is to forego its future growth. To replace a meadow with a shopping mall is to forego the meadow's natural beauty and wildlife habitat. To consume nonrenewable resources today is to forego their benefits tomorrow.

Environmental policymaking raises questions as challenging as they are important. How sustainable should our practices be? What values should policymakers place on benefits to be received by future inhabitants of the Earth? This chapter explains approaches to contemplated trade-offs that go beyond basic cost-benefit analysis. You will learn about dynamic efficiency, present-value calculations, and the discounting of future costs and benefits. The final section clarifies the differences between measures of economic growth and measures of well-being, and it explores prospects for economic growth that are consistent with environmental goals.

DOI: 10.4324/9781003428732-5

If we harvest too many fish today, there will be too few fish to catch tomorrow.

Trade-Offs Between Present and Future

Why Discount Future Benefits?

A dollar received today can be placed into a savings account where interest payments from the bank will make that account worth more than $1 a year from now. A dollar received a year from now is simply worth $1 a year from now. Clearly, it's better to have $1 now than a year from now. People also feel impatience, fear uncertainty and inflation, and have finite lifetimes, all of which makes it preferable to receive money and other benefits sooner rather than later. Consider the option to receive $1,000 today or in 10 years. When would you rather have the money? The rational response is today, in part because

- *Money received now can be invested to obtain even more money in the future.*

- *It is discomforting to delay gratification from money and expenditures.*

- *The further in the future a benefit arrives, the more likely it is that you will not live to receive it or that it will come when you are less able to enjoy it due to poor health.*

Because people prefer to receive benefits soon, 80 to 90 percent of lottery winners decide to receive a lump-sum payment of around half their earnings immediately rather than annual payments over several decades. If the carbon dioxide buildup in the oceans were curbed, annual revenues for the mollusk fishing industry alone would increase by between $75 million and $187 million over the coming decades, but the delay of benefits dampens interest in policies that would reduce carbon emissions. To prefer benefits now and place a lower value on benefits received later is to "discount" future benefits.

Why Discount Future Costs?

Costs to be paid later are preferable to costs paid up front. People appreciate the luxury of time when trying to assemble payments. On a global scale, the desire to pay for purchases later explains the accumulation of several trillion dollars of credit card debt. Wouldn't you prefer to pay $1,000 in 10 years rather than today? The option to put things off has several advantages:

- *It gives you more time to acquire the money needed.*

- *You could invest the money now and earn even more before the payment is due.*

- *Anything can happen in 10 years. You might be rich, or dead, before the payment must be made.*

Non-monetary costs are also more tolerable when delayed. Consider whether alcohol consumption would be moderated if hangovers hit with the first sip. Because the health and environmental costs of pollution come sometime after the activities that cause pollution, the costs weigh more lightly on the decisions to pollute than if the costs were immediate. To pay less attention to costs that come later rather than sooner, or to favor costs that come later, is to discount the value of future costs.

Our discounting of the future means that the timing of costs and benefits matters. The loss of a forest today is not justified by the gain of an equivalent forest in 1,000 years. It would therefore be inappropriate to evaluate policies by simply comparing amounts paid or received at different times, even if the amounts were adjusted for inflation. To make a valid comparison, future costs and benefits can be discounted to reflect their values in the present. The following sections explain how this discounting is performed and how policymakers approach the determination of discount rates.

Dynamic Efficiency

A **static** model depicts a situation in one time period. The analysis of static models is sufficient when decisions are independent across time periods. For example, individuals can decide to recycle today whether or not they recycled yesterday, and with no effect on tomorrow's decision to recycle. So recycling decisions over the period of a year can be seen as a series of static analyses of one-day situations. It may be that every day the individual weighs the costs and benefits of recycling and makes an independent decision one way or the other. In contrast, individuals cannot choose to tap rubber trees today on a plot of land that was burned yesterday for cattle grazing. *When today's decisions affect the choices available in the future, dynamic analysis is appropriate.*

Dynamic efficiency is achieved by maximizing the value today (the *present value*) of net benefits to be received over time. When deciding the fate of a plot of land in Brazil, for example, the owner would want to determine the net present value of each possible use of the land: cattle grazing, rubber tapping, commercial development, rental, sale, and so on. The timing of the stream of net benefits from these options matters a great deal. To simplify the story, assume there are only two periods: this year and next year. Suppose that immediate sale of the land would bring net benefits of $1,000 this year, rubber tapping would bring net benefits of $500 each year, and cattle grazing would yield $1,000 in profits upon the sale of the cattle next year. Despite the equality of the total payoffs, the preferred option of immediate sale is unambiguous. As we have seen, and as summarized by the principle of the **time value of money**, it is preferable to receive a given amount of money sooner rather than later. Monetary benefits of $1,000 received this year could be invested at any positive rate of return to yield more than $1,000 next year, making immediate sale the clear choice. If the choices were $950 this year, $500 each year, or $1,040 next year, the best choice would be unclear. In cases like that, satisfaction of the dynamic efficiency criterion requires present-value calculations and the determination of a discount rate.

Present-Value Calculations

South America's Amazon basin still holds some barter economies where goods and services are exchanged for the likes of 20-pound balls of smoked rubber and bushels of cocoa beans. This can lead to confusion when a canoe seller receives alternative offers of five balls of rubber, 150 pounds of cocoa beans, or six machetes. The relative value of differing options is uncertain in the absence of a common metric for comparison. The same is true when comparing monetary costs or benefits that come at different points in *time*, such as receiving $950 this year or $1,040 next year. In the case of international markets for goods, prices stated in a common currency simplify comparisons among the prices charged in different countries. Likewise, to simplify comparisons among streams of costs and benefits that differ in their timing, we can calculate the **present value**, which is the value today of amounts to be paid or received in the future.

Present-value calculations help us sort out trade-offs across time. Retail stores still compete with relatively inexpensive Internet e-tailers in part because consumers prefer to have their products today rather than tomorrow. But the decision to go to college reveals that the future benefits from an education are worth more to you than the money you could earn now by working instead. How much more you must receive in the future to compensate for a loss today depends on your *rate of time preference*. A **rate of time preference** is a discount rate applied to future benefits to determine their present value.

Here's a thought experiment: What is the smallest amount you would accept in one year in exchange for giving up $100 today? Your answer indicates your rate of time preference. If you would accept $100 in a year in exchange for $100 today, you do not discount the future at all, and your rate of time preference is zero. If you would accept as little as $110 a year from now in exchange for $100 today, your rate of time preference is 10 percent, because you would require 10 percent more in a year to be indifferent between the future benefit and $100 now.

For easier analysis, we can summarize the relationship between present values and future values with an equation. Continue to suppose your preferences make you indifferent between $110 in a year and $100 in the present. The future value of $110 that you would accept equals the present value of $100 plus 10 percent of $100. We can write this as $110 = $100 + 0.10($100), or equivalently,

$$\$110 = \$100(1 + 0.10).$$

More generally, with a future value of FV, a present value of PV, and a rate of time preference of R, the future value that compensates for a one-period delay is

$$FV = PV(1 + R).$$

Solving the equation for PV yields

$$PV = \frac{FV}{(1 + R)}.$$

Knowing any two of these variables, we can solve the present-value equation to determine the third.

Let's compare the three options presented to the Amazon landowner under the assumption that the rate of time preference is 10 percent. As a reminder, the options were to receive $950 in the first period, $500 each period, or $1,040 in the second period. The present value of the

$950 from an immediate sale would simply be $950. There is no need to discount money received in the present. The $500 received in the second period from rubber production would have a present value of

$$PV = \frac{\$500}{(1+0.10)} = \$454.50.$$

Combined with the $500 received for rubber in the first period, rubber tapping would yield a total of $500 + $454.50 = $954.50 in present value.
　　The cattle grazing option would bring in a net value of $1,040 in the second period, which has a present value of

$$PV = \frac{\$1,040}{(1+0.10)} = \$945.40.$$

So the harvest of rubber from the plot of land would yield the greatest present value ($954.50) and provide the dynamically efficient outcome.
　　It is no problem that reality involves more than two periods. The present value of a net benefit t periods in the future is

$$PV = \frac{FV}{(1+R)^t}$$

where t can be any value from zero to infinity. Solving for the future value yields

$$FV = PV(1+R)^t.$$

A similar apparatus can handle more complicated situations with little difficulty. To find the net present value of a stream of benefits, B, and costs, C, received once per period for n periods, sum the discounted net benefits for each period:[1]

$$PV = \sum_{t=0}^{n} \frac{(B_t - C_t)}{(1+R)^t}.$$

1　In case you have not seen this type of notation, the large Greek letter *sigma* (Σ) with a zero below it and an n above it indicates the summation for each period from period zero to period n. The term $(B_t - C_t)$ represents the net benefits in period t, and the term $(1 + R)^{-t}$ discounts the benefits to reflect present values.

When a net present value calculation accurately reflects all benefits and costs including external benefits, external costs, and opportunity costs, a positive net present value indicates that the project should be carried out. A negative net present value indicates that the project should not be pursued.

Discount Rates—Who's Got the Number?

Policymakers adjust anticipated future payments and receipts to reflect present values using estimates of the rate of time preference referred to simply as **discount rates**. Unfortunately, discount rate selection is not straightforward. The federal Office of Management and Budget (OMB) in the United States has used discount rates as high as 10 percent to calculate the present value of future regulatory costs and benefits. Since 2003, the OMB has advised government officials to use a discount rate of 7 percent for regulations that affect the allocation of capital and 3 percent for those that affect private consumption. In 2023, the OMB proposed adjusting the low end of the discount rates to 1.7 percent based on actual rates of time preference exhibited in the economy.

The selection of discount rates can strongly influence policy decisions. Table 5.1 illustrates the present values of each option for the Amazon land given the discount rates of 7 percent, 10 percent, and 12 percent. With a discount rate of 7 percent, the greatest net present value is achieved by clearing the land for cattle. With a discount rate of 10 percent, the trees should stand for rubber tapping. With a discount rate of 12 percent, the land should be sold. The importance of the discount rate is clear from its ability to determine which approach is adopted.

The rates of return on investments provide clues about appropriate discount rates. Suppose an investor personally discounts the future at an annual rate of 2 percent. She will forego alternative uses of her money to make a risk-free investment only if the investment offers a rate of return of 2 percent or more. This means that the return on risk-free investments that motivate large numbers of investors to forgo the use of their money offers a good approximation of the rate of time preference. Bonds issued by stable governments are an example of nearly risk-free investments and typically offer returns in the vicinity of

		Discount Rate ($)		
		0.07	0.10	0.12
	Sell	950.00	950.00	950.00
Land Use	Rubber	967.29	954.55	946.43
	Cattle	971.96	945.45	928.57

Table 5.1
The Effect of Discount Rate Selection

1–2 percent. However, there is more to consider. Four normative questions are central to discount rate debates:

1. **How self-serving should discount rates be?**

Impatient individuals naturally discount future benefits and costs, preferring to receive benefits now and pay later. We've already discussed how time preference is rational at a personal level due to investment opportunities, uncertainty about future health, and desires for immediate gratification. All these factors lead to relatively high discount rates.

American political theorist John Rawls argued against the influence of self-interest and felt that discounting should occur as if policymakers did not know the period in which they would live. This approach removes discounting based on impatience and time preference. However, this does not mean that a $1,000 investment today would be justified by the inflation-adjusted equivalent to a $1,000 benefit in 50 years. The real (inflation-adjusted) returns society could receive from alternative investments represent opportunity costs, with or without impatience and time preference. The Rawlsian approach would thus retain discounting of the benefits from an investment to reflect the social opportunity cost of the best forgone alternative investment.

In the process of policy evaluation, the returns can be compared directly with the returns available from alternative investments *if* the alternative returns would accrue with the same timing as the policy's benefits. If the alternative returns would not last as long or would fluctuate over time, the *shadow price* provides a more accurate measure of the opportunity cost of the policy. The **shadow price** is the present value of the returns from the best alternative investment. The shadow price should be compared with the present value of the policy's benefits to determine which is the best option.

As a simple example, suppose a community with a discount rate of 2 percent is considering a $10,000 expenditure on a community garden that would provide a harvest worth $5,500 1 year from now and again 2 years from now. The best alternative use of that $10,000 would be to expedite the repair of the community swimming pool, which would allow an additional $10,800 worth of pool memberships to be sold one year from now. Applying the formula for the present value $(PV = FV/(1+R)^t)$, we see that the present value of the garden is $5,500/(1.02) + $5,500/(1.02)^2 = $10,679$, and the shadow price is $10,800/(1.02) = $10,588$. So the benefit of the community garden exceeds the shadow price and is therefore the best choice.

2. **Should environmental benefits receive special treatment?**

The costs of environmental regulations generally come sooner than the benefits, meaning that the application of a single discount rate to costs and benefits diminishes the present value of the benefits more so than the costs. For this reason, cost-benefit analysis supports fewer regulations as the discount rate increases. Many economists argue for *dual-rate discounting*, saying that environmental values should be discounted at a lower rate than financial values.[2] There are several justifications for this. One is that putting off environmental benefits does not cause the same level of losses as putting off the use of money. Having a life or a wilderness area now does not enable investments that are forgone by having the life or the wilderness area next year instead. As a result, there is less reason to discount the value of an environmental asset gained next year rather than now. Another reason is that environmental benefits can have unrecognized value, the loss of which is irreversible. For example, suppose a regulation would save a tree species from future extinction. The possibility the species offers yet-to-be-discovered medicinal cures, along with the impossibility of getting the species back from extinction, may warrant a relatively low discount rate for its future value. A third reason for dual-rate discounting is that future generations are likely to have fewer animal species, wilderness areas, and other environmental assets than exist now. Increased scarcity will make those assets more valuable in the future, which warrants a lower discount rate for those particular assets as we consider policies that would protect them.

3. **How should we treat benefits and costs that extend to future generations?**
It is one thing to wait a year to pay a cost or receive a benefit. It's a bigger consideration if the cost or benefit goes to people in a future generation. It is only natural for self-interested decision-makers to place a higher discount rate on costs and benefits for unknown people yet to be born. The discount rate in this case depends on the extent to which people today feel connected to future generations. The more responsibility we feel to shield others from burdens, and the more utility we gain from helping others, the lower are the appropriate discount rates for future costs and benefits.

Selfishness is not the only reason to apply higher discount rates to values for future generations. There is a general trend for each generation to be wealthier than the one before it. If that trend continues, according to the *law of diminishing marginal utility* (as introduced in the Chapter 2 Appendix), the wealthier generations of the future would receive less benefit from an additional dollar than the current generation. In that case, relatively high discount rates would satisfy desires for intergenerational equality. The balancing effect comes from high rates discouraging sacrifices by current, less wealthy generations that would have benefited future, more wealthy generations.

4. **What is the appropriate treatment of uncertainty?**
When the benefits from a contemplated policy are clear and certain, the risk-free rate of return may be an appropriate guideline for the discount rate, as discussed earlier in this section. In the case of risky investments, relatively high expected rates of return are needed to attract investors who dislike uncertainty about gains and losses. For example, if stock market investments provided average returns no higher than the

---

2 See Polasky and Dampha (2021).

return on low-risk government bonds, there would be little incentive to purchase stocks. People invest in stocks despite the risk of low returns or losses because, on average, stock investments yield a higher return than low-risk bonds. The difference between the rate of return on a risk-free investment and the average rate of return on a risky investment is called the **risk premium**. The risk premium provides an incentive for investors to bear risk. If a risk-free investment offers a 2 percent rate of return and the average rate of return on stock investments is 8 percent, there is an 8–2 = 6 percent risk premium that attracts investors to the stock market.

Investments in environmental policies with uncertain returns may warrant the addition of a risk premium to the discount rate. This lowers the assessed present value of an investment to reflect disfavor in having a range of possible outcomes. For example, suppose a 2 percent discount rate would be appropriate for a risk-free project, but the project being assessed involves risks of failure similar in scale to the risks of stock market investments. Policymakers could add a risk premium of 6 percent to the discount rate when estimating the present value of the project. The appropriate size of the risk premium depends on both the risks involved and the investors' attitudes toward risk. There are three categories of risk preference among investors:

Risk-neutral investors care only about the expected value of the outcome. These investors have no preference between (1) receiving $1 with certainty and (2) a fair coin flip that will determine whether they receive $0 or $2, each with equal probability. Applying the expected value formula from Chapter 2, the expected value of the coin flip outcome is (0.5)($0) + (0.5)($2) = $1, and that value is all risk-neutral people care about.

Risk-averse investors care not only about the expected value of the outcome but also about the range of possible outcomes. They would pay extra to receive a certain $1 rather than being subjected to the uncertainty of a coin flip. Most people who purchase insurance pay more than the expected value of their losses from a house fire or an auto accident in exchange for certainty about having a house and a car at the end of the day regardless of their luck.

Risk-loving investors prefer a range of possible outcomes to a certain outcome with the same expected value. They would pay extra to receive the outcome of a coin flip for $0 or $2 rather than a certain $1. Similarly, gamblers typically pay more than the expected value of a lottery ticket for the small chance of winning big.

Risk-neutral investors need not attach a risk premium to their discount rate because risk doesn't bother them. Risk-loving investors would actually have a lower discount rate for uncertain future benefits than for certain future benefits, making their risk premium negative. The more risk averse an investor is, the larger the risk premium that should be added to the discount rate to accommodate uncertainty. Individual investors, including individual car owners and homeowners, are likely to be risk averse because they cannot bear the burden of a large loss. In a classic article, economists Kenneth Arrow and Robert Lind (1970) suggest that society at large should be relatively risk neutral when evaluating public investments because society can spread any losses far and wide. The implication is that a risk premium is unneeded for environmental policies when large segments of society share the costs and benefits.

Corrosion Proof Fittings v. EPA

The case of *Corrosion Proof Fittings v. EPA* (947 F.2d 1201) highlights the complexity and controversy behind discounting future costs and benefits and describes how the debate played out in a U.S. Circuit Court.

Asbestos is a naturally occurring fibrous material used in heat-resistant insulation, cement, building materials, clothing, and brake linings. Exposure to asbestos dust can result in mesothelioma, asbestosis, and lung cancer. The EPA issued a final rule under Section 6 of the Toxic Substances Control Act to ban asbestos in almost all products. They estimated this rule would save 202 lives, which had a present value of 148 lives when a discount rate was applied to human lives. The rule would cost between $450 million and $800 million, depending on the prices of substitutes for asbestos. A number of petitioners, including Corrosion Proof Fittings, contended that the EPA rule-making relied on flawed analyses of the necessary trade-offs between the environment and the economy. The U.S. Court of Appeals for the Fifth Circuit asked the EPA to reconsider the matter.

In the eyes of the Court, the EPA may have demonstrated that a complete ban of asbestos was preferable to the status quo, but "failed to show that there is not some intermediate state of regulation that would be superior to both the currently-regulated and the completely-banned world." In other words, the EPA compared the total costs and benefits of a ban without conducting sufficient marginal analysis of incremental policies, as described in Chapter 2.

While discounting the future monetary *costs* of an asbestos ban, the EPA presented alternative cost-benefit analyses that did and did not discount future *benefits*. Recognizing the ongoing dispute over discounting benefits measured in terms of human lives, the Court stated a preference for discounting everything, monetary or otherwise. The EPA chose to use a real discount rate of 3 percent. Citing historical real (inflation-adjusted) interest rates in the range of 2 to 4 percent, the Court found the 3 percent figure to be "not inaccurate."

The Court criticized the EPA for discounting health losses from the time of exposure to asbestos, rather than from the time when injury would result. If, without the ban, some individuals would be exposed to asbestos in 5 years and feel the adverse health effects in 10 years, the Court

(continued)

held that the health losses should be discounted by 10 years rather than 5. The Court also reprimanded the EPA for including only the lives saved over the next 13 years and counting lives saved after that time as simply "unquantified benefits." The EPA's mistakes came at an early stage in the application of cost-benefit analyses to human lives. More carefully crafted studies followed this valuable scrutiny.

What's Your Number?

The subjectivity and complexity of discount rate selection explain the broad range of values chosen. Studies find significant numbers of individuals favoring discount rates in every range, including negative and very high values.[3] Low and negative discount rates may be signs of altruism. Indeed, parental expenditures on education would seem to indicate negative discount rates, as they involve people effectively placing the next generation before themselves. Rational, selfish individuals should adopt a discount rate of at least the probability they will not live until the next period—around 1 percent for most age groups.

From a societal standpoint, there are again rational discount rate guidelines. Society's opportunity cost is an appropriate lower bound for discount rates. For example, if society can receive a 1 to 2 percent risk-free real rate of return on investments in government bonds, it would make no sense to discount future returns by less than that amount. To adopt a 0 or ½ percent discount rate would be to advocate expenditures on projects with lower net returns to society than an investment in bonds.

A thought experiment will provide perspective on the relatively subjective upper bound for *social discount rates*, meaning discount rates applied to costs and benefits for society rather than for a particular person or firm. Policies that curtail pollution save "statistical lives," meaning it is unknown who will be saved. Assume that a statistical (unidentified) life is worth $10 million today—a figure close to the estimates of many economists.[4] What amount would you say is the most society should spend on an environmental regulation that, with certainty, would save 1,000 lives 300 years from now? Continue reading only after you have decided on a maximum expenditure.

Table 5.2 indicates the maximum expenditure implied by several discount rates. If the discount rate is 10 percent, society would only be willing to spend a fraction of a cent on an environmental regulation that

---

3 See, for example, Attema et al. (2018).

4 See Banzhaf (2022) for an overview of value-of-life studies.

Discount Rate (%)	Maximum Expenditure ($)
10	0.004
9	0.06
8	0.94
7	15.31
6	256.00
5	4,397.54
4	77,624.39
3	1,408,745.65
2	26,299,559.32
1	505,344,874.52
0	10,000,000,000.00

Table 5.2
Discounting the Value of Future Lives

would certainly save 1,000 lives with a combined value of $10 billion in 300 years. If it were appropriate for society to forego 94 cents or more— less than the price of a can of soda—to save 1,000 lives in 300 years, then discount rates would have an upper bound of 8 percent, and the appropriate discount rate range would be from 0 to 8 percent. Find the maximum expenditure in Table 5.2 that is the closest to your own. The corresponding discount rate approximates your own discount rate.

The use of a single social discount rate to analyze environmental regulations would prevent agenda-driven manipulations from ushering particular programs in or out. Among economists, the broadest agreement on rates lies in the 2 to 3.5 percent range. For example, the *Green Book* that provides guidance to the British government recommends a social discount rate of 3.5. U.S. agencies including the General Accounting Office, the Environmental Protection Agency, and the Congressional Budget Office have adopted rates in the 2–3 percent range. When a discount rate is not stipulated, rate selection should hinge on the goals of the project. In general, low-risk projects intended to benefit society should receive low discount rates, while projects with more uncertainty and those meant to satisfy private goals should receive relatively high discount rates.

Trade-Offs Between Growth and the Environment

Every decision involves trade-offs. Economic growth offers a path to more products and jobs but can trample on health, the environment, free time, and other elements of our quality of life. What is the net effect of growth on social welfare? After defining relevant terms, this section explains how the virtues of economic growth can be measured with consideration of the associated problems, and how some of the perceived problems need not be associated with growth at all.

Growth Versus Welfare

Gross domestic product (GDP) is the final value of goods and services produced within a country in one year. The media tend to herald GDP growth as a uniformly splendid event. In reality, increases in GDP may or may not reflect welfare improvements. A better measure of social welfare would exclude expenditures that make no net contribution to welfare and add components of welfare not captured in GDP.

Defensive goods and services are purchased in response to pollution, congestion, work-related anxiety, and other unfortunate side effects of economic growth, or to recover from unwelcome events such as crime. Expenditures on cleanups to remedy environmental degradation indicate that society has suffered losses. Counselors and physical therapists are paid to help workers cope with the stress and strain of fast-paced jobs, but those payments reflect recovery and not advancements in social welfare. The same is true of expenditures on disease, war, natural disasters, and the replacement of stolen items. Spending on all these sorts of *regrettables* should be subtracted from GDP to form a better measure of social welfare.

Buildings, machines, and equipment are examples of **capital**—goods used to make other goods. **Capital depreciation** occurs whenever the value of capital erodes during the production process. Suppose that after the production of $10,000 worth of filing cabinets, $250 worth of metal stamping equipment must be replaced due to wear and tear. GDP would increase by the $10,250 value of the filing cabinets and new stamping equipment. However, the net gain for society would only be $10,000 because the $250 worth of stamping equipment simply replaces worn-out machinery. The U.S. government publishes net national product (NNP) figures that adjust GDP for the depreciation of (human-made) capital. NNP does not account for losses of natural capital such as forests, coal seams, and wetlands. A true measure of social welfare would account for the depreciation of natural capital.

It's tempting to look at measures of income as indications of welfare, but those have problems as well. Consider, **national income**, which is all the income earned by the factors of production owned by a nation's citizens. An increase in national income can be a good sign, typically reflecting an increase in production and spending power. However, as with the closely tied measure of GDP, national income rises with expenditures on defensive goods and services and fails to capture environmental effects among other determinants of the quality of life.

An overall increase in national income does not guarantee that most people are earning higher incomes; it may be that a small percentage of citizens are earning more, while the incomes of most people are stagnating. An ideal measure of social welfare would incorporate changes in income distribution.

Nonmarket production is the unpaid production of goods and services, typically to benefit the worker's household, family, or community. Examples include childcare, yard work, home cooking, housework, community activism, and various types of volunteer work. Nonmarket production provides benefits to society but is not counted in GDP, although the same activities are counted in GDP when performed by a paid professional. Unreported production in the illegal *underground* or *cash economy* is also not reported to the government and is not counted in GDP. It is important to include these goods and services to the extent possible in a true measure of social welfare.

Scholars have proposed numerous indicators of welfare and progress as alternatives to GDP. These include

- *Better Life Index (BLI)*

- *Canadian Index of Wellbeing (CIW)*

- *Genuine Progress Indicator (GPI)*

- *Human Development Index (HDI)*

- *Index of Sustainable Economic Welfare (ISEW)*

- *Net National Welfare (NNW)*

Many of these are considered *enlarged GDP indicators* because they are found by adjusting GDP to add omitted elements and subtract those that do not promote social welfare. For example, the formula for Net National Welfare is

NNW = GDP + Nonmarket Output

  − Externality Costs − Pollution Abatement and Cleanup Costs

  − Capital Depreciation (including human-made capital and
    natural capital).

Some of the measures are formulated from scratch rather than being based on GDP. For example, the Index of Sustainable Economic Welfare formula begins with the values of personal consumption and household labor, adds the values of beneficial public expenditures on things like health and highways, and subtracts the costs of environmental

Figure 5.1

*GDP, GPI,
and GPI
Environmental
Indicators for
Hawai'i*

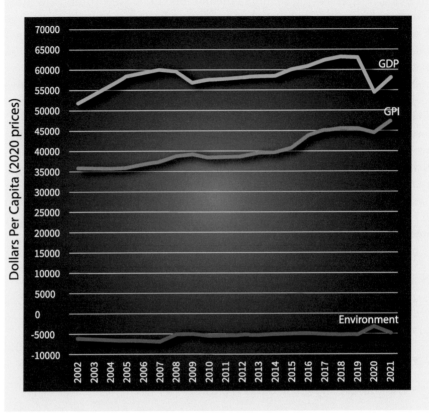

Source: Data from https://dbedt.hawaii.gov/economic/hawaii-genuine-progress-indicator-hi-gpi/.

degradation, auto accidents, and other problems. The Genuine Progress Indicator equation is very similar to the ISEW equation.

The Canadian Index of Wellbeing incorporates eight "domains": leisure and culture, democratic engagement, community vitality, education, environment, healthy populations, living standards, and time use.[5]

The Human Development Index is based on life expectancy at birth, literacy, school enrollment, and standard of living as measured by GDP per capita.

The indicators of welfare and progress tend to differ significantly from GDP. Figure 5.1 illustrates the GDP per capita, GPI per capita, and environmental indicators within the GPI for the state of Hawai'i. The GPI is well below the GDP and sometimes remains flat even as GDP rises. The environmental indicators are negative, relatively flat, and fall in 2021 just as the GDP and GPI turn upward. Figure 5.2 shows the contrasting changes in Canada's GDP per capita, the environment

5 See www.ciw.ca/.

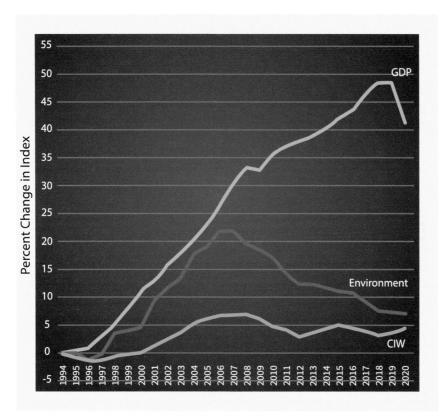

Figure 5.2

Changes in the Canadian GDP Per Capita, CIW for the Environment, and Overall CIW

Source: Data from: https://uwaterloo.ca/canadian-index-wellbeing/.

domain of the CIW, and the overall Canadian Index of Wellbeing. While GDP per capita has risen dramatically in Canada, the environment has had ups and downs, and well-being has been slow to rise. In many other places around the world, stagnating welfare and a faltering environment have likewise accompanied rising GDP.

Because GDP figures alone can be misleading, it is important for policymakers and the media to consider GDP in the context of other measures such as the Human Development Index. Table 5.3 shows the 38 countries with the highest HDI values.

"Green" Growth

Alfred Marshall, the father of neoclassical economics, described utility, not material accumulation, as the true standard of production and wealth. Utility and economic growth can be derived from many activities that tread lightly on the environment. It is therefore a fallacy to present economic growth as the antithesis of environmental goals. Some types of growth, particularly in the service industries, are not at odds with the

Table 5.3

The 38 Countries With the Highest Human Development Index Values

Rank	Country	HDI
1	Switzerland	0.962
2	Norway	0.961
3	Iceland	0.959
4	Hong Kong (SAR)	0.952
5	Australia	0.951
6	Denmark	0.948
7	Sweden	0.947
8	Ireland	0.945
9	Germany	0.942
10	Netherlands	0.941
11	Finland	0.940
12	Singapore	0.939
13	Belgium	0.937
14	New Zealand	0.937
15	Canada	0.936
16	Liechtenstein	0.935
17	Luxembourg	0.930
18	United Kingdom	0.929
19	Japan	0.925
20	South Korea	0.925
21	United States	0.921
22	Israel	0.919
23	Malta	0.918
24	Slovenia	0.918
25	Austria	0.916
26	United Arab Emirates	0.911
27	Spain	0.905
28	France	0.903
29	Cyprus	0.896
30	Italy	0.895
31	Estonia	0.890
32	Czechia	0.889
33	Greece	0.887
34	Poland	0.876
35	Bahrain	0.875
36	Lithuania	0.875
37	Saudi Arabia	0.875
38	Portugal	0.866

Source: The Human Development Index measures health, education, and standard of living.

https://hdr.undp.org/data-center/human-development-index.

environment. More broadly, the components of GDP can be divided into three categories according to their causes and effects:

- ***Defensive goods and services*** are purchased in response to undesirable occurrences, including crime, natural disasters, and pollution. Examples include security fences and cancer treatments.

- ***Utility-justified goods and services*** involve subjective trade-offs between the benefits of consumption and employment, and the costs of resource depletion and environmental degradation. Clothing, education, and most other goods and services fall into this category.

- ***Environmentally beneficial goods and services*** provide employment and utility with no net loss, or a net gain, to the environment. Solar panels, clotheslines (that take the place of dryers), and the planting of trees are examples.

You've already read that expenditures on defensive goods and services rise when society becomes worse off and that these expenditures are subtracted from GDP in the calculation of net national welfare. The tools of discounting and the criteria for sustainable development can be applied to potentially utility-justified goods to determine whether the trade-offs are indeed in the best interest of society. To see growth and the environment as strictly adversarial interests is to overlook the subset of goods and services with minimal environmental impact, as well as those that benefit the environment. Examples include safer pesticides, more efficient appliances, and toxic waste cleanups. The next section highlights more goods and services that require few environmental trade-offs.

Treading Lightly

The service occupations in Table 5.4 represent relatively low-impact sources of jobs and satisfaction. Of course, there is some sort of resource use associated with virtually all industries. Workers in the service industries typically use motorized transportation to get to work, and they perform their jobs in buildings whose climate control systems use considerable amounts of energy. But in other industries, these same needs are compounded with additional resource demands associated with production, processing, packaging, delivery, and use. Society can look to myriad service-oriented industries for economic growth with fewer externalities. Transportation and climate control systems can also be switched over to clean energy sources.

To make the manufacturing sector more sustainable, consumers and policymakers can support high-quality, multi-use products rather than inferior and single-use products. Ceramic mugs can last for centuries,

Is There Employment Outside of Manufacturing?		
Author	Therapist	Guard
Mechanic	Teacher	Entertainer
Dietician	Doctor	Firefighter
Police Officer	Masseur	Lawyer
Actor	Politician	Chef
Programmer	Athlete	Professor
Repairperson	Farmer	Secretary
Analyst	Scientist	Historian
Banker	Nurse	Coach
Economist	Ranger	Clergy
Insurance Agent	Broadcast Producer	Artist
Sociologist	Curator	Musician

Table 5.4

Service Occupations

unlike single-use Styrofoam cups, and a well-made bike that lasts 20 years is preferable to a series of five low-quality bikes that last 4 years each.

Environmentally beneficial goods and services provide win-win opportunities for employment and sustainability. For example, the U.S. Inflation Reduction Act introduced in Chapter 4 is projected to create 5.55 million jobs in clean energy, clean transportation, and environmental justice.[6] Producers of alternative fuels such as biodiesel, high-efficiency batteries, geothermal heating systems, and solar panels provide employment and a net environmental gain. The standard of living improves after expenditures on education, health care, communications, entertainment, and food service without substantial adverse consequences. With more focus on the efficient use of recycled materials, insulation, alternative energy sources, and water conservation, architects and builders can take great strides toward sustainable development.

If enough people purchase environmentally beneficial goods such as recycled paper, photovoltaic cells, wind turbines, and plug-in hybrid automobiles, producers will enjoy economies of scale and prices will fall. Because jobs and goods can tread lightly on the environment, and because low-impact goods are increasingly affordable, tensions between the economy and the environment can be mitigated by wise choices about growth and consumption.

Summary

Why don't people want to delay gratification? Impatience, opportunity cost, selfishness, and mortality cause us to prefer benefits sooner and costs later. Dilemmas over environmental policy arise when the choice is between a benefit now and a larger benefit later. Discount rates recast future benefits and costs in present-value terms to make it easier for policymakers to weigh the options. Although there is no consensus on a particular social discount rate for environmental benefits, most of the defensible rates are in the 1 to 8 percent range, and many economists agree that rates should fall between 1 and 3 percent.

Economic growth sometimes creates externalities that work against social and environmental goals. GDP, a common measure of economic growth, is a flawed measure of social welfare. GDP does not account for some good things, such as leisure, and it increases with expenditures on some "regrettables," including pollution and crime. Alternative measures such as the ISEW and the GPI are designed to increase if and only if society is truly better off.

6   See www.bluegreenalliance.org/site/9-million-good-jobs-from-climate-action-the-inflation-reduction-act/.

In the process of allocating scarce resources among competing ends, we encounter compelling reasons to make trade-offs between the environment and growth. The good news is that some of the imagined trade-offs are avoidable, and others can be minimized with the appropriate efforts. Production, employment, and consumption practices can occur within the guidelines of sustainability and dynamic efficiency. Even so, free lunches are rare. Decision-makers at every level must struggle with trade-offs between the present and the future and sometimes between growth and the environment. Economic tools can guide those endeavors.

• • • • • • • • • • • • • • •

Problems for Review

1. Under what circumstances is it appropriate to apply dynamic analysis rather than static analysis?

2. Kate is eager to make blueberry pie with berries from a bush in her backyard. Kate has 100 ripe blueberries today, but more berries will be ripe in a couple of days. Kate must decide whether to bake with the 100 blueberries today or wait 2 days for a larger harvest that will make a bigger pie. If Kate's rate of time preference for blueberries is 10 percent per day, at least how many blueberries must be available in 2 days to justify delaying her harvest?

3. Layers of clay, plastic, and sand help prevent landfills from leaking toxins. Suppose a $150,000 expenditure on protective layers would eliminate $1 million worth of healthcare costs 20 years from now. If the social discount rate were 10 percent, would society be willing to spend the $150,000 today?

4. Again consider the situation in Problem 3.

   a) How would your answer change if the social discount rate were 5 percent?

   b) What is the most society would spend to avoid the $1 million loss in 20 years with a 5 percent discount rate?

5. What discount rate would you apply if you were in charge of cost-benefit analysis for a new policy to save future lives by banning cars with high levels of toxic emissions? Write one paragraph to explain your answer using justifications from this chapter.

6. Provide one example (not given in this chapter) of each of the following:

   a) A defensive good or service

   b) A utility-justified good or service

   c) An environmentally beneficial good or service

7. Explain two reasons why environmental degradation need not accompany growth.

8. Explain two reasons why GDP is not a good measure of social welfare.

9. What is the difference between GDP and NNW?

10. In the opinion of the Fifth Circuit Court, what was one thing the EPA did right and one thing the EPA did wrong in their analysis of asbestos policies?

websurfer's challenge

1. Find detailed descriptions of at least two alternatives to GDP that may provide more accurate measures of national welfare.

2. Find a website that discusses trade-offs between economic growth and the environment. Do you agree with the arguments made on this website?

3. Find a website that describes the application of discount rates to a specific environmental cost or benefit in the future. Do you agree with the rate they have applied?

Key Terms

Capital
Capital depreciation
Defensive goods and services
Dynamic efficiency
Environmentally beneficial goods and services
Gross domestic product
National income
Nonmarket production
Rate of time preference

Present value
Risk averse
Risk loving
Risk neutral
Risk premium
Shadow price
Static
Time value of money
Utility-justified goods and services

Internet Resources

The customizable OECD Better Life Index:
www.oecdbetterlifeindex.org/

Institute of Wellbeing:
www.ciw.ca

Office for Management and the Budget: Guidance on Discount Rates:
www.whitehouse.gov/wp-content/uploads/legacy_drupal_files/omb/circulars/A94/a094.pdf

Resources for the Future:
www.rff.org/

Further Reading

Arrow, Kenneth J., and Robert C. Lind. "Uncertainty and the Evaluation of Public Investment Decisions." *American Economic Review 60*, no. *3* (1970): 364–378. Suggests that public investments should be evaluated from the perspective of a risk-neutral party.

Attema, Arthur E., Han Bleichrodt, Olivier L'Haridon, Patrick Peretti-Watel, and Valérie Seror. 2018. "Discounting Health and Money: New Evidence Using a More Robust Method." *Journal of Risk and Uncertainty 56*, no. *2* (2018): 117–140. https://doi.org/10.1007/s11166-018-9279-1. Compares the discount rates individuals place on health outcomes and financial gains.

Banzhaf, H. Spencer. "The Value of Statistical Life: A Meta-Analysis of Meta-Analyses." *Journal of Benefit-Cost Analysis 13*, no. *2* (2022): 182–197. https://doi.org/10.1017/bca.2022.9. A useful summary of estimates of the value of human lives and the underlying literature.

Baumol, William J. "On the Social Rate of Discount." *American Economic Review 58* (1968): 788. Argues for a discount rate that equals or exceeds the market interest rate.

Cobb, Clifford, Ted Halstead, and Jonathan Rowe. "If the GDP Is Up, Why Is America Down?" *The Atlantic Online* (October 1995). www.theatlantic.com/politics/ecbig/gdp.htm. As the subtitle says, "why we need new measures of progress, why we do not have them, and how they would change the social and political landscape."

Cooley, Sarah R., and Scott C. Doney. "Anticipating Ocean Acidification's Economic Consequences on Commercial Fisheries." *Environmental Research Letters 4*, no. *2* (2009). Quantifies the damage to the U.S. fishing industry from acidification caused by carbon dioxide buildup.

Council of Economic Advisers. "Discounting for Public Policy: Theory and Recent Evidence on the Merits of Updating the Discount Rate." Council of Economic Advisers Issue Brief (January 2017). Advocates for a decrease in the discount rates applied by government agencies. https://obamawhitehouse.archives.gov/sites/default/files/page/files/201701_cea_discounting_issue_brief.pdf

Daly, Herman E., and John B. Cobb, Jr. *For the Common Good: Redirecting the Economy Toward Community, the Environment, and a Sustainable Future.* Boston, MA: Beacon Press, 1989. Introduces the Index of Sustainable Economic Welfare and presents comparisons between it and the gross national product.

Howe, Charles W. "The Social Discount Rate." *Journal of Environmental Economics and Management 18* (March 1990): S1–S2. This edition of *JEEM* includes a collection of articles focusing on social/environmental discount rates.

Lind, Robert C., Kenneth J. Arrow, Gordon R. Corey, Partha Dasgupta, Amartya K. Sen, Thomas Stauffer, Joseph E. Stiglitz, and J. A. Stockfisch. *Discounting for Time and Risk in Energy Policy.* Washington, DC: Resources for the Future, 1982. A classic collection of essays on discount rates and their influence on policy.

Nordhaus, William, and James Tobin. "Is Growth Obsolete?" *Economic Growth Economic Research: Retrospect and Prospect 5* (1972). Presents the authors' measure of economic welfare.

Pigou, Arthur C. *The Economics of Welfare.* London: Macmillan, 1932. A classic in which Pigou questions human foresight.

Polasky, Stephen, and Nfamara K. Dampha. "Discounting and Global Environmental Change." *Annual Review of Environment and Resources 46* (2021): 691–717. A broad discussion of the factors to consider when establishing discount rates for environmental change.

Portney, Paul R., and John P. Weyant (eds). *Discounting and Intergenerational Equity.* Washington, DC: Resources for the Future, 1999. An updated collection of papers on discounting written by many of the foremost environmental economists, including Lind, who was behind the first such collection.

Rawls, John. *A Theory of Justice*. Cambridge, MA: Belknap Press, 1999. In this classic, Rawls explains his "veil of ignorance" from which to consider fair procedure.

United Kingdom. HM Treasury. *The Green Book*. Updated 2022. Accessed September 8, 2023. www.gov.uk/government/publications/the-green-book-appraisal-and-evaluation-in-central-governent/the-green-book-2020.

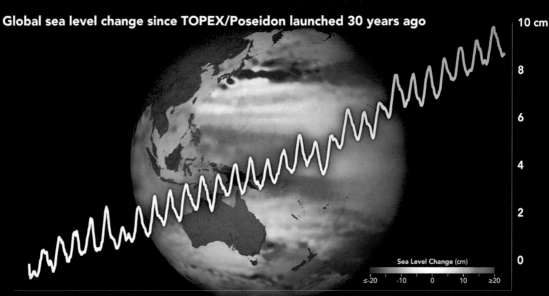

Imagery from the TOPEX/Poseidon Satellite reveals that as global climate change warmed the oceans and melted land ice between 1992 and 2022, the mean sea level rose by 10.1 centimeters (3.98 inches). Sea level rise is not uniform—the warmer the water, the higher the sea level. The rising seas are starting to destroy communities and ecosystems in coastal areas.

Source: NASA Earth Observatory.

6 Environmental Quality

*M*ore than 1,300 hazardous waste sites threatened environmental quality in the United States in 2023, and industries
released more than 3 billion pounds of toxic chemicals.[1] Every
individual, firm, and government faces decisions that affect environmental quality. Many of these decisions could be improved. The
appropriate size and scope of environmental initiatives depends on
the enormity of impending problems. Regrettably, views of environmental quality are clouded by misinformation and biased by personal agendas. Fervor for votes, profits, or environmental concerns
may blind some politicians, manufacturers, and environmentalists from moderate (and more accurate?) standpoints. Such dis-
tractions from the best available information are a problem with
global repercussions.

The United Nations Environment Program tracks countries'
pledges to cut greenhouse gas emissions. The goal is to limit global
warming to 2°Celsius in this century. The Program reports the
emissions gap between the projected level of emissions in 2030 and
the 42-gigaton emissions level needed to remain on target. In 2022,
the gap was about 15 gigatons, about 2 gigatons larger than the gap
reported for 2021. Amid alarm and apathy, dispassionate research
and objective quality measures offer a basis for convergence. This
chapter summarizes foundations for mutual understanding and
case studies about possible solutions.

1 See www.epa.gov/system/files/documents/2023-03/complete_2021_tri_national_analysis.pdf.

DOI: 10.4324/9781003428732-6

What Is the Quality of the Environment?

Terms of the Trade

A few key terms are useful in discussions of environmental quality. They are presented here in one place to help you compare and contrast terms and for convenience when you need to locate them again.

An **environmental sink** is a repository for potentially damaging by-products of human activity. As with a kitchen sink, we can send a certain amount of "gunk" into an environmental sink before problems occur. Limited amounts of some pollutants can be transformed, diluted, dispersed, or absorbed into the environment without causing any damage. Animals can store small traces of DDT pesticide in their fat tissues with few if any ill effects. Soil can receive moderate amounts of organic waste before harm is done, and by a process called **carbon sequestration**, the oceans and vegetation absorb carbon dioxide (CO_2), release oxygen (O), and store carbon (C).

Carbon sequestration is part of a set of reactions known as photosynthesis, in which solar energy, carbon dioxide from the atmosphere, and water combine to form carbohydrates (sugars) that are the building blocks of **biomass** (plant matter). Considering this, possible solutions to excessive carbon dioxide levels include reforestation and the placement of iron in the oceans to generate more carbon-dioxide-absorbing tiny organisms called phytoplankton. Researchers are studying numerous related sequestration concepts, including the injection of carbon dioxide into geologic formations or deep into the ocean, and chemically transforming it into various products. The U.S. Department of Energy is progressing with a process that pumps carbon dioxide gas into slurry made of water, salt, and magnesium silicate. The slurry is heated and then compressed into solid rock. The CO_2 is thereby transformed chemically into a mineral that will remain stable over millions of years.

The World Business Council for Sustainable Development estimates that 1 billion tires reach the end of their life each year. Too often, those tires create an unsightly confluence of nature and neglect.

For practical reasons, environmental regulations typically apply to the output of stationary pollution sources such as refineries and to the inputs into mobile pollution sources such as automobiles.

When biomass or fossil fuels (which long ago were biomass) are burned, the process is reversed. Oxygen from the atmosphere combines with released carbon to produce carbon dioxide once again.

Natural capital includes natural resources (forests, oceans, mineral deposits) and ecosystems that provide clean air and water. Natural capital is susceptible to resource degradation from the overharvesting of natural resources and from **pollution**, which results from the exhaustion of environmental sinks. The environment has little or no absorptive capacity for **stock pollutants**, which quickly overwhelm environmental sinks and accumulate in the environment over time. Examples of stock pollutants include polychlorinated biphenyls (PCBs), dioxin, DDT, and heavy metals. Pollutants for which the environment has moderate absorptive capacity are called **fund pollutants**. Examples of fund pollutants include carbon dioxide and organic waste. **Flow pollutants** can be initially damaging but are dissipated into environmental sinks with relative ease. Examples include light, noise, and heat pollution and smog.

Natural pollutants include harmful emissions from the Earth and its creatures. Volcanoes and bubbling springs emit substantial amounts of carbon dioxide. Decaying plants emit methane (CH_4). Oceans and bacteria in soil release nitrous oxide (N_2O). Animals, including humans, emit hydrogen methane (H_2), hydrogen sulfide (H_2S), nitrogen (N_2), and carbon dioxide in the processes of breathing, eating, and digestion. **Anthropogenic pollutants** are the undesired products of human activity. When humans use fossil fuels to manufacture goods, transport themselves and their possessions, heat and cool their homes, and turn on their lights and other electronics, they generate pollution. After goods are manufactured, transported, and consumed, their disposal creates further pollution. Garbage in landfills and sewage in treatment plants emit methane and nitrous oxide, among other pollutants, into the air, water, and soil.

A pollution source that remains in the same place, such as a power plant or an oil refinery, is called a

This pipe in India flows directly into the Arabian Sea. Point sources of pollution like this are identifiable and can be targeted for cleanup.

stationary source. A pollution source that moves around, such as a car or a ship, is called a mobile source. The source of pollution has implications for pollution-control policies. It is relatively easy to monitor the *output* of stationary sources. Regulators always know where smokestacks and other major stationary sources are. Mobile sources such as motor vehicles are much harder to monitor. A solution is to regulate the types of fuel that go *into* the mobile sources. For example, policies in many countries restrict the *output* of sulfur dioxide (SO_2) from manufacturing plants and prohibit the use of lead (Pb) as an *input* into gasoline.

Any identifiable source of pollution is considered a point source. Smokestacks, effluent pipes, and leaking oil tankers are all examples. Stationary point sources whose emissions are diffuse such as wildfires, and those too small to track individually such as homes, are called area sources.

Pollution that doesn't come from particular points at all, large or small, is nonpoint source (NPS) pollution. The degradation of water quality is generally associated with NPS pollution. The runoff of rain and melted snow picks up pollutants, including fertilizers, herbicides, insecticides, oil, road salt, and bacteria from human and livestock waste. The runoff then flows into waterways, wetlands, and underground sources of drinking water. Because of its unidentified sources, NPS pollution presents difficult challenges for policymakers. The U.S. Environmental Protection Agency tells the public, "*You* are the solution to NPS pollution."

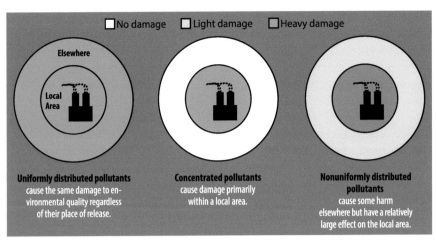

Figure 6.1
Types of Pollutants

No damage   Light damage   Heavy damage

Elsewhere

Local Area

Uniformly distributed pollutants
cause the same damage to environmental quality regardless of their place of release.

Concentrated pollutants
cause damage primarily within a local area.

Nonuniformly distributed pollutants
cause some harm elsewhere but have a relatively large effect on the local area.

Figure 6.1 illustrates how pollutants are categorized on the basis of their range and density. A **uniformly distributed pollutant** (sometimes called *uniformly mixed*) causes the same environmental damage regardless of where it is released. Examples include **greenhouse gases** such as carbon dioxide, methane, nitrous oxide, and hydrofluorocarbons, which cause the **greenhouse effect** by trapping heat in the Earth's atmosphere, no matter where they are released. Similarly, chlorofluorocarbons erode the protective ozone layer and allow more solar radiation to penetrate the atmosphere.

A **concentrated pollutant** causes damage primarily within a local area. For example, the activity within a city creates noise, light, and smog that are unlikely to cause much damage beyond the urban boundaries. Radiation levels from normally functioning nuclear power plants, X-ray equipment, and radon releases become insignificant outside the immediate areas of the sources.

A **nonuniformly distributed pollutant** (also called nonuniformly mixed) causes some harm elsewhere but has a relatively large effect on the local area. For example, arsenic, lead, and mercury emitted from coal-fired power plants may travel a considerable distance, but people living near the pollution source have more health problems and shorter lives than those living further away.[2]

Air Quality

Air quality is subject to both natural and anthropogenic influences. Natural air pollutants include gaseous emissions from the Earth, volcanic ash, pollen, and methane emitted by termites among other living

2   See https://ncmedicaljournal.com/article/54966.

creatures. Sources of anthropogenic air pollution include motor vehicles, power plants, and factories. Although there is controversy over the relative importance of natural air pollutants, it is clear that anthropogenic sources are more readily controllable by existing technologies. For that reason, anthropogenic pollution is the focus of this section.

Standards   **Ambient air** is the air that surrounds us. The ambient air quality in Beijing, China, is often so poor that school children and the elderly are forced to stay indoors. The record-high pollution levels have the Chinese media speaking of an "airpocalypse." In 1952, smog killed roughly 4,000 people in London, England, within only 4 days and led to passage of the U.K. Clean Air Act of 1956. In keeping with the U.S. Clean Air Act of 1970, the EPA established the National Ambient Air Quality Standards (NAAQS) for six **criteria air pollutants** considered harmful to public health, welfare, or the environment. Table 6.1 provides summary information for each pollutant.

Primary standards are designed to protect public health, including that of children, the elderly, asthmatics, and other particularly sensitive groups. For example, the EPA's primary standard for carbon monoxide is an average of 35 parts per million over a 1-hour period. **Secondary standards** are designed to protect public welfare, which is diminished by damage to plants, animals, buildings, and agriculture and by reduced visibility. For example, the EPA's secondary standard for sulfur dioxide is an average of 0.50 parts per million over a 3-hour period.

Table 6.1
Criteria Air Pollutants

Pollutant	Anthropogenic Sources	Health and Welfare Effects	Control Methods
Particulate Matter (PM$_{10}$, PM$_{2.5}$) Airborne particles less than 10 microns in diameter (PM$_{10}$) or less than 2.5 microns (PM$_{2.5}$) in diameter	• Power plant boilers • Steel mills • Chemical plants • Unpaved roads and parking lots • Wood-burning stoves and fireplaces • Motor vehicles	• Aggravates respiratory diseases • May cause lung and heart problems • May carry toxic materials deep into the respiratory system • Impairs visibility	• Filters • Scrubbers • Reduction in fuel combustion

Source: Adapted from information compiled by Ken Larson (www.co.broward.fl.us/aqi02700.htm) and the EPA (www.epa.gov/reg5oair/emission/critpllt.htm).

Table 6.1

(continued)

Pollutant	Anthropogenic Sources	Health and Welfare Effects	Control Methods
Sulfur Dioxide (SO_2) Colorless, nonflammable gas	• Power plant boilers • Sulfuric acid plants • Petroleum refineries • Smelters • Paper mills • Fuel combustion in diesel engines	• Respiratory irritant • Aggravates lung and heart problems • Can damage marble, iron, steel, crops, and vegetation • Impairs visibility • Precursor to acid rain	• Use of low-sulfur fuel • Energy conservation (reduces power plant emissions) • Pollution-control equipment
Carbon Monoxide (CO) Odorless, colorless gas	• Incomplete combustion of carbon-based fuels in motor vehicle and industrial boilers	• Impairs the delivery of oxygen to vital tissues • Impairs vision • Can cause dizziness, unconsciousness, or death	• Transportation planning • Vehicle emissions testing • Efficient combustion techniques • Energy conservation

(continued)

147

Table 6.1
(continued)

Pollutant	Anthropogenic Sources	Health and Welfare Effects	Control Methods
Ozone (O_3) Colorless or bluish gas (smog) formed from volatile organic compounds (VOC) and nitrogen oxides	• Fuel combustion in motor vehicles • Gasoline storage and transport • Solvents and paints • Landfills	• Irritates mucous membranes • Aggravates lung and heart problems • Damages rubber, some textiles, and dyes • Damages plants • Reduces crop yield	• Use of low-VOC solvents • Evaporative controls • Vehicle emissions testing • Pollution-control equipment
Nitrogen Dioxide (NO_2) Reddish-brown gas	• Fuel combustion in motor vehicles and industrial sources	• Respiratory irritant • Aggravates lung and heart problems • Precursor to ozone and acid rain • Causes brown discoloration of atmosphere	• Exhaust gas recirculation • Reduction of combustion temperatures • Energy conservation • Pollution-control equipment

Table 6.1

(continued)

Pollutant	Anthropogenic Sources	Health and Welfare Effects	Control Methods
Lead (P$_b$) Toxic heavy metal	• Smelters • Lead-acid battery man-ufacturing • Electric arc furnaces • Incineration of garbage containing lead • Use of leaded gasoline	• Toxic to the nervous system, organs, and most lev-els of body function	• Phaseout of leaded gasoline • Use of pollution-control equipment in industrial plants

In addition to the criteria air pollutants, the Clean Air Act amend-ments of 1990 added 189 **hazardous air pollutants** (HAPs), also called *toxic air pollutants* or *air toxics*.[3] These pollutants are known or suspected to cause serious health problems as well.

Actual Levels   Over 4,000 air quality monitoring stations across the United States measure pollutant concentrations on an hourly or daily basis. States submit their air-monitoring data monthly for inclusion in the EPA's Aerometric Information Reporting System (AIRS) database. Table 6.2 indicates the number of people living in counties with crite-ria air pollutant concentrations in excess of the National Ambient Air Quality Standards in 2022.

Criteria Pollutant	People Living in Noncompliant Counties
One or more pollutant	85,000,000
O$_3$	78,300,000
PM$_{10}$	19,600,000
PM$_{2.5}$	20,200,000
SO$_2$	900,000

Table 6.2

Population Exposed to Noncompliant Pollution Levels in the United States

Source: EPA, available at www.epa.gov/air-trends/air-quality-national-summary.

3   A list of HAPs from Section 112(b)(1) of the Clean Air Act is available at www.epa.gov/ttn/atw/188polls.html.

Typical incentives for compliance with air quality standards include *offset sanctions* and the withholding of federal highway funds for noncompliant counties. When expansion or new development causes a firm in a noncompliant county to increase its emissions by some amount, a **2:1 offset sanction** requires the firm to reduce existing emissions by twice the amount of the increase. This reduction can be accomplished with pollution-control equipment or by purchasing tradable pollution permits as described later. For example, suppose a factory expansion causes a firm to emit 10 additional tons of lead per year. The firm would have to reduce lead emissions from another part of its operations by 20 tons. The idea is that nonattainment counties must decrease their emissions to reach the acceptable standard, and a 1:1 offset (meaning each additional ton of emissions in the new facilities would be offset by one fewer ton of emissions elsewhere) would simply maintain the current level of pollution. If firms must reduce their emissions by twice the amount created by the new facilities, total emissions will fall to a level closer to the desired standard. Over the past two decades, the EPA has sent out hundreds of formal notifications of intent to apply sanctions. Most counties remedy the problem within the 18-month period between formal notification and the imposition of sanctions.

Acid Deposition   Acidity is measured on a pH scale, with lower numbers meaning higher acidity. Wet **acid deposition**, commonly referred to as acid rain, occurs when rainwater, fog, or snow has a pH level lower than 5. Pure water has a pH of 7.0. Rain has a slight natural acidity caused by atmospheric carbon dioxide, giving it a pH of about 5.5. Acid rain generally has pH levels between 4.2 and 4.4. Acid rain is deadly to forests, aquaculture, and humans and damaging to automobiles and buildings. The ability for trees to absorb nutrients is dependent on soil acidity, and increases in acidity can starve trees and other vegetation. Beyond its direct toxicity, acid releases copper, aluminum, and mercury stored in rocks and pipes.

The primary cause of acid rain is sulfur dioxide (SO_2), which dissolves in water to form sulfuric acid (H_2SO_4). Various oxides of nitrogen, collectively called NO_x, and hydrochloric acid are contributing causes. Coal-fired power plants are the largest controllable source of SO_2 and NO_x, followed by industry and road transportation. The Clean Air Act amendments in 1990 substantially reduced acid rain caused by U.S. emissions over the following decades. China is among the countries facing continuing problems with acid rain.[4]

Acid deposition can affect wildlife great distances from the causal sources of air pollution. Lifeless trees and lakes in Canada and Norway have been attributed to sulfur dioxide emissions in the United States

4   See https://doi.org/10.5814/j.issn.1674-764x.2021.05.002.

Source: NASA Earth Observatory.

Warmer oceans fuel bigger hurricanes. Hurricane Ian, shown here as photographed from the International Space Station, caused more than $113 billion worth of damage in 2022.

and central Europe. Among the financial costs, Norwegians spend more than $12 million each year adding lime to rivers and lakes in efforts to protect fish stocks from acidification.[5] The EPA reports that their cap of 8.95 million tons of SO_2 emissions decreased mortality, hospital admissions, and emergency room visits in the United States and brought the annual health benefits from the Acid Rain Program to $50 billion.[6]

As the most accessible supplies of low-sulfur coal are depleted, power companies turn to coal with sulfur levels up to ten times higher. Scrubbers that remove SO_2 and NO_x from power plant emissions have price tags in the hundreds of millions of dollars. On the bright side, an innovative approach to acid rain involving tradable pollution allowances has reduced annual sulfur dioxide emissions in the United States by about 10 million tons. The economics of scrubbers and emissions trading is explained in greater detail in the last section of this chapter and in Chapter 12.

Global Climate Change   British naturalist David Attenborough is among the experts who consider climate change the greatest challenge modern humans have ever encountered. Pollution contributes to collections of greenhouse gases that hold heat within Earth's atmosphere and keep it about 33°C (59°F) warmer than it would otherwise be.

---

5 See www.environment.no/topics/air-pollution/acid-rain/.
6 See www.epa.gov/sites/production/files/2016-03/documents/clearingtheair.pdf.

Table 6.3

Indications of Climate Change

Greenhouse Gases

- U.S. Emissions
- Global Emissions
- Atmospheric Concentrations
- Climate Forcing

Weather and Climate

- U.S. and Global Temperature
- High and Low Temperatures
- U.S. and Global Precipitation
- Heavy Precipitation
- Drought
- Tropical Cyclone Activity

Oceans

- Ocean Heat
- Sea Surface Temperature
- Sea Level
- Ocean Acidity

Snow and Ice

- Arctic Sea Ice
- Glaciers
- Lake Ice
- Snowfall
- Snow Cover
- Snowpack

Society and Ecosystems

- Streamflow
- Ragweed Pollen Season
- Length of Growing Season
- Leaf and Bloom Dates
- Bird Wintering Ranges
- Heat-Related Deaths

Note: On the basis of these indicators, the EPA concludes that climate change now affects ecosystems and society, and that it is linked to fossil fuel combustion.

Atmospheric concentrations of the most prevalent greenhouse gas, carbon dioxide, have risen by nearly 30 percent since the industrial revolution. Methane concentrations have more than doubled, and nitrous oxide concentrations have risen by about 15 percent. These increases, and other human and natural trends, contribute to the 1.06°C (1.90°F) increase in global temperatures since the end of the nineteenth century.[7] The ten warmest years since record keeping began in 1850 have all occurred since 2010. The warming trend causes more powerful hurricanes (also called cyclones), a thinning of ice over the Arctic Ocean, more frequent extreme rainfall, and a 6- to 8-inch increase in the sea level over the past century.[8]

The EPA tracks the 26 indicators of global climate change shown in Table 6.3. The indicators are selected based on the quality of available data. The conclusion is that climate change is already affecting society and ecosystems. The EPA reports that

Scientists are confident that many of the observed changes in the climate can be linked to the increase in greenhouse gases in the atmosphere, caused largely by people burning fossil fuels to generate electricity, heat and cool buildings, and power vehicles.[9]

7 See www.climate.gov/news-features/understanding-climate/climate-change-global-temperature.

8 See https://sealevel.nasa.gov/faq/13/how-long-have-sea-levels-been-rising-how-does-recent-sea-level-rise-compare-to-that-over-the-previous/.

9 See www.epa.gov/climatechange/science/indicators/index.html.

According to the Intergovernmental Panel on Climate Change:*

- Warming of the climate system is unequivocal, as is now evident from observed increases in the global average air and ocean temperatures, widespread melting of snow and ice, and the rising global average sea level.

- Global atmospheric concentrations of carbon dioxide, methane, and nitrous oxide have increased markedly as a result of human activities since 1750 and now far exceed pre-industrial values determined from ice cores spanning many thousands of years.

- The global increases in carbon dioxide concentration are due primarily to fossil fuel use and land-use change, while those of methane and nitrous oxide are primarily due to agriculture.

- Improvements in the understanding of anthropogenic warming and cooling influences on climate have led to very high confidence that the global average net effect of human activities since 1750 has been one of warming.

- Paleoclimatic information supports the interpretation that the warmth of the last half century is unusual considering at least the previous 1,300 years.

- The last time the polar regions were significantly warmer than they are now for an extended period—about 125,000 years ago—reductions in polar ice volume led to a 4- to 6-meter rise in the global sea level.

- For the next two decades, a warming of about 0.2°C per decade is projected for a range of emission scenarios.

- Most of the observed increase in global average temperatures since the mid-twentieth century is very likely due to the observed increase in anthropogenic greenhouse gas concentrations.

- Discernible human influences now extend to other aspects of climate, including ocean warming, continental-average temperatures, temperature extremes, and wind patterns.

- It is very likely that hot extremes, heat waves, and heavy precipitation events will continue to become more frequent.

Source: *See www.ipcc.ch/.

Figure 6.2 Global atmospheric carbon dioxide compared to annual emissions (1751–2022)

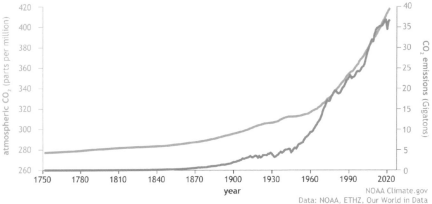

Carbon Dioxide: Annual Emissions and Atmospheric Levels

Source: NOAA Climate.gov. Data: NOAA, ETHZ, Our World in Data.

Figure 6.2 shows U.S. National Oceanic and Atmospheric Administration estimates of carbon dioxide levels in the atmosphere (the blue line) and carbon dioxide emissions by humans (the gray line). Humans emit more than environmental sinks can remove, so our emissions directly increase atmospheric levels. That is problematic given our level of emissions. Scientists working for the United Nations Environment Program report that with current trends in emissions and emission-reduction efforts, we are on track for temperature increases in excess of 2°C by the end of this century. Continuing changes in climate, soil moisture, sea level, and weather patterns threaten vital ecosystems and agricultural production. Experts have stressed to the United Nations Security Council that climate change will raise seas above cities and cause forced migration and competition for fresh water, land, and other critical resources.[10]

Water Quality

Surface water is water that is open to the Earth's atmosphere. It includes the rivers, lakes, oceans, and streams that cover 70 percent of the planet. *Groundwater* is fresh water located primarily in large aquifers beneath the Earth's surface. Over 95 percent of the world's water supply is salt water in oceans and inland seas. Groundwater, ice caps, and glaciers make up almost all the fresh water. Fresh surface water makes up only 0.01 percent of the total water supply. None of these sources is impervious to pollution.

Today's older generations may remember when bottled water was not commonplace. People no longer have the same trust in their water

10 See https://press.un.org/en/2023/sc15199.doc.htm.

supplies. In the more distant past, those canoeing the great rivers could push a paddle into the water and then lift the paddle above their heads to drink water as it ran off the handle. Don't try this on your next canoe trip. Like air, water is vulnerable as a common property resource for which few property rights exist. As detailed in the Reality Check, pollutants are found in virtually every corner of the world. In 1969, the surface of the badly polluted Cuyahoga River in Ohio caught fire and burned for several days. Still today, there are advisories warning consumers about high levels of mercury, PCBs, or dioxin in certain types of fish from waterways in Ohio among many other states and regions all over the world. Swimming also poses health risks in many waterways.

reality check

Even in the Arctic: PCBs and Hermaphrodite Polar Bears

PCBs are a group of synthetic industrial chemicals used in everything from coolants and casting wax to pesticides and plastics. There are no known natural sources of PCBs. Due to their extreme toxicity, the Toxic Substances Control Act of 1976 directed the EPA to ban the manufacture of PCBs and regulate their use and disposal, which the EPA did in 1979. Most industrialized nations followed suit. The Stockholm Convention on Persistent Organic Pollutants banned the production of PCBs in 2004 and aimed to phase out the use of PCBs by 2025. Nonetheless, as evidence of the difficulty of reversing environmental problems, PCBs are still present in and around many waterways, landfills, and industrial spills. Like all persistent organic pollutants (POPs), PCBs are considered *forever chemicals* because they are highly resistant to breaking down. Other forever chemicals include the widely banned dichlorodiphenyltrichloroethane (DDT) used to fight mosquitos and the perfluoroalkyl substances (PFAs) still found in consumer products such as some food packaging and stain-resistant coatings.

These chemicals are as widespread as they are persistent. PCBs and other POPs including DDT and PFAs show up in the most remote regions of the Earth. On the Arctic island of Svalbard, PCBs are thought to be the cause of hermaphrodite polar bears with both male and

(continued)

female reproductive organs. Analysis of the DDT components found in the Arctic indicates that some of the culprit chemicals are only weeks old when they arrive. This is evidence that the banned substances are still in production and that their damage spans the globe.

The PCB levels Bytingsvik et al. (2012) found in polar bear cubs were 100 times the levels known to affect thyroid hormones in human babies. Other problems associated with PCBs in humans include reproductive disorders, skin ailments, liver disease, and cancer. PCBs accumulate in body fat and are passed through the food chain. The U.S. Food and Drug Administration has issued an advisory against eating fish with more than 2 parts per million of PCBs. Fishers, human and otherwise, take heed.

The abuse of river resources is particularly tempting because effluents released into flowing water immediately become the problem of someone else downstream. Of course, that is the short-run story. What goes around comes around. Rivers flow into oceans, which provide food for the masses and support critical ecosystems for the planet as a whole.[11]

The Value of Water Quality   Astronomers are ever alert for evidence of water on other planets because it is a precursor for all life forms. Some amount of non-toxic water must pass through all living things during the *hydrological cycle* that takes water down from the clouds as precipitation, to (and below) the Earth's surface, up through evaporation, and back down as precipitation again. The human body is about 60 percent water. Beyond that, we rely on varying degrees of water quality for hygiene, aquatic life support, recreation, agriculture, many facets of production, and sustenance of the natural environment more broadly.

Table 6.4 provides examples of the high value of clean water. Notice that most of these studies estimate the value of water quality to specific groups like homeowners or swimmers. Many other individuals value clean water for its existence and availability, now and in the future, for humans and wildlife. Chapter 10 discusses the specific methods of placing values on natural capital and provides additional examples of estimated values placed on wildlife.

Losses from oil spills provide striking examples of the costs of polluted water. The 2010 *Deepwater Horizon* spill at an offshore drilling rig released 210 million gallons of oil into the Gulf of Mexico and caused nearly $10 billion in lost earnings in the seafood industry, medical costs, and property damage. In 2023, the *MT Princess Empress* tanker

---

11 A sobering anecdote: In 2001, an economist from Wisconsin (one of the northernmost tributaries to the Mississippi River) passed away after eating toxic seafood from Louisiana (where the Mississippi flows into the sea).

Table 6.4

The Value of Clean Water

Study	Quality Issue	Group Studied	Estimated Value
Austin et al. (2007)	A 20 percent reduction in beach closings due to pollution	Swimmers in the Great Lakes region of the United States	$2.5 billion per year
Georgiou et al. (2000)	Clean water at bathing beaches	Bathers in England	$53/bather per year
Guha (2007)	Availability of drinkable water	435,000 in Calcutta, India	$4.5 million per year
Jalilov (2018)	Water clean enough to swim and fish in	Residents of Manila, Philippines	$190 million per year
Mamun et al. (2023)	A 10 percent improvement in lake water quality	Property owners in the United States	$6–9 billion appreciation in property values
Michael et al. (2000)	Lake water clarity	Homeowners in Maine	$5,000–$10,000 per home
von Haefen et al. (2023)	Improved water quality in a river	Residents in North Carolina	$54 million

carrying 211,338 gallons of oil spilled its cargo off the Philippines. The oil-blackened mangroves and coral reefs, sidelined 18,000 people in the fishing industry, dried up tourism, and threatened food chains for most living things in the area.

Polluted water causes death and disease among humans and wildlife alike. Fuller et al. (2022) estimate that water pollution causes 1.36 million deaths and that pollution of all types causes 9 million deaths globally each year. As another indication of the value of clean water to humans worldwide, consumers purchased about 475 billion liters of bottled water in 2023 at a cost of more than $340 billion.[12] Indeed, as an essential ingredient to life, the value of potable (fit to drink) water can be no less than the value of life.[13]

Threats to Water Quality   The United Kingdom's Environmental Agency reported 44 "serious" pollution incidents involving water companies in 2022, down from 62 the previous year. Spills that are not

12 See www.statista.com/outlook/cmo/non-alcoholic-drinks/bottled-water/worldwide.

13 Specific estimates of the values of human and animal lives are discussed in Chapter 10.

Pipelines are vulnerable to spills and contamination.

Fish advisories signal dangerous levels of water pollution.

considered serious are far more common. Tens of thousands of accidental releases of hazardous substances are reported to the U.S. National Response Center each year. Among the policy responses to past oil spills, the Oil Pollution Act of 1990 required vessels operating in U.S. waters to have protective double hulls by 2015. The International Convention for the Prevention of Pollution from Ships (MARPOL) required all tankers to be double-hulled by 2023.

Unlike the air, almost anything can be spilled or dumped into water. Despite the thousands of identified oil spills and effluent releases every year, nonpoint source pollution is the leading cause of water pollution. The U.S. Clean Water Act requires states to conduct biennial water quality inventories. Over half of the rivers, streams, and lakes in the United States are considered impaired due to excess pollution. Nonpoint source pollution such as excessive nutrients (nitrogen and phosphorus), metals (primarily mercury), bacteria, siltation, oxygen-depleting substances, and pesticides are among the leading pollutants. Figure 6.3 provides an example of waterways designated as polluted in the state of Colorado. In China, where roughly 70 percent of lakes and rivers are unsafe for human use, water pollution causes an estimated 100,000 deaths and $1.5 trillion in financial losses.[14]

Water Quality Standards Most countries and all U.S. states set quality standards for water. Among the common bases for these standards are the maintenance of existing water quality, goals for

14 See www.pnas.org/doi/10.1073/pnas.2015175118.

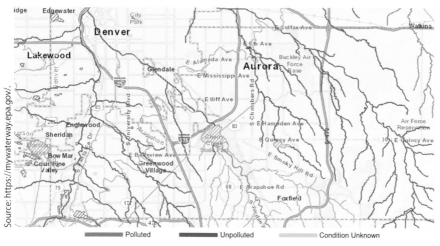

Source: https://mywaterway.epa.gov/.

Figure 6.3

Impaired U.S. Waters

Polluted   Unpolluted   Condition Unknown

Pollution threatens waterways around the world. Even places like Colorado, which is famous for its natural beauty, struggle with water impairments that limit aquatic life and recreational opportunities.

intended uses of the water, and criteria for protecting the water. Water quality criteria include keeping pollutants at levels that are unlikely to harm plants, animals, and human health. There are also biological water quality criteria that use the presence and condition of algae, fish, and other aquatic organisms as indicators of the health of a body of water.

The quality and vulnerability of a region's water depends largely on its *watershed*, which is a land area that catches rain and snow and sheds (drains) it into a body of water. Watersheds carry nonpoint pollutants into the flow of consumable water. The EPA tracks the attainment of various water quality standards separately in rivers and streams, lakes and reservoirs, and bays and estuaries. For example, for the entire United States, they find that 32 percent of lakes and reservoirs meet the standards for fish consumption and 29 percent of rivers and streams meet the standards for drinking water.[15] If you live in the United States, you can learn the specifics of your own watershed at www.epa.gov/waterdata/hows-my-waterway.

Under the Clean Water Act, states face federal sanctions if they fail to restore impaired bodies of water. States often delegate standard setting, implementation, and enforcement duties to branches of local government or designated conservation districts. Standards are applied by an overlapping set of *general discharge prohibitions*, agriculture laws, forestry laws, fish and game laws, nuisance prohibitions, land-use planning and regulatory laws, and criminal laws. The **general discharge prohibitions** either require a permit for the release of designated

15 See https://environmentalintegrity.org/wp-content/uploads/2022/03/CWA-report-3.23.22-FINAL.pdf.

substances or prohibit emissions that cause (or help cause) water quality levels to violate established standards.

Unfortunately, it is often difficult to prove a direct link between specific releases and impaired water quality. The requirement of a *discharge permit* removes the need to demonstrate any causal link between pollution and the violator's behavior and thereby lessens the burden of enforcement. The polluters are asked to file for a permit, which means they must identify themselves and, in some cases, subject themselves to inspection. The number of permits can be limited as a check on pollution levels. And the application process makes it easy to communicate cautions and restrictions to polluters. Polluters who do not obtain a permit face sanctions upon discovery, regardless of any demonstrable association between their releases and subsequent damage. Many of the more targeted laws are enforced as petty criminal offenses. These include laws that limit the percentage of vegetation that may be removed near a waterway, special rules for timber operations near wetlands, and sediment control laws that require construction companies to prevent eroded soil from damaging nearby bodies of water.

Noise and Light Pollution

Problems on Land   Noise and light are good examples of flow pollutants because they do not accumulate in the environment. Even so, they can cause considerable damage. The World Medical Association warns that sound levels produced by some industrial sources, transportation systems, and audio systems may be emotionally disturbing and lead to permanent hearing loss.[16]

The detrimental effects are not limited to humans. Chapter 2 discussed how noise disrupts the hibernation of bears. Noise pollution also harms living things ranging from insects and crustaceans to birds and mammals.[17] The repercussions include reduced food intake, separations of parents and offspring, temporary deafness, reduced thyroid function, and loss of reproductive fitness.

Problems Under the Sea   Noise pollution is also a problem underwater, where sources include ships, oil exploration equipment, military sonar, pingers, explosives, and dredgers. Unnatural noises have serious effects on marine life. At close range, loud noises can rupture lungs and ears (McCarthy, 2001). Even distant sounds can mask the echolocation pulses marine mammals use to navigate and feed in dark or murky water. And noise pollution can mask communications with offspring and mates, while giving predators a deadly advantage. When the COVID

16 See www.wma.net/policies-post/wma-statement-on-noise-pollution/.
17 See Sordello et al. (2020) for a meta-analysis of noise's influence on wildlife.

pandemic shut down ferry traffic around Hong Kong, the quieted waters brought back the rare Indo-Pacific humpback dolphin.[18]

The European Parliament and the U.S. Congress are among many legislatures around the world that have considered noise restrictions in places that include national parks and marine reserves. While good ideas abound, it is challenging to enact and enforce noise pollution standards. The common property aspects of the oceans and skies are further complicated by the international scope of ocean vessels, the noise associated with military operations, and the many mobile, nonpoint sources of noise.

Problems in the Sky   Noise pollution isn't much to look at, but the NASA photograph sheds light on another flow pollutant. Excess light that spills into the atmosphere represents wasted energy. Note the uneven distribution of this waste across densely populated areas with differing levels of wealth and development. For example, India has four times the population of the United States but wastes less light. As discussed in the next chapter, the production of energy for lighting among other uses is a primary source of environmental damage worldwide. Light pollution also causes stress for wildlife just as noise pollution does, and its mere existence can render otherwise suitable wildlife habitat uninhabitable.

Contemporary light fixtures tend to spread light outward and upward where it is not needed. Unneeded lighting in the United States wastes an estimated $6.3 billion worth of energy annually, which amounts to 60 billion kWh and 23 billion pounds of CO_2 emissions.[19] Solutions include

- *The use of "fully shielded" or "full-cutoff" outdoor lighting that points light downward and inward on the subject to be illuminated rather than outward and upward*

- *The elimination of unnecessary light fixtures*

- *The transition from all-night illumination to illumination only when facilities are in use (with the exception of security lighting)*

- *The reduction of bulb wattages*

- *The replacement of incandescent bulbs with fluorescent bulbs, or even better, LED bulbs.*

No place knows light bulbs like Las Vegas, where several hotels boast multimillion bulb light displays that shine all night long. Hotels owned by MGM Resorts have replaced over a million light bulbs with LED

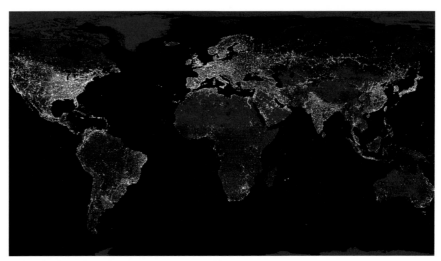

Excess light that spills into the atmosphere represents wasted energy and light pollution that can disrupt wildlife and disturb humans.

bulbs, and the city of Las Vegas converted 42,000 streetlights to LED lighting. By what kind of light do you read?

Standards   The United States offers a case study of noise pollution approaches and pitfalls. The Noise Control Act of 1972 requires the EPA to regulate major sources of noise. The Federal Highway Administration enforces EPA noise emission standards for motor vehicles used in interstate commerce; the Occupational Safety and Health Administration regulates workplace noise; the Federal Aviation Administration regulates aircraft noise emissions; and the National Environmental Policy Act requires agencies to assess noise impact as part of environmental impact statements.

Although noise regulations are in place[20] and many are effectively enforced, political wrangling can prevent their full implementation. For example, the Office of Management and Budget (OMB) under President Ronald Reagan defunded the Office of Noise Abatement and Control (ONAC) in 1981. ONAC was the EPA's source of funding to enforce Noise Control Act regulations. So to this day, the EPA is responsible for enforcing the Act's noise regulations but receives no funding so do so. To make matters worse, the Noise Control Act preempts many state and local noise regulations, making it difficult for local governments to fulfill the goals of unfunded federal programs.[21]

To accommodate the differing needs and interests of diverse communities, light pollution is generally regulated at the local level. As an

20 For example, the Boeing Corporation operates a website about airport noise regulations around the world.
   See www.boeing.com/commercial/noise/.

21 See www.acus.gov/recommendation/implementation-noise-control-act.

Homes in Sedona, Arizona, where local policies restrict light and sound pollution and require buildings to be painted colors that blend in with the natural environment.

example, the Keep Sedona Beautiful Dark Skies committee in Sedona, Arizona, helps the city establish and enforce its lighting ordinances. Targets include accident-causing glare, light trespass from neighboring homes and businesses, energy waste, and sky glow. The committee works explicitly to balance commercial needs for signage and safety lighting with community needs for public health, safety, and welfare. Careful attention to the environmental issues of lighting, noise, water quality, and low-impact development has also improved local tourism and property values.

Where Do We Go from Here? A Brief Look

Efficiency dictates that the optimal quantity of pollution is not zero. Chapter 2 explained the desirability of continuing all activities until the social marginal benefit no longer exceeds the social marginal cost. The social marginal benefit from the first ton of SO_2 emissions might be that ambulances can rush accident victims to hospitals. At the same time, a ton of SO_2 spread around the atmosphere might not have any appreciable social marginal cost. Given that the social marginal benefit of activities that cause pollution starts out high and the social marginal cost starts out low, some positive level of pollution is efficient and permissible in the name of human welfare, as illustrated in Figure 6.4. The social marginal benefit decreases after the most valuable polluting activities are carried out, and the social marginal cost increases as environmental sinks are exhausted and contamination levels become unsafe.

When pollution levels exceed the efficient level, the task is to identify which pollution avoidance measures provide benefits that exceed the costs. There are ongoing debates over the efficiency of measures to thwart global warming, regulations on nuclear power, and

Figure 6.4

*The Efficient
Level of
Pollution*

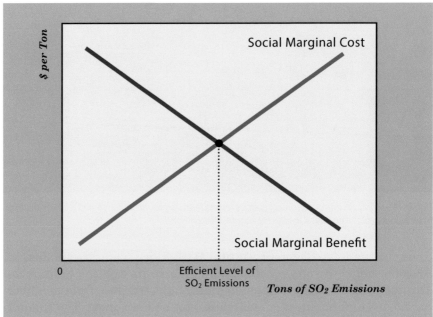

The efficient level of pollution is not zero. It is the level that equates the social marginal cost and the social marginal benefit.

emissions standards for automobiles. The advisability of other measures is more clear-cut. Low-wattage light bulbs, low-volume-flush toilets, and improved home insulation can all cut costs and pollution at the same time, as can turning out the lights when leaving a room. Similarly, Chapter 7 discusses alternative energy sources with few environmental costs. This section begins the conversation about progress toward efficient levels of pollution.

Policy

Point sources of pollution are easier to monitor than nonpoint sources, making point sources a common target for emissions standards and emissions charges. For example, planes landing at the Bern-Belp Airport in Switzerland pay a landing fee based on the level of NO_x emissions and the noise level generated by the particular type of aircraft.[22] Because nonpoint sources are harder to monitor, it is preferable to regulate the manufacturers of products that cause nonpoint pollution, such as the makers of pesticides or lumber with chemical preservatives, rather than regulating the pollution itself.

22 See https://plonesaas.devel4cph.eea.europa.eu/epanet/reports-letters/reports-and-letters/ig-noise_aircaft-noise-abatement.pdf.

Unlike flow pollutants, stock pollutants affect present and future generations and necessitate the selection of discount rates for the analysis of future costs and benefits. Paradoxically, the most damaging pollutants are often the most difficult to regulate. While concentrated pollutants are more likely to be internalized and controlled locally, distributed pollutants can extend beyond national jurisdictions. The control of uniformly distributed pollutants, such as greenhouse gases, and nonuniformly distributed pollutants, such as SO_2, may require multinational accords established at meetings like the UN Climate Change Conference in Poland.

International policymaking is tricky due to power struggles, income disparities, and pollution that is created in one place to serve customers in another place. When leaders meet to address global environmental issues, the solutions typically involve specific reductions in the pollution from each country. But if pollution is created in one country to satisfy the needs and wants of consumers in another country, it is arguable that the pollution should be attributed to the consumers' country, rather than to the producers' country. Moran and Kanemoto (2016) trace emissions to final consumers and discuss the prospect of shared responsibility for export-driven pollution.

Education

Many people don't understand the environmental repercussions of their material consumption, municipal solid waste generation, or energy use. The average household in the United Kingdom wastes $260 worth of energy each year with such habits as overfilling teakettles and washing clothes in warmer water than is necessary.[23] Surveys in the United States reveal that many adults do not know how their energy is generated. The outpouring of money to save identified victims, such as beached whales and oil-covered otters, suggests that if more of the victims of pollution were identified (particularly big and fuzzy ones), we would observe greater sacrifices in favor of the environment.

Habits can be broken. For example, when 75,000 U.S. residents received information on how much energy their neighbors used, those with relatively high levels of use reduced their consumption by between 1.2 and 6 percent on average.[24] Like information, education makes a clear difference when carried out well. The most successful programs in environmental education convey knowledge of relevant environmental concepts, provide issue analysis, and involve participants in the resolution of environmental problems. Programs that last several months or

---

23 See https://www.msn.com/en-gb/money/other/the-energy-wasting-teatime-habits-adding-205-to-your-bill/ar-AA1mpylW?ocid=finance-verthp-feeds.

24 See https://eric.ed.gov/?q=ed320761&id=ED320761.

longer are more effective than brief or one-time activities.[25] Education also increases wealth, which can lead to more consumption of material goods, but also more spending on environmental protection. Many wealthy nations achieve relatively high air and water standards and spend more on clean energy, environmental regulations, and environmental agencies.

New Technology

Automobile gas mileage has improved dramatically over the past 25 years with the wide availability of *hybrid electric vehicles* (HEVs). The gas-and-electric engines in HEVs reduce pollution by nearly doubling the gas mileage of their all-gasoline counterparts. A hybrid engine runs on electricity at low speeds when gasoline engines are the least efficient and reverts to gasoline at higher speeds for improved power and range. The batteries in HEVs are charged by the gasoline engine. *Electric vehicles* (EVs) run entirely on electricity, and their batteries are charged with a plug fed by any external electricity source. EVs reduce the environmental impact of transportation *if* the electricity is produced with clean energy (as from solar, wind, hydropower, tidal power, and so on).

Plug-in hybrid electric vehicles (PHEVs) can be powered by electricity from an external source or by gasoline. Their larger batteries allow PHEVs to surpass the all-electric range of HEVs. The option to use gasoline eliminates the "range anxiety" that EV drivers can experience on long trips, although battery technology is rapidly extending the range of EVs. Other energy-saving technology for cars includes *integrated starter-generator* (ISG) systems that prevent idling by automatically shutting the engine down at stoplights and restarting it when the brake is released. *Regenerative braking systems* capture energy from braking to help recharge the battery. And fuel economy improves with design changes that lower the weight and height of vehicles and improve their aerodynamics.

Advancements in clean energy include solar panels that are more efficient, longer lasting, and more recyclable. Solar panels are being integrated into roof shingles and windows, making every rooftop and every window a potential source of solar power. Other innovations include underwater wave-energy harvesters and bladeless wind-power generators.[26] With growing acceptance, economies of scale can make many types of environmentally responsible technology more affordable. The next section describes a successful means of encouraging the use of clean energy.

25 See www.ed.gov/databases/ERIC_Digests/ed320761.html.
26 See www.cnet.com/science/this-underwater-buoy-could-power-homes-by-capturing-the-oceans-power/
and https://vortexbladeless.com/.

Market-Based Incentives

Market-based policies create incentives for pollution reduction by influencing market prices. Common mechanisms for this influence include taxes, subsidies, and *emissions trading programs*. Market-based policies generally provide polluters flexibility in how they respond to the incentives. Having discussed taxes and subsidies in previous chapters, we will introduce emissions trading here and follow up on these and other policies in later chapters. One type of emissions trading, *allowance trading* or *cap and trade*, was adopted in the United States to address the acid rain problem in 1995. Like most **cap-and-trade programs**, the U.S. Acid Rain Program sets a limit or "cap" on the volume of a particular pollutant (SO_2) that can be released from regulated sources. Each existing source then receives **allowances** to emit a set amount of a pollutant. Some number of allowances may be provided to polluting firms for free. Firms have the option to purchase more allowances at the market price, but the total number of allowances remains capped. When the cap is set below existing emissions levels, the cap-and-trade program reduces the total volume of emissions. The European Commission developed the first international emissions trading system in 2005 to limit CO_2 releases. Similar programs are in place for many substances around the world.

Emissions trading programs allow for flexibility and creativity. Permissible emissions levels can be achieved via conservation, new technology, alternative energy sources, lower-sulfur coal, or the purchase of allowances—whichever is the most feasible and affordable on a case-by-case basis. Older pollution sources that could only reduce emissions at great expense can purchase allowances from newer sources that achieve reductions relatively inexpensively.

Under a cap-and-trade program, even firms with enough allowances to cover their current emissions levels are incentivized to find ways to emit less. As long as reductions cost less than the going price of allowances, further reductions allow firms to sell extra allowances to other firms for a profit. The applications of this approach are limitless. For example, tradable permits are used to allocate fish in New Zealand and tradable fuel economy credits have been proposed for automakers.[27] Tradable emissions permits are among the policies discussed in greater detail in Chapter 12.

In contrast to market-based incentives, **command-and-control policies** require uniform reductions or specific technology applications for every pollution source of a particular type. A command-and-control policy could require all coal-fired power plants to implement a cement kiln flue gas recovery scrubber.[28] Such a requirement may achieve the

27 See www.rff.org/files/document/file/RFF-Rpt-AutoCreditTrading.pdf.

28 This is a system that cleans pollutants out of the emissions of coal plants.

targeted pollution level, but it does not allow for flexibility and creativity. The rigidity of command-and-control policies is problematic when the polluters vary in their ability to comply with policies. For example, requirements for state-of-the-art smokestack scrubbers may be easy for new power plants to obey but difficult for others, depending on their age and the adaptability of their equipment.

Summary

During the early stages of child development, infants watch objects while they are visible but do not pursue them behind a blanket, as if the objects cease to exist when they cannot be seen. In some contexts, this out-of-sight, out-of-mind mentality continues into adulthood. Some people litter as if the trash does not exist when they cannot see it. Worldwide, consumption decisions lead to the release of more than 45 billion metric tons of carbon dioxide, as if what goes around does not come around. Are we polluting too much? And if so, how important is the problem? The answers to these questions determine the extent to which individuals and policymakers should pursue private and public environmental efforts.

Air pollution can impair the heart, lungs, and nervous system of humans and threatens wildlife, buildings, and the global climate. Rising ocean temperatures intensify hurricanes and rising seas submerge seaside communities. Water pollution contaminates drinking water, poisons the habitat of endangered species, and causes mutations even in the farthest reaches of the planet. Impaired waterways also threaten food supplies, not to mention the world's most popular participation sport: fishing. Noise pollution causes hearing loss, injurious stress, and mammalian deaths. Beyond being a nuisance to humans, light pollution causes exorbitant energy consumption, animal fright, habitat loss, and the erosion of property values.

The marginal cost curve intersects the marginal benefit curve at a positive level of pollution. Determining this efficient level of pollution is easier said than done, and it is a moving target. Innovative products and processes are lowering the marginal benefit from pollution while the exhaustion of environmental sinks increases the marginal cost. There are several possible solutions. Some members of society respond to educational programs that raise awareness of problems and alternative paths. Even decision-makers who are entirely self-interested respond to market-based incentives. And emissions trading programs have decreased pollution at the regional, national, and international levels. Above all, an understanding of the state of our environment is the first step toward success in the optimization of environmental practices and policies.

Problems for Review

1. Categorize each of the following pollutants as uniformly distributed, concentrated, or nonuniformly distributed:

   a) Methane (a greenhouse gas) emissions from livestock

   b) Light pollution from a football stadium

   c) Heavy metals emitted from a coal-fired power plant

2. Classify the following as a stock pollutant, a fund pollutant, or a flow pollutant:

   a) Manure from cattle farming, reasonable amounts of which can be absorbed into the environment within a few months

   b) Noise pollution from fireworks

   c) Used automobile tires, which take a half century or longer to decompose

3. According to the chapter, what is the difference between the preferable pollution-control policy for point sources and for nonpoint sources?

4. This chapter highlights the value of clean water.

   a) Table 6.4 reveals that billions of dollars of value can be gained by improving water quality by as little as what percent?

   b) What is the most you would be willing to pay per year to be able to drink water safely out of the tap rather than filtering water or purchasing bottled water?

   c) Multiply your answer to part (a) by the population of your country to determine what safe drinking water is or would be worth to the citizens of your country if everyone valued it like you do.

5. According to the Intergovernmental Panel on Climate Change, global increases in methane and nitrous oxide concentrations are primarily due to which of the following?

   a) Volcanic activity

   b) Transportation systems

   c) Agriculture

   d) Manufacturing

6. The chapter explains the use of offset sanctions.

   a) In what particular situation are offset sanctions applied?

   b) Suppose a factory expansion led to an increase of 1 ton of particulate matter emissions per day. If an offset sanction required the firm to decrease particulate matter emissions by 3 tons per day elsewhere in its operation, what specific offset sanction was applied?

7. According to the chapter, what is one way education helps the environment and one way it hurts the environment?

8. Recall the discussion of discharge permits in the chapter.

   a) What is a specific advantage of requiring discharge permits when it comes to enforcement?

b) *Explain a pollution problem in your area for which a discharge-permit requirement might help protect the environment.*

9. A thorough cost-benefit analysis of motor vehicles with fuel combustion engines and lead-acid batteries would include estimates of the costs of emitting which of the criteria air pollutants? (Hint: See Table 6.1.)

10. Suppose that to reduce smog, the city where you live is contemplating either (1) requiring everyone to ride a moped to work or (2) providing only enough gasoline for each family to receive 10 gallons per week, which can be bought or sold among families.

a) *Categorize each of these policies as either command-and-control or market-based.*

b) *If you were asked to recommend one of these policies, which would you choose? Explain one advantage of your chosen policy and one problem with the alternative policy.*

websurfer's challenge

1. Find one map that indicates levels of a particular type of pollution in the area where you live.

2. Find one website that agrees and one website that disagrees that global warming is a serious problem. Evaluate the relative quality of the arguments.

Key Terms

2:1 offset sanction
Acid deposition
Allowances
Ambient air
Anthropogenic pollutants
Area source
Biomass
Cap-and-trade programs
Carbon sequestration

Command-and-control policies
Concentrated pollutant
Criteria air pollutants
Environmental sink
Flow pollutants
Fund pollutants
General discharge prohibitions
Greenhouse effect
Greenhouse gases

Hazardous air pollutants
Market-based policies
Mobile source
Natural capital
Natural pollutants
Nonpoint source
Nonuniformly distributed pollutant

Point source
Pollution
Primary standards
Secondary standards
Stationary source
Stock pollutants
Uniformly distributed pollutant

Internet Resources

Toxic chemicals released by industries this year:
www.worldometers.info/view/toxchem/

The Indoor Air Quality Association:
www.iaqa.org/

The Noise Pollution Clearinghouse:
www.nonoise.org/

The United Nations Environmental Program:
www.unep.org

United Nations Framework Convention on Climate Change:
http://unfccc.int/

World Forum for Acoustic Ecology:
www.wfae.net/

ENVIRONMENTAL QUALITY

Messages painted on the street urge residents in Hawai'i not to use storm drains as waste receptacles.

Further Reading

Austin, J. C., S. Anderson, P. N. Courant, and R. E. Litan. *America's North Coast. A Benefit Cost Analysis of a Program to Protect and Restore the Great Lakes.* Washington, DC: Brookings Institution, 2007. A review of studies on the value of clean water to swimmers, anglers, and other users of the Great Lakes.

Bytingsvik, J., E. Lie, J. Aars, A. E. Derocher, O. Wiig, and B. M. Jenssen. "PCBs and OH-PCBs in Polar Bear Mother-Cub Pairs: A Comparative Study Based on Plasma Levels in 1998 and 2008." *Science of the Total Environment 417–418* (2012): 117–128. Research on the lingering effects of illegally produced persistent organic pollutants on polar bears in the Canadian Arctic.

Fletcher, J. L. "Review of Noise and Terrestrial Species: 1983–1988." In *Noise as a Public Health Problem Vol. 5: New Advances in Noise Research Part II*, edited by B. Berglund and T. Lindvall. Stockholm: Swedish Council for Building Research, 1990. A summary of 1980s research on noise pollution and its effects on the environment.

Fuller, Richard, Philip J. Landrigan, Kalpana Balakrishnan, Glynda Bathan, Stephan Bose-O'Reilly, Michael Brauer, Jack Caravanos, Tom Chiles, Aaron Cohen, Lilian Corra, Maureen Cropper, Greg Ferraro, Jill Hanna, David Hanrahan, Howard Hu, David Hunter, Gloria Janata, Rachael Kupka, Bruce Lanphear, Maureen Lichtveld, Keith Martin, Adetoun Mustapha, Ernesto Sanchez-Triana, Karti Sandilya, Laura Schaefli, Joseph Shaw, Jessica Seddon, William Suk, Martha María Téllez-Rojo, and Chonghuai Yan. "Pollution and Health: A Progress Update." *The Lancet Planetary Health 6*, no. 6 (2022): e535–e547. https://doi.org/10.1016/S2542-5196(22)00090-0. An overview of the number of deaths caused by various types of pollution.

Georgiou, S., I. Bateman, I. Langford, and R. Day. "Coastal Bathing Water Health Risks: Assessing the Adequacy of Proposals to Amend the 1976 EC Directive." *Risk, Decision and Policy 5* (2000): 49–68. A "contingent valuation" (CVM) study of the value of clean water to bathers in England.

Guha, Shion. "Valuation of Clean Water Supply by Willingness to Pay Method in a Developing Nation: A Case Study in Calcutta, India." *Journal of Young Investigators* (October 2007). www.jyi.org/issue/valuation-of-clean-water-supply-by-willingness-to-pay-method-in-a-developing-nation-a-case-study-in-calcutta-india/. A CVM study that finds that the value of providing clean water in Calcutta would exceed the cost.

Jalilov, Shokhrukh-Mirzo. "Value of Clean Water Resources: Estimating the Water Quality Improvement in Metro Manila, Philippines." *Resources 7*, no. *1* (2018). https://doi.org/10.3390/resources7010001. A contingent valuation study of the value of clean water for purposes that include swimming and fishing.

Kremer, Dario, Till Strunge, Jan Skocek, Samuel Schabel, Melanie Kostka, Christian Hopmann, and Hermann Wotruba. "Separation of Reaction Products from Ex-Situ Mineral Carbonation and Utilization as a Substitute in Cement, Paper, and Rubber Applications." *Journal of CO2 Utilization 62* (2022): 102067. https://doi.org/10.1016/j.jcou.2022.102067. A study of the feasibility of sequestering carbon in commercial products.

Mamun, Saleh, Adriana Castillo-Castillo, Kristen Swedberg, Jiarui Zhang, Kevin J. Boyle, Diego Cardoso, Catherine L. Kling, Christophe Nolte, Michael Papenfus, Daniel Phaneuf, and Stephen Polasky. "Valuing Water Quality in the United States Using a National Dataset on Property Values." *Proceedings of the National Academy of Sciences 120*, no. *15* (2023): e2210417120. https://doi.org/10.1073/pnas.2210417120.

McCarthy, Elena M. "International Regulation of Transboundary Pollutants: The Emerging Challenge of Ocean Noise." *Ocean and Coastal Law Journal 6*, no. *2* (2001): 257–292. http://digitalcommons.mainelaw.maine.edu/oclj/vol6/iss2/2. This article highlights the international problem of underwater noise pollution and the challenges of trying to regulate its sources.

Michael, H. J., K. J. Boyle, and R. Bouchard. "Does the Measurement of Environmental Quality Affect Implicit Prices Estimated from Hedonic Models?" *Land Economics 76*, no. *2* (2000): 283–298. This study looks at the determinants of housing prices in Maine, including water clarity, which turns out to make a sizable difference.

Moran, Daniel, and Keiichiro Kanemoto. "Tracing Global Supply Chains to Air Pollution Hotspots." *Environmental Research Letters 11*, no. *9* (2016). https://doi.org/10.1088/1748-9326/11/9/094017. Discusses the shared responsibility for pollution caused by exported goods.

Natural Resources Defense Council. "Sounding the Depths: Supertankers, Sonar, and the Rise of Undersea Noise", 1999. www.nrdc.org/

wildlife/marine/sound/sdinx.asp. A comprehensive assessment of noise pollution under the sea.

Pascal, C. *Global Warring: Environmental Change and the Looming Economic, Political and Security Crisis.* Toronto: Key Porter Books, 2010. Discusses the social volatility that could accompany climate change.

Smith, J. B., S. H. Schneider, M. Oppenheimer, and J. van Ypersele. "Assessing Dangerous Climate Change Through an Update of the Intergovernmental Panel on Climate Change (IPCC) 'Reasons for Concern'." *Proceedings of the National Academy of Sciences USA 106* (2009): 4133–4137. Evaluates reasons for concern about global climate change.

Smith, V. Kerry. "Social Benefits of Education: Feedback Effects and Environmental Resources." In *The Social Benefits of Education*, edited by Jere R. Behrman and Nevzer Stacey. Ann Arbor, MI: The University of Michigan Press, 1997, 175–218. This chapter examines why wealthier nations with higher levels of educational attainment generally have higher levels of environmental quality.

Sordello, Romain, Ophelie Ratel, Frederique Flamerie De Lachapelle, Clement Leger, Alexis Dambry, and Sylvie Vanpeene. "Evidence of the Impact of Noise Pollution on Biodiversity: A Systematic Map." *Environmental Evidence, 9*, no. 20 (2020). https://doi.org/10.1186/s13750-020-00202-y. This article maps evidence of the effects of all types of anthropogenic noises on all wild species and ecosystems.

U.S. Bureau of the Census. *Statistical Abstract of the United States: 2000.* Washington, DC, 2000. This publication is filled with data on everything from abrasives to zinc. The chapters on the environment and parks are particularly relevant.

von Haefen, Roger H., George Van Houtven, Alexandra Naumenko, Daniel R. Obenour, Jonathan W. Miller, Melissa A. Kenney, Michael D. Gerst, and Hillary Waters. "Estimating the Benefits of Stream Water Quality Improvements in Urbanizing Watersheds: An Ecological Production Function Approach." *Proceedings of the National Academy of Sciences of the United States of America 120*, no. *18* (2023): e2120252120. https://doi.org/10.1073/pnas.2120252120. A study of residents' willingness to pay for improved water quality in a river.

What energy source keeps your lights on?

Energy

The confluence of environmental concerns and profit-seeking creates particularly muddied waters in the realm of energy production. At the core of the problem is ongoing reliance on energy from nonrenewable sources. Since Thomas Edison harnessed electricity and Nikolaus A. Otto popularized the internal combustion engine, energy derived from fossil fuels has received a central role in commerce and everyday life. The health and environmental costs of burning fossil fuels arise at every stage of production and consumption. Invasive exploration for new supplies of coal and oil brings noise, roads, and heavy equipment into vulnerable wilderness areas. Fossil fuel mining and drilling operations extend human influence into even the most remote natural habitats. The transportation of fossil fuels to their processing and distribution sites necessitates truck, train, and ship emissions, and involves thousands of accidents and spills each year. The refinement of oil into fuel is a source of toxic emissions into the air, wastewater discharges, and hazardous waste. And the combustion of fuels in power plants and engines pollutes communities around the globe.

Fossil fuels also provide benefits. The U.S. petroleum and coal industries generate more than $1 trillion in sales annually, and fuel activities such as commercial air travel that would be difficult to power with green energy. Oil and coal provide 60 percent of the marketed energy humans use to attain their current standard of living.

Controversy surrounds policies to support alternatives to fossil fuels. What is certain is that the tremendous profits being made in

DOI: 10.4324/9781003428732-7

the energy industry make energy policy a focus of interest and a target for influence. This chapter details the primary energy sources and describes policies designed to provide ample energy while protecting health and the environment.

Energy Sources

Wise decisions about energy sources hinge on the available options and necessary trade-offs. Awareness of viable substitutes removes the need to endure externalities from the most damaging energy sources. A look at options also reveals prospects for energy indepen-

A coal-fired power plant like this one creates many benefits and many costs.

dence, perils of pursuing fossil fuels in environmentally sensitive areas, and possibilities for transitions to new sources in response to changing prices or attitudes. This section provides a critical base of information on energy sources and raises pertinent economic issues along the way.

Energy Terminology

Energy is the amount of work a physical system is capable of performing. Strictly speaking, energy cannot be created, consumed, or destroyed. The catch is that *entropy*—the amount of energy not available to perform useful work—increases every time someone mows a lawn or lifts a finger because energy is dissipated into the universe as heat. Energy can also be converted or transferred into different forms. *Kinetic energy* is the energy of motion. The kinetic energy of moving air molecules can be converted into *rotational energy* by the rotor of a wind turbine and then converted into *electrical energy* by a wind turbine generator. Since the real use of a wind turbine is to convert one type of energy into another, purists sometimes call them *wind energy converters* (WECs). With each conversion, part of the energy from the source is converted into *heat energy*. The *thermal efficiency* (TE) of an energy source indicates the percentage of its energy that can be used directly in the next link of the energy conversion system, rather than being converted into heat. A coal-fired power plant is about 33 percent efficient, meaning that it uses 3,000 megawatts of energy stored in coal to generate 1,000 megawatts of electricity. The other 2,000 megawatts are lost as heat energy. The TE of a typical car is about 26 percent. Some racing engines have a TE of 34 percent. The photovoltaic cells in solar panels are 7 to 44 percent efficient.

Energy is measured in joules (J) or kilowatt hours (kWh). One *calorie* equals 4.18 joules, the energy needed to raise the temperature of one gram of water by 1°Celsius. A *British thermal unit* (BTU) equals 1,055 joules, the energy needed to raise one pound of water 1°Fahrenheit. **Power** is the rate of energy transfer per unit of time. Power is measured in watts or horsepower. A watt is one joule per second, so a 60-watt light bulb uses 60 joules of energy per second, converting the electric energy into light energy and heat energy. One *horsepower* equals 746 watts. One *kilowatt* (kW) equals 1,000 watts, and one *megawatt* (mW) equals one million watts.

Mentions of energy "production" or "generation" refer to the conversion of energy into a more usable form. Energy "loss" means energy is dissipated as heat or for some other reason becomes unavailable for useful work. The fundamental task of most commercial electricity production is to make turbines spin and thereby create an electrical current. Both nuclear fission and coal combustion create steam that turns turbines. The kinetic energy from wind and falling water turns turbines directly to generate electricity. You can generate electricity yourself by turning a properly rigged bicycle wheel or the crank on some emergency radios. A photovoltaic (solar) cell, in contrast, converts light energy into direct-current electricity with no moving parts and no emissions.

Global energy consumers receive 36 percent of their electricity and 27 percent of all marketed energy from coal. Oil and other liquids (including biofuels) are the largest source of marketed energy, followed by coal, natural gas, renewable resources (other than biofuels), and nuclear power, as shown in Figure 7.1. The following section explains relevant information about each of these sources of energy.

Fossil Fuels

Coal Coal is the compressed remains of tropical and subtropical plants, predominantly from the Carboniferous and Permian periods 225 million to 345 million years ago. Coal includes varying levels of carbon, hydrogen, sulfur, and nitrogen. Coals are classified according to their carbon content, from the lowest to the highest, as lignite, subbituminous, bituminous, and anthracite. Coals with higher concentrations of carbon are harder and hold more energy. North America has the world's largest coal

Figure 7.1
*World
Consumption
of Marketed
Energy*

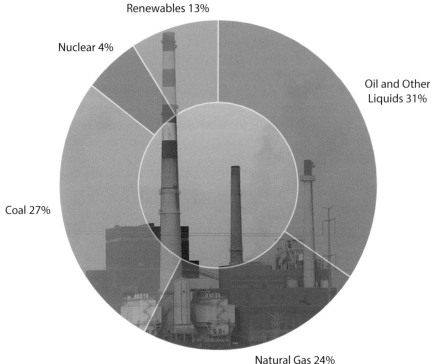

Renewables 13%

Nuclear 4%

Oil and Other
Liquids 31%

Coal 27%

Natural Gas 24%

Source: Statistical Review of World Energy, 2022.

reserves, followed by the former USSR, China, India, and the European Union. Current major producers of coal also include Australia, Indonesia, and South Africa. Lignite, with the least energy available to burn, is the most abundant in the United States; high-energy anthracite is relatively rare. In general, coals from the eastern and midwestern United States are bituminous, with high heat values but high levels of acid-rain-causing sulfur as well. Coals from the western U.S. states are largely subbituminous or lignite, with low heat value and low sulfur content.

Coal was used for fuel in ancient China and the Roman Empire thousands of years ago and by the Aztecs in the Americas more than 500 years ago. Coal-powered steam engines were vital to the Industrial Revolution of the late eighteenth and early nineteenth centuries. And still today, coal is burned to produce heat for the steel, concrete, and paper industries, among others, and to produce large amounts of electricity.

Environmental hazards arise from coal mining and combustion. Mines upend wildlife habitats and threaten groundwater contamination. Exposure to coal-mine dust is associated with deadly diseases that include chronic obstructive pulmonary disease, lung cancer, and coal workers' pneumoconiosis (black lung disease).[1] Upon combustion, coal

1   See https://blogs.cdc.gov/niosh-science-blog/2023/02/27/mining-lung-disease/.

releases sulfur dioxide, nitrous oxides, particulates, carbon dioxide, ash, and heavy metals including mercury. Those emissions collectively cause acid deposition, smog, respiratory illnesses, lung disease, climate change, and neurological and developmental damage in humans and other animals. New clean coal technologies (CCTs) and land reclamation programs are reducing the environmental impact of coal. Critics argue that greater benefits would come from expenditures on emission-free, renewable energy sources.

Oil **Crude oil** or *petroleum* is a formation of aliphatic hydrocarbons. Hydrogen and carbon atoms link together to form molecular chains of all lengths. Shorter chains are liquid at room temperature;[2] longer chains form solids. In the refinery, distillation separates the hydrocarbon chains into "fractions" that can be blended to produce fuels with desired characteristics. Hydrocarbons with between 7 and 11 carbon atoms per molecule form gasoline. Related processes form kerosene, diesel fuel, fuel oil, lubricating oil, and asphalt.

The scarcity of oil forces heavy users to rely on a limited number of primary providers. Most of the world's oil is produced in Russia, Saudi Arabia, the United States, Iraq, China, Canada, the United Arab Emirates, Kuwait, and Venezuela. Beyond the political and financial risks of oil dependency, the extraction, transportation, refinement, and combustion of petroleum products are major sources of pollution and environmental degradation. Concerns about energy security and environmental health prompt interest in gasoline-ethanol blends among other renewable, domestically produced, and cleaner-burning fuels. Former World Bank President Robert Zoellick and every U.S. president from Bill Clinton to Joe Biden are among supporters of subsidies to help ethanol blends compete with gasoline. However, ethanol production is energy intensive and therefore not a clear choice above oil from an environmental standpoint, as discussed in the later section on ethanol.

As the current source of about one-third of marketed energy, oil conveys power and wealth. Long ago labeled "black gold," oil is still the font of envy and international conflict. The Organization of the Petroleum Exporting Countries (OPEC) oil cartel provides about 40 percent of the world's oil and has a strong influence on oil prices. The private costs of petroleum products such as gasoline (petrol) vary considerably across nations due in large part to differences in tax and subsidy policies. For social efficiency, it is ideal for a Pigouvian tax to raise the private marginal cost to the level of the social marginal cost as discussed in Chapter 4. That is, a tax at the appropriate level can bring users to internalize the negative externalities of oil use and lead to the socially optimal quantity level of consumption.

---

2 This applies to chains as large as $C_{18}H_{32}$, meaning that they contain 18 carbon atoms and 32 hydrogen atoms.

Figure 7.2

The Oil Market

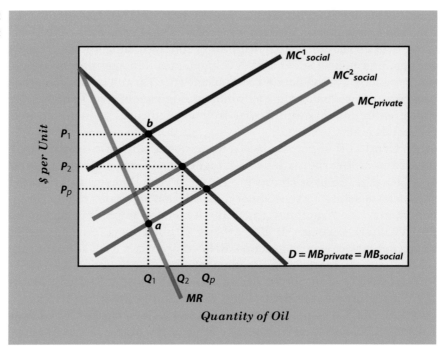

A competitive oil market would produce quantity Q_p at price P_p. A cartel would try to behave like a monopoly and maximize profits by restricting quantity to Q_1 and charging P_1. The same quantity and price, Q_1 and P_1, would also result if competitive sellers paid a tax represented by the vertical distance between points a and b. For goods such as oil that cause negative externalities, quantity restrictions associated with cartels, monopolies, and taxes can lead to more efficient outcomes. If the social marginal cost is MC^1_{social}, the monopoly quantity Q_1 is efficient. If the social marginal cost is MC^2_{social}, the monopoly quantity Q_1 is inefficiently low.

Market structure has interesting implications for the efficiency of oil use as well. Consider the oil market depicted in Figure 7.2 and assume the marginal cost curve would be the same whether the market is competitive, monopolistic, or a cartel. In a competitive market, economic theory foretells a price equal to private marginal cost, resulting in the production of quantity Q_p at price P_p. This quantity is excessive because it does not take into account the negative externalities of oil use. A monopoly would produce the quantity Q_1, which equates marginal revenue with private marginal cost, and charge the highest price that consumers would pay for that quantity, P_1. A cartel like OPEC would try to reap monopoly profits and divide them among the member firms. With success in quelling incentives to cheat, the cartel would produce Q_1 and charge P_1 just like a monopoly.

If the social marginal cost were at the level of the dark purple line labeled MC^1_{social}, the competitive outcome would be excessive and the monopoly/cartel outcome would be efficient. A tax equal to the marginal external cost (the vertical distance between points a and b) would be one way to bring a competitive oil market to produce the efficient quantity Q_1. If the social marginal cost were MC^2_{social}, the monopoly/cartel production level would be too low, and the competitive outcome would again be too high. The "Market Structure and Price Controls" section of this chapter discusses further implications of differing market structures. For a review of graphs involving externalities, see Chapter 3.

Natural Gas   Like coal and oil, **natural gas** has its origins in the swamps of past geologic periods. Natural gas is primarily methane (CH_4) with a mixture of other hydrocarbons. It is obtained from gas wells and as a by-product of crude oil extraction. Although it is a fossil fuel, it is also considered an "alternative" fuel because of its clean-burning qualities and relative abundance. Unlike most alternative fuels, natural gas has a broad infrastructure available to transport it, making it relatively accessible and affordable. Natural gas provides about a quarter of the marketed energy consumed and powers more than 23 million vehicles worldwide.

The world's major natural gas producers include Russia, the United States, Canada, Qatar, and Iran. Natural gas is produced in most regions of North America and transported through pipelines in liquid or gaseous form to every mainland state. Pure methane is a highly flammable, odorless gas with high marks for safety and limited emissions. Although leaks of unburned gas can ignite and explode, gas companies have inserted foul-smelling additives into natural gas to make leaks detectable ever since a natural gas explosion killed 293 people in a New London, Texas, school in 1937. Explosions along the world's millions of miles of natural gas pipelines are still a problem.[3]

Oil and natural gas are sometimes trapped within shale rock. A procedure called **hydraulic fracturing** or *fracking* releases the fuel by forcing various combinations of sand, water, and chemicals into the rock. The fracking process was first implemented a half century ago. New applications that involve drilling horizontally into shallower rock are increasingly common and controversial. Among potential threats to health and the environment, fracking requires large volumes of water, it can release methane—a greenhouse gas—into the air, and the chemicals injected into the ground may contaminate groundwater supplies. The EPA reports that fracking can severely affect drinking water.[4]

---

3 See Anderson (2020).

4 See https://cfpub.epa.gov/ncea/hfstudy/recordisplay.cfm?deid=332990.

Nuclear Energy

Fission is the splitting of atoms to release energy. **Fusion** is the combination of atoms, again with the release of energy. Nuclear power can come from the fission of uranium, plutonium, or thorium or from the fusion of hydrogen into helium. Convenience and accessibility make uranium the fuel of choice in today's reactors. Nuclear power has appeal because the fission of a uranium atom produces 10 million times the energy produced by the combustion of a carbon atom in coal. Excluding naval power reactors, there are 93 active nuclear reactors in the United States and about 410 in the world. These reactors satisfy 18 percent of U.S. electricity consumption and 4 percent of global energy use.

The greatest environmental threats from nuclear power involve the storage of spent radioactive fuel from reactors and the possibility of accidents. The worst accident in a commercial reactor occurred on April 26, 1986, when a reactor in Chernobyl, Ukraine, melted down and released at least 3 percent of its core radiation.[5] The incident killed 32 local workers immediately and thousands more in the following years. The plume of radiation reached Asia, Europe, and parts of the United States. Radioiodine and radiocesium contaminated food supplies, increasing the incidence of thyroid cancer, mutations, and associated health effects. Hundreds of threatened animal species live in the evacuation area, including deformed fish living in the reactor's cooling ponds. Crops grown in the areas still register dangerous levels of radioactivity, and it may be 24,000 years before humans can safely live there again.[6]

In the United States, there are currently over 750,000 metric tons of depleted uranium fuel stored in dozens of states. Most of the nuclear waste is stored at power plants because, after prolonged efforts to create a national repository at Yucca Mountain, Nevada, and elsewhere, the Department of Energy could not overcome health and safety issues. Both the storage of radiated waste and its transportation across the country have met with violent opposition. Proponents of nuclear energy assert that nuclear power plants cause less than 1 percent of U.S. radiation exposure.

Alternative Fuels

Alternative fuels are generally renewable, available domestically, and less toxic than their mainstream counterparts. Beyond their environmental benefits, the use of alternative fuels provides energy security by reducing reliance on foreign sources of fossil fuels. One trade-off is that many alternative fuels are not cheap. The direct costs of ethanol and liquefied

5   There are some claims that as much as 80 percent of the core was released. See www.ratical.com/radiation/inetSeries/ChernyThyrd.html.

6   See www.mentalfloss.com/article/646387/chernobyl-animals-thriving.

natural gas are higher than those of gasoline, although improved technology and economies of scale via mass production could make any of the alternative fuels economically viable. Energy collected from the sun, wind, or moving water generally requires a sizable initial investment and then the marginal cost of producing electricity becomes very small if not zero.

Biodiesel   **Biodiesel** is a relatively clean-burning diesel fuel made from renewable plant oils or animal fats. It can run in current diesel engines and is often blended with petroleum diesel fuel. B20 is a common biodiesel blend of 20 percent biodiesel and 80 percent petroleum diesel. Biodiesel emissions are sulfur-dioxide-free and contain 75 to 90 percent lower levels of unburned hydrocarbons, aromatic hydrocarbons, and carbon dioxide than diesel emissions. Biodiesel is also less flammable than petroleum diesel, reducing the risk of vehicular fires. However, biodiesel has a relatively high private cost. For example, soy-based biodiesel costs about three times as much as petroleum diesel. Researchers hope to increase the oil content of soybeans as one way of lowering the overall biodiesel cost.

Electric Fuel   *Electric vehicles* (EVs), a subset of **zero-emission vehicles** or ZEVs, require no tailpipe and lose no fuel to evaporation. Their electric power can come from solar panels on the owner's home or be transported from power plants to homes and industries through the same power grids that provide electricity for your computer. Of course, unless the power source itself uses an alternative fuel, the problem of pollution from the power plant remains. Some EVs are expensive, although high start-up costs can be balanced by low fuel costs. The U.S. Department of Energy estimates the average EV driven 1,124 miles will cost $60 to charge, while the average gasoline vehicle would cost $129 to fuel for the same distance.[7] Subsidies for EVs ameliorate car-price concerns in many countries, including China, Spain, Norway, Germany, France, the United Kingdom, and the United States.

Ethanol   **Ethanol** (CH_3CH_2OH), also called ethyl alcohol or grain alcohol, is a clear, colorless biofuel. Ethanol is distilled from a mash of renewable sources that can include corn, wheat, sugarcane, switchgrass, and waste accumulated from wood and paper processing. Ethanol and animal feed can be produced simultaneously from the same grains.[8] Before advocating or subsidizing ethanol produced from a crop, it is important to make sure the energy obtained from the ethanol and any by-products exceeds the energy that went into the farm operation, fertilizers, irrigation, and transportation of the crop. For example, Hill et al. (2006) estimated corn-based ethanol production creates 25 percent more energy than it uses, thanks mostly to the energy value of the animal-feed by-products.

7 See www.energy.gov/energysaver/cost-charge-electric-vehicle-explained.
8 See www.wardlab.com/ethanol-co-products-for-animal-feed/.

Weighing in its favor, ethanol combustion releases lower levels of greenhouse gases than gasoline combustion, and the growth of ethanol feedstocks draws carbon dioxide from the atmosphere. Ethanol is already popular as a fuel in Brazil and serves as a common octane-enhancing additive to gasoline in Canada, Europe, Thailand, and the United States. The most common mixture, containing 5 to 10 percent ethanol, is called *gasohol*. Gasoline-ethanol blends made up of 85 percent and 95 percent ethanol, called E-85 and E-95, respectively, are currently available for specially modified "flexible-fuel" vehicles such as the Ford F-150 truck.

Hydrogen **Hydrogen gas** (H_2) can power combustion engines and fuel cell electric vehicles.[9] Electricity generated from the reaction of hydrogen fuel and oxygen from the air provides fuel cells with clean, quiet power. The only emissions are pure water and heat. A steam reformation process splits hydrogen atoms from carbon atoms in natural gas (primarily methane, CH_4). Electrolysis or extreme heat (2,800°C) can also split hydrogen atoms from the oxygen atoms in water. Fuel cells and hydrogen-fueled internal combustion engines already power cars and buses; trains and submarines are in the works.[10]

Hydrogen is the primary propellant used in space flight. Hydrogen fuel cells on spacecraft power life-support systems and computers, returning drinkable water as a by-product. Back on Earth, most major automakers are experimenting with hydrogen fuel cell vehicles. One of the challenges at this point is to safely distribute and store the hydrogen, which is an explosive gas at room temperature. Scientists are exploring more manageable forms, including compressed hydrogen, liquid hydrogen, and chemical bonding between hydrogen and metal hydrides. Little infrastructure is in place to transport hydrogen. If fuel cells become popular, a growing network of hydrogen pipelines may replace the current canister and tanker truck distribution systems.

Methanol **Methanol** (CH_3OH) serves as a gasoline additive, a replacement for gasoline and diesel fuels, and a source of hydrogen for fuel cells. Methanol sources include natural gas and many renewable resources containing carbon, including seaweed, waste wood, and garbage. Relative to gasoline, methanol offers lower emissions of hydrocarbons, nitrogen oxides, and particulate matter. Methanol's fuel efficiency is currently half that of gasoline, although that may improve with ongoing research. Methanol's high-octane performance and low flammability make it the only fuel used in Indianapolis 500 race cars.

Propane **Liquefied petroleum gas** (LPG) is primarily propane and contains a mixture of hydrocarbons, including butane, butylene, and propylene. LPG is a by-product of petroleum refinement and natural

9 See https://afdc.energy.gov/vehicles/fuel_cell.html.

10 The military is particularly interested in fuel-cell vehicles because their relatively cool and quiet engines are more difficult to detect. For more on hydrogen, see the Fuel Cell & Hydrogen Energy Association website: www.fchea.org.

gas processing. While it is gaseous at room temperature, propane can be delivered to vehicles as a gas or a liquid. Propane is 270 times more compact as a liquid than as a gas. Propane fuels cause less carbon buildup than gasoline or diesel, allowing spark plugs and engines to outlast their counterparts in gasoline and diesel engines. Propane has been used for lighting and cooking for over 100 years. Today, there are about 8 million propane-powered vehicles in Europe and 60,000 in the United States. Propane vehicles are popular for corporate fleets and offer lower emissions and fuel prices than gasoline-powered vehicles.

Solar Fuel   The sun emits energy at a rate of 1.56×10^{18} kWh per year. All humans combined require less than 0.1 percent of this amount of energy. Collecting solar energy is the hard part, but it's not that hard. *Solar thermal* applications use the sun's energy directly to heat air or liquid, primarily for residential use. *Photoelectric* methods use semiconductors and photovoltaic cells to convert solar energy into electricity. The most efficient commercially available photovoltaic cells can convert 22.8 percent of light energy into electrical energy.[11] Once thermal or photoelectric applications are in place, the use of solar energy creates no emissions and no noise. Worthwhile amounts of solar energy can be collected even under cloudy conditions, although energy is not produced at night, and more energy is collected in sunny places than in cloudy places. Battery storage can make energy collected from the sun available when and where there is no sunlight. Many countries in Europe, Asia, North America, and elsewhere provide subsidies for solar energy that have contributed to the proliferation of this alternative fuel.

Wind   Wind whips some regions of the Earth at speeds exceeding 200 mph. Small wind turbines can create energy using a minimum of 8 mph winds; large turbines require 13 mph winds to achieve efficiency. Wind turbines use long blades called rotors that spin by the force of the wind and create electricity. Wind turbines offer a sustainable source of kinetic energy and create no emissions. A disadvantage is that the turbines are currently expensive. A field of turbines, sometimes called a *wind farm*, is needed to power a town of significant size, and sufficient wind energy is only available in certain areas. However, the capital cost of wind turbines has fallen rapidly over the past two decades, making wind energy increasingly competitive with energy from traditional fuels.

Wind turbines are a sustainable source of clean energy. In light of their ability to supplant negative externalities from fossil fuels, many countries subsidize wind energy. The countries with the largest wind energy capacity are China, the United States, Germany, India, Spain, and the United Kingdom. While there are concerns about birds colliding

11 For more, see the Department of Energy's photovoltaic site: www.energy.gov/eere/solar/photovoltaics.

The winner of a 2,500-mile solar-powered car race. The sun is an abundant source of energy.

with wind turbines, the problem is reportedly negligible,[12] especially compared with the threat to birds of climate change.[13]

Geothermal   **Geothermal energy** is heat held beneath the Earth's crust. This energy can be brought up to the surface to heat buildings or icy sidewalks. It can also be converted into electricity for broader applications. Countries with enough installed geothermal capacity to serve millions of customers include the United States, Indonesia, the Philippines, Türkiye (the country formerly known as Turkey), New Zealand, and Mexico. Even in areas with no hydrothermal reservoirs, liquid antifreeze can be cycled underground to pick up the Earth's relatively constant ground temperature. In the winter, *geothermal heat pumps* (GHPs) transfer heat from the ground into homes and buildings. In the summer, GHPs transfer indoor heat into the ground and draw up the relatively cool temperatures from below. Approximately 85,000 GHPs are installed in the United States each year, including one in the home of this author. Ball State University in Indiana has one of the world's largest GHPs, which heats and cools 47 buildings and saves the school roughly $2 million per year. *Direct-use geothermal systems* in Iceland pump hot water from underground directly into buildings and greenhouses for heating and below city streets for effortless snow removal. Likewise, residents of Klamath, Oregon, cherish

12 See www.carbontrust.com/news-and-insights/news/pioneering-study-finds-seabirds-avoid-offshore-wind-turbines-much-more-than-previously-predicted.

13 See www.audubon.org/climate/survivalbydegrees/county.

These wind turbines in Hawai'i provide enough electricity for more than 10,000 homes.

the safety of their snow-free streets and sidewalks, all cleared by naturally hot water piped beneath them.

Energy Policy

Efficient Source Selection

Ongoing transitions in the transportation industry exemplify the importance of energy source selection. While traditional combustion engines run mostly on petroleum-based fuels, increasingly popular EVs run on batteries that can be charged with any energy source. The challenge, then, is to determine the appropriate amount of energy to draw from each source. Figure 7.3 illustrates how to solve this puzzle based on the social marginal costs of three representative sources. The social marginal cost of energy from nonrenewable resources such as oil and coal includes extraction costs and the opportunity cost of future generations not being able to use the resources. As more of a nonrenewable resource is depleted, future uses of greater importance are forgone, and the social marginal cost of use rises. The social marginal cost of renewable energy sources such as solar and wind is relatively small because there is no extraction, other variable costs are minimal, and future uses are not forgone. The social marginal cost curve for an energy source with a social marginal cost of *zero* would lie along the horizontal axis and, if available in ample quantities, render alternative sources obsolete.

The first three graphs in Figure 7.3 indicate the social marginal cost of producing electricity from solar energy, oil, and coal. The last graph shows the social marginal benefit of each level of electricity and the market social marginal cost curve, found as the horizontal sum of the social marginal cost curves for the three available sources. To

Figure 7.3

The Energy Market and Representative Components

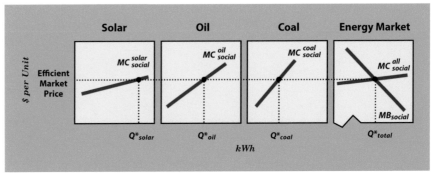

The market social marginal cost (MC^{all}_{social}) curve for energy is found by summing the MC^{all}_{social} curves for each energy source horizontally. The efficient quantity of energy to draw from a particular source is found at the intersection of the efficient market price and the social marginal cost curve for that source.

find the horizontal sum at any particular cost level, simply add up the distance from the vertical axis to the social marginal cost curve for each energy source. For example, suppose you wanted to find the point on the market social marginal cost curve at a height of 20 cents. If that cost level intersected the MC_{social} curve for solar energy at 2 billion kWh, oil at 1 billion kWh, and coal at 0.5 kWh, then the horizontal sum would be $2 + 1 + 0.5 = 3.5$ billion kWh. This number of kWh could be produced in the combined energy market at a social marginal cost of 20 cents.

Energy should be produced until the social marginal cost equals the social marginal benefit. A market price determined by the equilibrium of MC_{social} and MB_{social} yields production and consumption at the efficient level, Q^*_{total}. Remember that the private marginal benefit curve is equivalent to the demand curve and that the private marginal cost curve (above average variable cost) is equivalent to the supply curve in a competitive industry. Thus, in a competitive market with externalities either absent or internalized (making $MC_{private} = MC_{social}$ and $MB_{private} = MB_{social}$), the market equilibrium between supply and demand is efficient.

The current energy markets are not competitive and do involve externalities, although targeted policies offer assistance. The next section describes how price controls can bring monopolies to behave more like competitive firms. Chapter 3 explained how taxes and subsidies can cause externalities to be internalized.

Figure 7.3 illustrates efficiency in the energy market. The market price is at the efficient level that equates MC_{social} and MB_{social} as shown in the graph on the far right. The efficient levels of solar-, oil-, and coal-sourced energy, labeled Q^* in each graph, are found where the line representing the efficient market price intersects the social marginal cost of each source. The sum of the quantities of each source equals the socially efficient quantity of energy, Q^*_{total}.

If the social marginal cost of one energy source were higher than that of another, any given quantity of energy could be produced at a lower cost to society by using more of the source with the lower social marginal cost and less of the more costly source. As that adjustment occurs, the social marginal costs of the sources come into equilibrium. Why? Because of what happens to marginal cost as quantity changes: Marginal cost curves typically slope upward, so using more of the less costly source causes its social marginal cost to increase, and using less of the more costly resource causes its social marginal cost to decrease. That is, the adjustment in use levels causes the social marginal costs of the two sources to gradually converge. This adjustment should continue until the social marginal cost of producing energy with each source is equal, at which point the cost of energy to society is minimized.

To better understand the possible efficiency gains when social marginal costs differ and too much energy is used overall, we'll consider a situation with just two energy sources: wind and coal. In Figure 7.4, we also drop the assumption that externalities are absent or internalized. This means the social marginal cost curve for coal is above the private marginal cost curve by the amount of the marginal external cost of coal as discussed in Chapters 3 and 4.

As discussed in Chapter 3, inefficient levels of production can result from externalities or market power among other sources of market failure. With consumers not feeling the negative externalities from coal, the intersection of the red private marginal cost (and supply) curve and the blue social marginal benefit (and demand) curve at point G in Figure 7.4 determines the market price, *Market Price$_1$*. The intersection of the price line and the

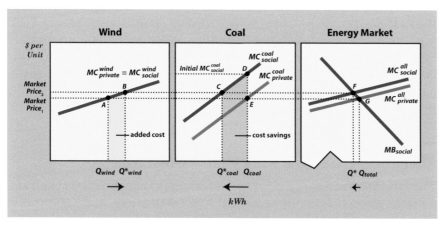

Figure 7.4
The Gains from Switching

When there are two energy sources with different social marginal costs, society can benefit from producing less energy with the relatively expensive source and more with the relatively inexpensive source. In this example, the yellow area represents the added cost of increasing solar energy production from Q_{wind} to Q^*_{wind}, and the lavender area represents the savings from reducing coal energy production from Q_{coal} to Q^*_{coal}. The net benefit to society of equating the dollars per kWh of wind and coal energy is the lavender area minus the yellow area.

private marginal cost curves for wind and coal at points A and E respectively determine the quantities of energy from each source: Q_{wind} and Q_{coal}.

There is an opportunity to reduce society's energy cost because at quantities Q_{wind} and Q_{coal}, the social marginal cost of energy from coal, shown at point D, exceeds the social marginal cost of energy from wind, shown at point A. We can actually see the gains from switching sources and reaching the socially optimal quantity. The area under an MC_{social} curve along any section of the quantity axis indicates the additional cost to society of producing the quantity represented by that section. The yellow area in Figure 7.4 represents the additional cost of increasing wind-sourced energy production from Q_{wind} to Q^*_{wind}. The lavender area represents the cost savings from reducing coal-sourced energy production from Q_{coal} to Q^*_{coal}. By making that change and producing at points B and C, the net benefit to society is the large lavender area minus the smaller yellow area.

How could that net gain be achieved? A Pigouvian tax on producers equal to the marginal external cost of coal would bring the private marginal cost of coal up to equal the social marginal cost of coal. In the market graph, that would raise the private marginal cost of all energy up to the social marginal cost. Market equilibrium would be at point F, and the market price would rise to *Market Price$_2$*, at which producers would supply Q^*_{wind} kWh of wind energy and Q^*_{coal} kWh of coal energy.

While taxes and subsidies can counteract the inefficiency of negative and positive externalities respectively, sometimes the incentive effects of these tools go awry. The taxation of goods causing positive externalities or the subsidization of goods causing negative externalities only exacerbates the problem. A tax on wind energy would shift the $MC_{private}$ curve further to the left and reduce wind energy production. A coal subsidy would bring the private marginal cost further to the right and worsen the over-production problem. In fact, the International Monetary Fund reports that global fossil fuel subsidies exceed $7 trillion annually.[14] Subsidies that increase the divergence between private and social costs or benefits are called **perverse subsidies**. If a good creates no externalities, inefficiency results from any tax or subsidy that separates the private marginal cost from the social marginal cost tax or separates the private marginal benefit from the social marginal benefit.

Market Structure and Price Controls

The structure of energy markets determines their efficiency and the policies needed to achieve socially optimal prices and quantities. Large-scale power plants are generally *natural monopolies* because the enormous cost of building one prevents competing firms from entering the same local market. Competition is another real possibility in the energy market. Most homes could have solar panels on their

14 See www.imf.org/en/Blogs/Articles/2023/08/24/fossil-fuel-subsidies-surged-to-record-7-trillion.

roofs, and wind turbines could sit on most farms, representing millions of small energy production facilities. Indeed, the 1978 Public Utilities Regulatory Policies Act (PURPA) established the right of independent power producers to sell electricity to regulated utilities when they create more energy than they need for their own purposes. And just as a penny saved is a penny earned, a kWh saved is effectively a kWh produced. In that sense, the ability of each individual to "produce" (save) energy can help to keep prices in check. Since the 1970s, energy made available via savings has exceeded energy made available by new energy sources.

Rocky Mountain Institute founder Amory Lovins describes two paths for energy dependence: a **soft path** of many renewable energy sources such as rooftop solar panels and a **hard path** of centralized fossil fuel production. Currently, large, centralized producers generate about 90 percent of energy in the United States; we are still largely on the hard path. Path-dependence theory suggests that once a particular path is taken, it becomes increasingly difficult to change course. Investments in research and development, infrastructure, training, and end-use systems create commitments to stay the course. The enormous fixed costs of entry into the traditional energy market create monopolies. However, governments can require monopolies to produce at relatively efficient levels and charge prices that approximate competitive market prices. This section discusses the use of price controls for that purpose.

Figure 7.5 illustrates three possible structures for the energy market. The first graph depicts a competitive market, as would result from a soft-path approach of millions of solar power sources. The second is a typical monopoly graph. This approximates the market structure for energy sources with barriers to entry and moderate fixed costs, such as a small or medium-sized hydroelectric power plant. The last graph depicts a natural monopoly, characterized by high fixed costs and decreasing average costs over the relevant range of output levels. Coal-fired electric utilities and nuclear power plants are natural monopolies because they necessitate fixed costs of several hundred million dollars and enjoy tremendous economies of scale.

The natural monopoly graph in Figure 7.5 shows that, as the firm increases production, its average cost falls because the large fixed cost is spread over more units of output. The falling average cost in this situation allows one firm to serve the entire market at a lower cost than two or more smaller firms that can't spread their costs over as many units of output.

In a competitive market with no externalities, the equilibrium of supply and demand will yield the efficient price and quantity, as shown in the top panel of Figure 7.5 by $P_{competition}$ and $Q_{competition}$. The imposition of a price ceiling in such a market would reduce the quantity supplied and increase the quantity demanded, causing a shortage. Production would occur at level Q^*, which is below the efficient quantity $Q_{competition}$. Now suppose there are negative externalities that cause the efficient quantity (where $MC_{social} = MB_{social}$ as in Figure 7.4) to be Q^*, not $Q_{competition}$. In that case, a price ceiling set to intersect the supply curve at the socially efficient quantity as shown achieves the efficient quantity. With or without externalities, the low price created by the price ceiling causes a shortage that necessitates some form of energy rationing.

A profit-maximizing monopoly chooses the quantity that corresponds with the intersection of marginal revenue and marginal cost ($Q_{monopoly}$ in the middle panel of Figure 7.5) and finds the highest price consumers are willing to pay on the demand curve above that quantity ($P_{monopoly}$). The monopoly charges more and produces less than would a competitive firm with the same costs.

If the monopoly depicted in the middle graph were instead a competitive market, the intersection of marginal cost and demand would determine the price and quantity. Remember that the marginal revenue curve is below the demand curve for a monopoly because it must lower its price to sell more units. If a price ceiling is imposed on the monopoly at the price a competitive firm would charge, as shown by $P_{ceiling}$, the firm will be able to sell up to Q^* units. In that case, the line representing the price ceiling becomes the firm's marginal revenue curve for quantities up to Q^* because each unit sold simply adds the price set by the price ceiling to total revenue.[15] That new marginal revenue will equal marginal cost at the competitive (and efficient) price and quantity, $P_{ceiling}$ and Q^*, respectively. Such a price ceiling provides one solution to a monopoly with an inefficiently high price and low quantity.

The natural monopoly depicted in the bottom graph of Figure 7.5 will select its price and quantity just as the monopoly did. However, if a price ceiling were set where the marginal cost curve crosses the demand curve, equating price and marginal cost as in a competitive market, the natural monopoly would incur losses because its price would be below its average cost. On the other hand, a price ceiling set where demand equals average cost, as shown, would cover all costs, and the quantity would be as large as is sustainable (without price discrimination, as discussed in the next paragraph). This makes a price ceiling at the intersection of demand and average cost an appealing choice for policymakers.

A second policy option is to permit **price discrimination**, whereby different customers pay different prices. Under **perfect price discrimination**, each customer is charged the most she or he is willing to pay. For instance, business customers and wealthy individuals would pay higher prices than people with relatively small budgets. Although consumers

---

15 Marginal revenue is the additional revenue received from selling one more unit. If a firm must lower its price from $1.00 per gallon of P-series to $0.90 per gallon in order to increase its sales from 5 gallons to 6 gallons, its total revenue increases from $1 × 5 = $5.00 to $0.90 × 6 = $5.40 and its marginal (additional) revenue from the sixth gallon is $0.40. On the other hand, if it can sell up to 10 gallons at a price ceiling of $0.70, then it can sell 5 gallons for $0.70 per gallon and it can sell 6 gallons for $0.70 per gallon. Total revenue is $0.70 × 5 = $3.50 for 5 gallons and $0.70 × 6 = $4.20 for 6 gallons, and the marginal revenue from the sixth gallon equals the price ceiling of $0.70.

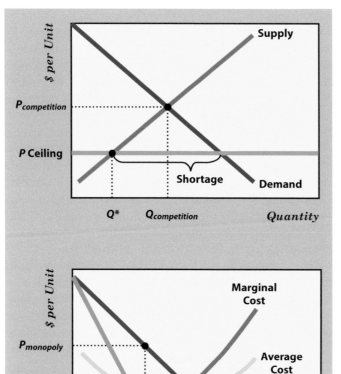

Perfect Competition

A price ceiling is a competitive market causes a decrease in the quantity supplied and a shortage.

Figure 7.5

Market Structure and the Influence of Price Ceilings

Monopoly

A price ceiling in a monopoly market can increase the quantity supplied while lowering the price and eliminating deadweight loss.

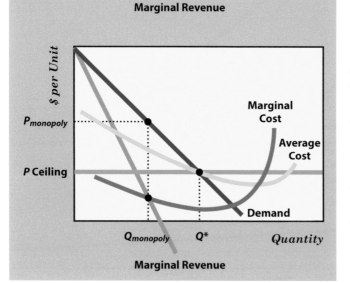

Natural Monopoly

A natural monopoly cannot operate under competitive conditions, but average-cost pricing can lead to reasonable prices for consumers and "normal" returns for the firm.

receive no consumer surplus when the price is the most they would be willing to pay, the outcome is efficient unless there are externalities. To explore why, first note that under perfect price discrimination, the height of the demand curve at any given quantity represents both a customer's willingness to pay for that unit and the additional revenue gained by serving that customer. That is, the demand curve is also the marginal revenue curve. For example, in Figure 7.6, a consumer would pay at most P_1 for the first unit and is charged P_1 for it. The second unit is valued at P_2 and the firm receives P_2 in revenue for it, without lowering the price on the first unit, and so forth. The other part of the explanation is that perfect price-discriminating firms maximize profits by producing where the demand/marginal revenue curve intersects the marginal cost curve. So a perfect price-discriminating firm produces where the firm's marginal cost equals the consumers' marginal benefit (willingness to pay)—the criterion for efficiency in the absence of externalities.

To achieve efficiency under perfect price discriminators without granting all of the net gains to firms, the resulting profit can be diminished using a lump-sum tax and redistributed as desired. In practice, price discrimination is rarely perfect due to the difficulty of determining each consumer's willingness to pay. However, multitiered pricing plans that charge different consumers different prices are common. Examples

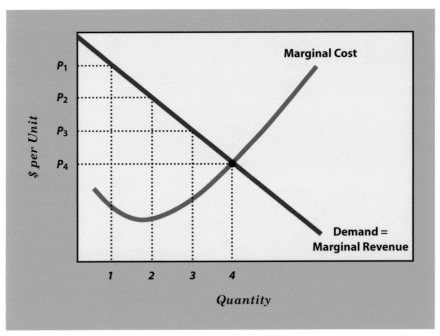

Figure 7.6
Perfect Price-Discriminating Monopoly

A perfect price-discriminating monopolist charges each customer the most she or he would be willing to pay as indicated by the height of the demand curve. Since the monopolist does not have to lower its price in order to sell additional units, its marginal revenue curve corresponds with the demand curve. The monopolist maximizes profits by producing until marginal cost equals marginal revenue at 4 units. This production level is efficient in the absence of externalities.

include different pricing for residential and business customers, seniors, students, children, coupon holders, and high-volume customers.

Deregulation

Beyond price controls, government policy influences the very structure of the electricity industry. As a means of deterring national conglomerates from dominating the electricity market, the 1935 Public Utility Holding Company Act (PUHCA) effectively gave U.S. electric utilities a monopoly in their geographic area but prevented them from expanding into other regions. The 2005 Energy Act repealed PUHCA. The intent of this deregulation was to allow customers a choice among energy providers, although a majority of U.S. households currently have no real choice. India, Thailand, Sri Lanka, and Nepal are among many Asian countries that are deregulated, and all of the European Union countries are at least partially deregulated.

Current technology allows electricity produced in one region to be sold in another, meaning that deregulation could introduce competition from other areas. It could also spawn dominant conglomerates. Suppose major power utilities A and B operate in Region 1 and Region 2, respectively. Regulation prevents Utility A from selling its power to Region 2 and Utility B from selling to Region 1. A stated intent of deregulation is to reduce rates for consumers by allowing utilities like A and B to compete as service providers in both Region 1 and Region 2. (Earlier legislation already allowed for some competition.) Some worry that rather than increasing competition, deregulation defeats the purpose of PUHCA by allowing Utility A and Utility B to be replaced by deregulated Electric Company C, which will dominate the energy market in both regions.

The economics of deregulation cannot be separated from the politics involved. Parties on all sides of the energy deregulation debate have funded political contributions, lobbyists, and advertisements in publications aimed at legislators. Before becoming embroiled in controversy for allegedly concealing over $1 billion in debt, large-scale energy trading pioneer Enron was a major contributor to pro-deregulation members of the U.S. Congress and presidential candidates. Although there is no consensus on the true impact of deregulation, the story in the Reality Check exemplifies the mixed results of partial deregulation.

Politics Rears Its Ugly Head Again: Oil and Automobiles

Political battles pit "big business" against environmental protection. Of concern is the possibility that political and financial forces may impede efficiency within the markets for fuel, automobiles, and other goods that affect the environment. Worldwide, biofuel subsidies amount to tens of billions of dollars annually. It would be naïve to suppose that

California's Energy Woes

California brought us the hula-hoop, the skateboard, the cable car, and the Frisbee. But we would not want to follow all the past trends of the Golden State. In 2001, California experienced rolling *brownouts* (periods of low voltage in power lines that cause lights to dim and equipment to malfunction) and *blackouts* (complete power failures due to storms or, in California's case, overburdened power grids). Stores were closed, traffic lights went out, cable cars halted, and elevators jammed. Energy prices in the state increased by anywhere from 12 to 55 percent.

Behind this calamity were natural and human errors and political and economic motives. Wholesale energy production in California was deregulated in 1996. The state's power utilities sold power plants to a handful of private energy wholesalers. The utilities purchased power from the deregulated wholesalers and sold it to customers at regulated retail prices. Increased energy demand and decreased supply caused the wholesale prices to exceed the fixed retail prices, leading to energy shortages and multi-billion-dollar debt among the utilities. Compounding these problems were increasing temperatures that caused more precipitation to fall as rain rather than snow, leaving less snowmelt to fuel hydroelectric power stations. Cold snaps in the winter and excessive heat in the summer increased energy demands. During the same period, a number of power plants were under maintenance and could not produce at full power.

Some have pointed their fingers at the power companies, which may have restricted supply to force prices and profits upward. In the long run, the higher prices may have served the opposite purpose.

Californians, already leaders in the use of alternative-fuel vehicles, have embraced solar and wind power among other means of gaining independence from the electricity grid. The number of energy-efficient homes made of used tires and straw bales is on the rise. Perhaps those trends will fly east like the Frisbee.

the economics of fuel subsidies have only to do with the internalization of external benefits. Ethanol producer Archer Daniel Midland Co. spent $2.7 million on lobbying in 2023, and at least 23 oil and gas companies spent more than $1 million each on lobbying in the same year. London-based petroleum giant BP announced it would end political contributions in 2002 because the legitimacy of the political process is "crucial both for society and for us as a company working in that society."[16] Nonetheless, BP affiliates spent $4.5 million on lobbying in 2023.[17]

Influence is also found in political ties. The wife of British Prime Minister Rishi Sunak is a large stakeholder in a company whose top clients include British oil and gas company Shell.[18] Before pushing for oil drilling in the Arctic National Wildlife Refuge as vice president of the United States, Dick Cheney earned over $5 million as the CEO of Halliburton Company, a diversified energy services company. In 2018, U.S. President Donald Trump appointed a former lobbyist for the coal industry to be the Acting Administrator of the EPA. And there was an uproar in 2023 when the head of an oil company was appointed as the president of the United Nations Climate Change Conference known as COP 28.[19] The power of incentives makes it important to have policymakers with no conflicts of interest.

Is environmental and natural resource economics in a political quagmire? Read on. The plight of emissions and *corporate average fuel economy* (CAFE) standards in the United States further exemplifies the struggle between short-run political interests and the long-run well-being of society.

CAFE Standards and Emissions Caps

Most modern automobiles receive their power from fuel combustion within their engines. Efficient combustion would turn the fuel's hydrogen and carbon atoms into water and carbon dioxide.[20] In reality, combustion is less than efficient, and by-products include unburned hydrocarbons, nitrogen oxides, and carbon monoxide. *Hydrocarbons* (HC) are the result of partially burned fuel molecules. They react with sunlight and nitrogen oxides (NO_x) to create ground-level ozone, a major component of smog. *Carbon monoxide* (CO) is created when fuel is only partially oxidized in the engine. When breathed, CO limits the flow of oxygen to the bloodstream and can cause severe illness or death.

Fuel evaporation adds to the vehicle emissions problem. *Running losses* of fuel occur when a vehicle's operating temperature causes fuel

16 See www.nytimes.com/2002/02/28/us/bp-s-chief-says-the-oil-company-is-ending-its-political-donations.html.

17 See www.opensecrets.org.

18 See www.bylinesupplement.com/p/how-rishi-sunaks-family-profited.

19 See www.cnbc.com/2023/01/12/cop28-uae-sparks-backlash-by-appointing-oil-chief-as-president.html.

20 For a discussion of why this isn't the case, see www.howstuffworks.com/question407.htm.

to vaporize. *Hot soak* is the vaporization of fuel when a car has been turned off but is still hot. *Refueling losses* occur when fuel is being transferred from a stationary tank into a vehicle. *Diurnal evaporation* occurs as the temperature rises, causing fuel tanks to heat and vent vapors.

The governments of many countries regulate emissions by controlling both what goes into the fuel and what comes out of the average tailpipe. Some countries and U.S. states also require each automobile to pass emissions tests to monitor exhaust problems that come about as cars age. The ups, downs, and stagnation of CAFE standards illustrate the political struggle between corporate and environmental interests in the United States, where fuel economy standards are lower than those in the European Union and several Asian countries but higher than in countries such as Saudi Arabia.

The CAFE standard for U.S. passenger cars started at 18 miles per gallon (mpg) in 1978 and rose gradually to 27.5 mpg in 1985. The standard then dropped back to 26 mpg in 1986, rose to 26.5 mpg in 1989, and was held at 27.5 mpg from 1990 to 2010. Over the next decade it rose to 42.4 mpg, and for 2026, the standard is 49 mpg. The standard was clearly driven by political influence rather than the limits of technology. Similar power games are played in the arenas of offshore oil drilling rules, waste disposal policies, and energy deregulation.

Politics and environmental economics are irrevocably intermeshed. When seeking progress on the environmental front, politicians and economists who look only within their own field and disregard the importance of the other run the risk of accomplishing little of value in either area.

Summary

As economic efficiency requires, we tolerate the climate-changing, health-degrading, environment-impairing costs imposed by the fossil fuels that turn the wheels of industry and agriculture. The question is: Are we tolerating too much? A host of renewable fuels are poised to provide far more clean energy than they do today. Solar and wind energy draw from virtually infinite sources and generate emissions-free electricity. Fuel cells create only pure water and heat but lack supporting research into low-cost hydrogen sources. Biofuel mixtures reduce the emissions from gasoline and come from carbon-sequestering vegetation. Nonetheless, we have not yet fully embraced a substitute for fossil fuels. We face the puzzle of needing popularity for mass production and needing mass production for the economies of scale and low prices required for popularity.

Political, economic, and scientific hurdles perpetuate the status quo in energy production. Strife in oil-exporting countries and the vulnerability of centralized energy producers to blackouts and hurricanes may motivate fresh approaches. Only with an understanding of our energy alternatives can society allocate resources efficiently between fossil fuels, renewable sources, and the research and development needed for the next generation of options. This chapter is intended to foster such an understanding and to explain the implications of various market structures that energy policy can create or control.

• • • • • • • • • • • • • • • • • • •

Problems for Review

1. Identify a strength and a weakness of each of the following fuels:

   a) *Natural Gas*

   b) *Hydrogen*

   c) *Coal*

   d) *Solar Fuel*

2. Explain how energy deregulation was meant to lower energy prices.

3. Suppose that any amount of solar energy could be supplied at a constant MC_{social} just below the preexisting market price in Figure 7.3. Draw a set of graphs similar to those in Figure 7.3 with the appropriate adjustments in the solar energy graph and the market MC_{social} curve. Indicate the appropriate level of energy production from each source. Repeat these steps under the new assumption that the constant MC_{social} for solar energy is almost zero, which is less than the MC_{social} for the first unit of oil or coal.

4. The introduction to this chapter discusses the environmental costs of fossil fuel consumption. Omitted are the environmental costs of manufacturing the trucks, oil rigs, bulldozers, and other equipment used in fossil

fuel production. Choose one of those items and list at least five stages of its creation that cause externalities. (*Hint*: Think about the components of the item. How is steel made? Where does iron ore come from? and so on.)

5. Suppose the wind and coal industries are producing at the levels Q_{wind} and Q_{coal} in Figure 7.4. Describe two specific policies for each source that could cause the production levels to change to the socially efficient production levels.

6. Suppose the supply curve in the top panel of Figure 7.5 represents $MC_{private}$ and that the demand curve represents MB_{social}. Assuming negative externalities exist, redraw that graph and add an MC_{social} curve such that Q^* is the efficient level of output. Besides a price ceiling, what other policy could reduce output to Q^*?

7. Draw a monopoly graph and include a price ceiling that would *decrease* the quantity of output to a level below the monopoly level of output.

8. If you had $10,000 to invest in an alternative fuel, which one

would you pick? Explain your answer.

9. Suppose the Green City Hydropower Company (GCHC) is a perfect price-discriminating monopolist.

a) *How much consumer surplus will GCHC's customers earn? Explain your answer.*

b) *If GCHC produces no externalities, how will its production level*

compare to the socially optimal production level? Explain your answer.

10. Draw hypothetical *marginal cost, marginal revenue, average cost, demand, quantity,* and *price* lines for

a) *A nuclear power plant*

b) *One of the energy providers in a country that has taken the "soft" energy path*

websurfer's challenge

1. Find a website that offers updated information on corporate fuel economy standards in your country.

2. Find the website of a major newspaper that includes a story about energy and read it.

3. Find a picture of a wind farm.

Key Terms

Biodiesel	Hydrogen gas
Coal	Liquefied petroleum gas
Crude oil	Methanol
Energy	Natural gas
Ethanol	Perfect price discrimination
Fission	Perverse subsidies
Fusion	Power
Geothermal energy	Price discrimination
Hard path	Soft path
Hydraulic fracturing	Zero-emissions vehicles

Internet Resources

American Solar Energy Society:
www.ases.org/

Energy Information Agency:
www.eia.doe.gov/

Geothermal Education Office:
http://geothermal.marin.org/

International Energy Agency:
www.iea.org/

National Renewable Energy Lab:
www.nrel.gov/

International Renewable Energy
Agency:
www.irena.org

Solar Energy International:
www.solarenergy.org/

Rocky Mountain Institute:
www.rmi.org/

World Nuclear Association
Radioactive Waste Management Site:
https://world-nuclear.org/information-library/nuclear-fuel-cycle/nuclear-wastes/radioactive-waste-management.aspx

Further Reading

Anderson, David A. "Natural Gas Transmission Pipelines: Risks and Remedies for Host Communities," *Energies 13*, no. *8* (2020): 1873. https://doi.org/10.3390/en13081873. An investigation of the determinants of severity for natural gas pipeline explosions.

Baumol, William J., John C. Panzar, and Robert D. Willig. *Contestable Markets and the Theory of Market Structure.* New York: Harcourt Brace Jovanovich, 1982. A seminal work that argues that natural monopolies cannot result from fixed costs alone.

Coady, David, Ian Parry, Louis Sears, and Baoping Shang. "How Large Are Global Fossil Fuel Subsidies?" *World Development 91* (March 2017): 11–27. An overview of the extent of subsidies for oil and gas companies.

Hill, Jason, Erik Nelson, David Tilman, Stephan Polasky, and Douglas Tiffany. "Environmental, Economic, and Energetic Costs and Benefits of Biodiesel and Ethanol Biofuels." *Proceedings of the National Academy of Sciences 103*, no. *30* (2006): 11206–11210. A report on the net energy and pollution created by corn-based ethanol and soybean-based biodiesel, with comparisons to the fossil fuels they displace.

Hubbard, Harold M. "The Real Cost of Energy." *Scientific American 264* (April 1991): 36–40. A commentary on the costs of energy that are not paid by consumers at the pump.

Koplow, Douglas. *A Boon to Bad Biofuels.* Cambridge, MA and Washington, DC: Earth Track and Friends of the Earth, 2009. A report on federal tax credits and mandates for biofuels that create negative externalities.

Koplow, Douglas, and Aaron Martin. "Global Warming: Federal Subsidies to Oil in the United States, Industrial Economics Incorporated, a Report for Greenpeace." 2001. www.greenpeace.org/~climate/oil/fdsub.html. A thorough report breaking down direct and indirect subsidies, and credits that benefit oil production.

Lovins, Amory. "Energy Strategy: The Road Not Taken?." *Foreign Affairs 55*, no. *1* (1976): 65–96. Sets forth the concepts of "hard" and "soft" paths for energy production.

Mehta, Rahul Tongia, and Vikram Singh. "Making Renewable Power Sustainable in India: Blowing Hard or Shining Bright?" *Brookings Institute* (January 8, 2015). www.brookings.edu/research/making-renewable-power-sustainable-in-india/. A look at the wind and solar energy industries in India, with applications elsewhere.

Piore, Adam. "A Nuke Train Gets Ready to Roll." *Newsweek* (July 30, 2001): 26–28. A story of policy and protest relating to the transport and storage of spent uranium fuel.

U.S. Environmental Protection Agency. *Hydraulic Fracturing for Oil and Gas: Impacts from the Hydraulic Fracturing Water Cycle on Drinking Water Resources in the United States*, EPA/600/R-16/236F. Washington, DC: U.S. Environmental Protection Agency, 2016. A detailed study that reports the possibility that fracking can severely damage drinking water supplies.

Watkiss, Jeffrey P., and Douglas W. Smith. "The Energy Policy Act of 1992—A Watershed for Competition in the Wholesale Power Market." *Yale Journal of Regulation 10*, no. *2* (Summer 1993): 447–482. Discusses avenues for small-scale competitors in the market for electricity.

Yergin, Daniel. *The Prize: The Epic Quest for Oil, Money, and Power.* New York: Simon & Schuster, 1993. A colorful history of the world petroleum industry.

Your author at the helm of a power plant. The next Homer Simpson?

ENERGY

SUSTAINABILITY

WEAK **STRONG**   THEN NOW

PHYSICAL CAPITAL

F
U   NATURAL CAPITAL
PRESENT T
U   OBLIGATIONS
R
E   CIRCULAR ECONOMY

RECYCLE

VEIL OF IGNORANCE

ENVIRONMENTAL SOCIAL HUMAN ECOLOGICAL

ECOLOGICAL RESILENCE

SURVIVABIL

*I
T
Y*

NATURAL CAPITAL DEPLETION TAX

ECOLOGICAL TARIFFS

SAFE MINIMUM STANDARDS

PRECAUTIONARY PRINCIPLE

WALK THE WALK

8 Sustainability

*I*n the small coastal town of Arcata, California, the Marsh Commons Cohousing Community adjoins a state wildlife refuge. The 13 housing units are built largely of recycled lumber, environmentally friendly SmartWood, Trex brand recycled-material decking, recycled paint, and carpets made from recycled plastic bottles. The inhabitants share a common garden, laundry room, shop, guest room, meeting and dining area, children's play area, workroom, and craft area. Among the primary goals of this and many similar cohousing communities worldwide is sustainability. By sharing common spaces and avoiding the use of virgin, depletable materials, the communities have a relatively small effect on the environment and on the options available to future generations.

Individuals, organizations, and governments share the goal of sustainability for differing reasons. People who wish to leave a legacy of health and environmental wealth for future generations seek to use resources in ways that can continue indefinitely. Ecological ethics bring many to tread lightly. Members of organizations that espouse sustainability, such as the National Wildlife Federation and participants in the student sustainability coalitions at many universities, see sustainability as a vital component of long-term social welfare. Elements of sustainability—energy conservation, recycling, and the sustainable use of raw materials—appeal to the profit motives of firms as well. When the National Cancer Institute gave Bristol-Myers Squibb the rights to produce the anticancer drug Taxol, the drug was made from the bark of a yew tree found in the U.S. Pacific Northwest. Those forests are home to endangered species including the spotted owl. Bristol-Myers Squibb

DOI: 10.4324/9781003428732-8

found that twigs and needles served as a sustainable substitute for bark as their essential ingredient, without harming the trees and with minimal impact on the ecosystem at large.

As a prescription for the allocation of scarce resources, sustainable development is closely tied to the world of economics. Like many environmental and resource economics issues, decisions about sustainable development force normative judgments about the welfare of present and future generations that encroach into ethics, as discussed in Chapter 16. This chapter explains several versions of the sustainability criterion, explores their implications, and discusses policies that support sustainability.

Sustainability Criteria

An activity with an unlimited time horizon is **sustainable**. The concept of sustainability is nebulous but attractive, with a guiding question for every activity: Can this go on? This general criterion for sustainability can be applied to everything from production and consumption to the flows of natural, physical, and human capital. The Bruntland Report issued by the World Commission on Environment and Development described **sustainable development** as "development that meets the needs of the present without compromising the ability of the future to meet their own."

The concept of sustainability complements the human instinct to endure and leave a legacy. However, there is disagreement over the appropriate target. Aside from debates over discount rates and the treatment of future generations as discussed in Chapter 5, there is controversy over what substitutions are allowed when we say something is being "sustained." And if we favor nondecreasing welfare over time, how is that best achieved, and how can we measure our level of achievement?

Weak Sustainability

At the heart of long-lived controversies over sustainability is the substitutability of natural capital (fossil fuels, biomass, the ozone layer, and so on) and physical capital (man-made capital such as machines, buildings, and roads). Advocates of **weak sustainability** argue that physical and natural capital are acceptable *substitutes*, making the relevant question: Is our accumulation of physical capital sufficient to make up for our loss of natural capital? The criterion for weak sustainability is the maintenance of the total capital stock, represented by the horizontal line across the top of Figure 8.1.

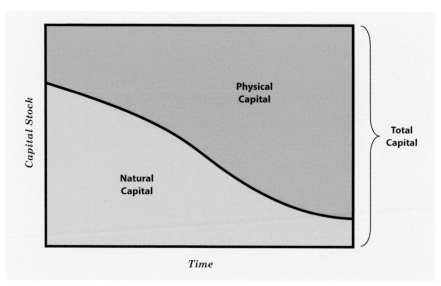

Advocates of weak sustainability argue that physical and natural capital are substitutes. As levels of natural capital decline, they believe that technological innovations will allow physical capital to take its place. Thus, the key to sustained welfare is maintenance of the total capital stock.

Neoclassical economists John Hartwick and Robert Solow are among those who have argued that decreases in natural capital over time will not be problematic if balanced by increases in physical capital. According to **Hartwick's rule**, *with no population growth and no depreciation of physical capital, consumption levels can remain constant from one generation to the next.* The stipulation is that exhaustible natural resources are never consumed but rather are turned into physical capital. The total stock of productive capital is thus sustained because the current generation converts natural capital into physical capital and "lives off" the output of physical and human capital.

David Pearce and Giles Atkinson present an index of weak sustainability. They begin with the assumption that human capital will not depreciate because it has public-good aspects and can be passed from one generation to another. Fitting with the assumptions of weak sustainability, they also assume that savings are invested in physical capital, which they treat as a perfect substitute for natural capital. The maintenance of the total capital stock, then, depends on a national savings rate (savings/GDP) that is at least as great as the combined depreciation rate of natural and physical capital. The relevant condition for sustainability is thus

$$Z = \frac{\text{Savings}}{\text{GDP}} - \frac{\left(\begin{array}{c}\text{Depreciation of} \\ \text{Natural Capital}\end{array} + \begin{array}{c}\text{Depreciation of} \\ \text{Physical Capital}\end{array}\right)}{\text{GDP}} \geq 0.$$

Figure 8.2

The Strong
Sustainability
Criterion:
Maintain the
Level of Natural
Capital

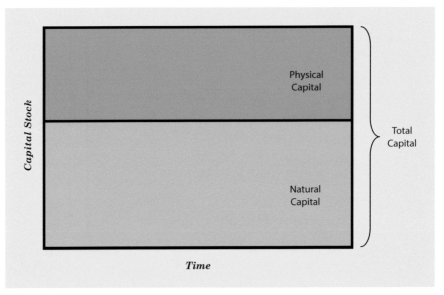

Advocates of strong sustainability prefer not to rely on uncertain technological innovations. They see natural and physical capital as complements and expect limited amounts of natural capital to form the binding constraint on future welfare. The strong sustainability criterion is the maintenance of sustainable levels of natural capital.

Pearce and Atkinson suggest the value of Z as an index of sustainability. They calculated index values for 22 countries, finding that countries including Brazil, Japan, and the United States have sustainable economies; Mexico, the Philippines, and the United Kingdom are marginally sustainable; and Ethiopia, Mali, and Nigeria are among countries with unsustainable economies.

Strong Sustainability

Ecological economist Herman Daly popularized a less forgiving view of sustainability. The **strong sustainability** criterion calls for the maintenance of natural capital as shown by the middle horizontal line in Figure 8.2. Advocates of strong sustainability see natural and physical capital as complements, not substitutes, with natural capital being an essential ingredient in production, consumption, and welfare. They perceive too many uncertainties and feel it is too dangerous to assume that physical capital can take the place of biodiversity or the life-sustaining cycles of oxygen, carbon, nitrogen, and water, for example. In their view, neoclassical economists give inadequate attention to environmental degradation and pollution, the presence of which can curtail social welfare regardless of the maintenance of total capital levels.

Several available measures of strong sustainability reflect the emphasis of this criterion. For example, Daly's and Cobb's Index of Sustainable

Figure 8.3

*The Course of
Intergenerational
Welfare*

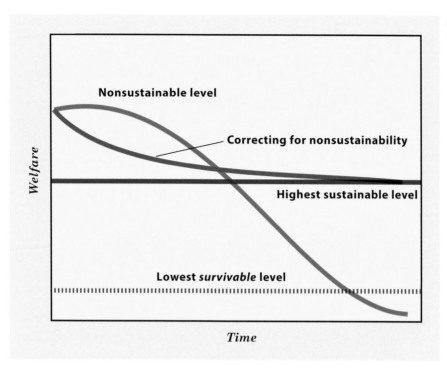

If our current levels of resource use and welfare are not sustainable, we will have to make sacrifices now to reach the highest sustainable level of welfare. Delays in this correction may lead to decreases in the highest level that can be sustained. As indicated by the red line that starts upward and then falls to the horizontal axis, some economists believe that continued consumption at current levels could cause severe environmental losses and bring welfare below the survivable level in the future.

Economic Welfare, as detailed in Chapter 5, explicitly adjusts GDP for pollution and losses in natural capital. The strong sustainability version of the Pearce-Atkinson sustainability rule is

$$\frac{\text{Natural Capital Depreciation}}{\text{GDP}} \leq 0.$$

This simply requires natural capital depreciation to be zero, or negative, which would cause an appreciation in natural capital. None of the 22 countries Pearce and Atkinson examined meet this criterion, although the Netherlands, Japan, and Finland come the closest.

Given that natural capital is indeed decreasing as in Figure 8.1, the achievement of strong sustainability would require changes that end the net loss of natural capital. This might mean fewer material goods and greater recycling efforts, among other sacrifices. Figure 8.3 illustrates this story as it relates to social welfare. If the goal is a sustainable level of welfare as indicated by the purple horizontal line, ecological economists advise a gradual adjustment of resource use to achieve a sustainable level of welfare starting now, as illustrated by the green line. If we continue to deplete natural resources at current rates, they say, we could sustain or improve welfare in the short run as shown by the red line, but we will face irreversible environmental damage and a corresponding loss of welfare in the future.

Strong sustainability requires the maintenance of aggregate levels of natural capital. It does *not* stipulate the sustenance of specific components of natural capital or specific physical flows of resources. The harvest of every fishery and forest at a sustainable rate would be a *sufficient* but unnecessary condition for strong sustainability. Depletable resources like fossil fuels can be used if they are replaced by renewable resources like switchgrass for ethanol production so that the total stock of natural capital is maintained.

Avoiding Mistakes

What if we choose the wrong criterion? If we limit our use of natural capital and it turns out that physical capital could have served the same purposes, then the reduction in social welfare depicted in Figure 8.3 will have been unnecessary. This raises the question of who should make the sacrifices that may or may not turn out to be necessary. Daly suggests that sustainable development should begin within the developed countries because they could reduce resource use while maintaining a relatively high level of comfort. Others fear it is inevitable that the brunt of conservation would be felt by the developing countries, which could be denied the luxuries of their rich neighbors.

Sustainable practices are a deliberate attempt to improve intergenerational equity. If this interest in equity across time carried over to equity across people, policymakers would allocate the sustainability burden by placing constraints on developed nations while allowing developing nations improve their conditions. In practice, any actual decrease in resource use would most likely occur in developed countries; developing nations have relatively little to sacrifice. A policy of holding resource use steady without mechanisms for redistribution would be a more direct threat to developing countries, which would have to compete with stronger economies for now-scarcer resources. In the end, policy and ethics will guide who sacrifices what.

The scarcity of resources such as water, land, and fossil fuels already limits human activity. If natural capital and not total capital turns out to be the binding constraint on output and we allow the stock of natural capital to decrease substantially, social welfare may be irreversibly diminished. If disruptions in food chains, environmental sinks, and ecosystems cannot be repaired with physical capital, the repercussions will be severe. If laboratories can't synthesize the medicinal cures that extinct species would have provided, human lives will be jeopardized.

Scientists study ecological resilience at Biosphere 2 in Arizona, an indoor facility with models of desert, ocean, wetland, rainforest, and savanna ecosystems.

The potential irreversibility of losses in natural capital has led some economists to call for **safe minimum standards** that set a lower bound on the level of critical natural resources. These standards could include the maintenance of viable populations of unique species and limits on the damage done to the air, water, and climate. Critics say that safe minimum standards represent an abandonment of cost-benefit analysis because the *benefits* of further loss are ignored. Advocates argue that if falling below a safe minimum standard means a high risk of catastrophic loss, it is not that benefits are ignored but that the cost will almost certainly exceed the benefit. Safe minimum standards can also guide public policies affecting the distant future, for which estimates of costs and benefits may be too tenuous to compare or too heavily discounted to garner attention.

To sidestep the worst possible mistakes, policies and actions must be chosen with attention to **ecosystem resilience**, also known as *ecological robustness*, which is the adaptability of ecosystems to disturbances such as chemical releases, floods, fires, invasive species, radiation, pollution, and climate change. Researchers examine temperature thresholds for marine life, pollution thresholds for trees, season-length thresholds for crops, and noise thresholds for birds, among other measures of resilience. With knowledge of what wildlife can and cannot withstand, decision-makers can assess the trade-offs inherent in government regulations and personal choices. Current uncertainty about ecosystem resilience motivates ongoing research at places like Biosphere 2 in Arizona, which houses small-scale ecosystems including a desert, an ocean, a wetland, a rainforest, and a savanna. There, is it possible to simulate global warming and study, for example, which species of coral are resilient to the changes caused by global climate change.

Other Types of Sustainability

Although the strong and weak sustainability criteria have received considerable attention among economists, a number of other criteria have been proposed. One that has already been mentioned is a target of sustenance for every specific component of natural capital and every flow of particular natural resources. This concept is called "very strong" sustainability or **environmental sustainability** and is associated with the deep ecology movement discussed in Chapter 16. Strictly interpreted, this criterion would seem impossible to meet, given the natural changes in biodiversity and other natural capital. In practice, deep ecologists would generally apply their criterion by respecting the environment and prioritizing ecological concerns over economic development, recognizing that some things will naturally change.

The term *environmental sustainability* shows up in many contexts with somewhat differing interpretations, as does the term *ecological sustainability*. A group of scientists commissioned by the U.S. Department of Agriculture advanced the meaning of **ecological sustainability** as "maintaining the composition, structure, and processes of an ecological system." The *composition* of an ecological system refers to its biodiversity, including the diversity of its genetics, species, and landscape. The *structure* refers to characteristics with biogenic origin, such as fallen trees, bogs, and coral reefs, as well as physical attributes, including navigable rivers, mountains, valleys, and wetlands. Natural *processes* include hydrologic and nutrient cycles, photosynthesis, disturbances like floods and fires, and the stages of succession within forests, prairies, ponds, and other wilderness areas.

Implicit in the goal of intergenerational equity are at least two other forms of sustainability: human sustainability and social sustainability. **Human sustainability** entails sustenance of the human capital needed to maintain levels of health, wealth, production, and therefore welfare. **Social sustainability** is the maintenance of social capital, which is made up of citizens' ethics, discipline, tolerance, trust, and other building blocks of social welfare. The least ambitious bottom line is **survivability**, which would allow social welfare to fall as long as consumption remains above the subsistence level. In Figure 8.3, a light blue dotted line represents the lowest survivable level of welfare. As discussed in Chapter 9, some economists expect current trends in population and consumption to take us on a welfare path below the highest sustainable level but above the collapse of human civilization, permitting survivability at an inferior level of welfare.

Sustainability and Efficiency

Economists generally focus on efficiency. The issue of sustainability provides an exception, for it involves vital resource-allocation questions wrapped up in human values and equity concerns. By providing an infinite time horizon for ecosystem services, sustainability provides welfare equity over time. This meets popular standards for fairness, including the Rawlsian *veil of ignorance*, under which decisions are made as if the deciders did not know in which period they would live. But achieving welfare equity over time is different from achieving dynamic efficiency, which involves discounting future values and maximizing the present value of net gains as explained in Chapter 5. Sustainability does not imply efficiency, nor does efficiency imply sustainability. The following example illustrates the distinction and a special case of compatibility between efficiency and equity in the context of sustainability.

We will examine the problem with a two-period model. The simplification of life into two periods permits a useful perspective on activities in the present (Period 1) and the future (Period 2). The implications carry over to relatively complex models with a more realistic number of periods. Consumers in each period are assumed to have the same preferences and receive diminishing marginal benefits, net of any associated costs, from a depletable resource. This resource might be coal, oil, clean water, lithium, or any other resource with a limited supply. The "consumption" described in this example refers to resource expenditures for the sake of utility in the period of consumption, not investments in the first period that would provide utility in the second period.

Suppose the world has 30 million tons of lithium that can either be used in Period 1 or Period 2. Figure 8.4 models the two-period allocation dilemma. The green arrows below the graph represent Period 1 consumption and the purple arrows below the graph represent Period 2 consumption. Similarly, the green *MB* curve represents the marginal benefit (net of any costs) in Period 1, and the purple *MB* curve represents the (net) marginal benefit in Period 2. Consumption in Period 1 starts at zero on the left side of the graph and increases to the right. Consumption in Period 2 starts at zero on the right side and increases to the left. The fixed quantity of lithium available means that if the total of 30 million tons is to be used, an increase in use in one period necessitates a decrease in use in the other period. The sustainable level of use is half of the resource in Period 1, which leaves the other half for Period 2.

To achieve dynamic efficiency, each unit of the resource would be allocated to the period in which it provides the largest present value of its marginal benefit. In the graph, that allocation occurs at the intersection of the *MB in Period* 1 curve, which reflects present values because those benefits are received in the present, and the light purple *Present Value in Period 1 of MB in Period 2* curve, which is discounted with the relevant discount rate to indicate the value of those benefits in period 1. Recall from Chapter 5 that this discounting simply amounts to dividing the marginal benefit by one plus the discount rate.

A positive discount rate decreases the present value of marginal benefits in Period 2, as from the dark purple line to the light purple line. The dynamically efficient consumption level of 18 in Period 1 and 12 in Period 2 then exceeds the sustainable level 15 in each period, as indicated by the intersection of the green and light purple *MB* curves. With any other allocation, some of the units received in one period would be worth more in the other period and, therefore, should be reallocated to the other period. For example, suppose the division is at point A, with

Figure 8.4

Equity Versus Efficiency in a Two-Period Model

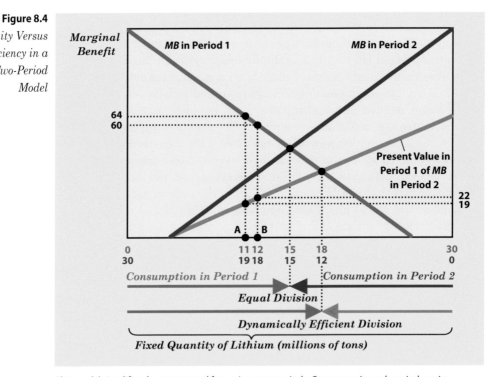

This model simplifies the present and future into two periods. Consumers in each period receive diminishing (net) marginal benefits from a depletable resource, which in this case is lithium. Consumption in Period 1 starts at zero on the left side of the graph and increases to the right. Consumption in Period 2 starts at zero on the right side and increases to the left. The sustainable level of use is half of the lithium in each period. A dynamically efficient allocation maximizes the present value of benefits by dividing the resource at the intersection point of the green and light purple curves that measure the present value of marginal benefits in each period. An equal division would only be dynamically efficient if the discount rate were zero. A positive discount rate decreases the present value of marginal benefits in Period 2, as from the dark purple line to the light blue purple, and the dynamically efficient consumption level in Period 1 exceeds the sustainable level.

Period 1 receiving 11 units and Period 2 receiving 19. By allocating an eleventh unit to Period 1 and moving to point B, the gain in Period 1 is $60 (the height of the *MB* in Period 1 curve above the unit gained), and the loss in Period 2 is only $19 (the height of the Present Value in Period 1 of *MB* in Period 2 curve above the unit lost).

An equal division is dynamically efficient only if the discount rate placed on benefits in the second period is zero. In that case, the appropriate allocation occurs at the intersection of the green and dark purple *MB* curves, which are mirror images of each other and cross in the middle of the graph. With any positive discount rate, equity and dynamic efficiency are at odds, and the distinction between equitable and efficient outcomes grows larger as the discount rate grows.

Walking the Walk

Sustainability has been the talk of many world conferences, documentaries, articles, and books. Actually living a more sustainable lifestyle is easier discussed than done. Consider the case of cotton, the pleasingly soft fabric in the T-shirts we wear to celebrate trips, colleges, 5K races, and sports teams. Traditional cotton farming involves the intensive use of toxic chemicals that can include insecticides, herbicides, fertilizers, and fungicides. It takes up to 2,700 liters of water to grow the cotton for one T-shirt. Patagonia is among several companies selling clothing made with organically grown cotton, but it isn't cheap. Walking the walk of sustainability means making efforts at the individual, corporate, and government levels to moderate the use of products made in unsustainable ways. The necessary sacrifices are often reduced by the availability of more sustainable substitutes. In the case of cotton, and depending on the production process employed, these include fabrics made of recycled fibers, bamboo, flax, or hemp.

Efforts toward sustainability are nothing new. Greece had laws to protect forests more than 2,000 years ago, and several European societies limited timber and game harvests more than

Efforts toward sustainability require moderation in the use of products with a large environmental footprint. This includes the traditional cotton T-shirts bought to celebrate trips, colleges, and sports teams.

The United Nations General Assembly's 2030 Agenda for Sustainable Development includes these 17 goals to guide national strategies for sustainable development around the world. See www.un.org/ sustainabledevelopment/sustainable-development-goals/. The content of this publication has not been approved by the United Nations and does not reflect the views of the United Nations or its officials or Member States.

1000 years ago. In the modern era, noting the "profound impact of man's activity" on the environment, the National Environmental Policy Act of 1969 made it the policy of every level of the U.S. government to "use all practicable means and measures . . . to create and maintain conditions under which man and nature can exist in productive harmony, and fulfill the social, economic, and other requirements of present and future generations of Americans." This nod to sustainability was followed by the National Forest Management Act of 1976 (NFMA), which requires the U.S. Secretary of Agriculture to provide for, among other things, sustainable timber harvests and diversity of plant and animal communities. A committee of scientists, asked by the U.S. Department of Agriculture to recommend management practices for national forests and grasslands in the new century, stated that the NFMA standards should be interpreted to mean ecological sustainability as defined in this chapter.

Agenda 21 of the 1992 United Nations (UN) Conference on environment and Development asked all participating countries to introduce national strategies for sustainable development (NSSDs). In 2015, the UN General Assembly adopted the 2030 Agenda for Sustainable Development, which included 17 sustainable development goals to guide countries' NSSDs. As shown in the illustration, the goals include sustainable cities, climate action, responsible consumption, and sustainable use of land and water. The 2022 UN Climate Change Conference, known as COP 27, launched new funds to protect land and water in Africa and to improve resilience to climate change in many developing countries.

Other specific policies and practices that support sustainability include advancements in renewable energy, tradable emissions permits, CAFE standards, brownfield (abandoned commercial site) reclamation, product substitution (for example, substituting hybrid cars for combustion engines and services for goods), municipal solid waste reduction, scrubbers and other "clean" technology, and "green" architecture. The remainder of this section discusses recycling, which can help make resource use sustainable.

Recycling

Historically, those who salvaged trash have been looked down upon. To thwart trash pickers, late-nineteenth-century Parisian prefect Eugène Poubelle required residents to set their trash out only minutes before collection time. Colonel George E. Waring, New York City's commissioner of streets during the same period, despised the many low-income individuals who picked garbage from the streets, and he saw them as a threat to sanitation. Still today, at dumpsites and trash receptacles, informal recyclers are often considered a nuisance.

To **recycle** is to take used goods and make them into new goods. For example, 36 used polyethylene terephthalate (PET) water bottles can be recycled into one square yard of carpet. Most of the plastic, metal, glass, and paper removed from the waste stream will be recycled into new goods. Some items can simply be reused as *alternatives* to newly manufactured goods. Thrift stores, flea markets, garage sales, and eBay visitors share the mission of making one person's junk another's treasure. Recognition of the private and public benefits of reusing and recycling has swayed some opinions. Even Poubelle gained greater appreciation for the benefits side of the equation and relaxed his Parisian policy after 30,000 beneficiaries of recycling rioted in the streets. In Hanoi, Vietnam, not unlike other developing urban areas, scavengers reuse or recycle about 30 percent of the city's refuse. Recycling's reputed contributions to environmental, natural resource, and sustainable development goals have made it the practice of soccer moms and ragpickers alike.

Is It Efficient?

Recycling rates have stagnated at around one-third in the United States since 2010. This includes items that are *composted*, meaning that organic waste such as food scraps and leaves are recycled into fertilizer. Germany leads the world by recycling about two-thirds of its municipal solid waste. The broad range of possible recycling efforts raises the question: Is it worthwhile? Naturally, the efficiency of recycling is case specific and changes with societal needs and technology. It is efficient to water plants with recycled dishwater rather than new water drawn from the tap—the distance the waterer must travel is the same either way, and the dishwater otherwise goes down the drain. It would be inefficient to recycle chalk dust—there is no shortage of the raw material, and the used resource is difficult to collect and reconstitute. The challenge is to draw the line between items to recycle and items to send to the landfill. Fluctuations in the supply and demand of recyclable materials, and for the products made from them, make it impossible to create a definitive list of materials that should or should not be recycled. However, there are relevant approaches and important points to address.

Critics of recycling contend it is a ritual that would not survive without undue guilt or public coercion. Mike Munger wrote, "If recycling were efficient, someone would pay you to do it."[1] Advocates find the recycling ritual makes people more aware of their resource expenditures and helps to replace a linear flow of resources—from extraction to use to disposal—with a circular flow in which old products become new products. That recapture of resources is fundamental to the broader goal of a *circular economy* in which products, businesses, and systems are designed to eliminate waste and keep materials in circulation for as long as possible.[2] For example, single-use shopping bags can be replaced with reusable shopping bags made of recycled fibers. Many cities and states already prohibit or tax single-use bags to incentivize more sustainable alternatives.

Chapter 3 explained the pitfall of market failure due to externalities. Although there are markets in which recycled products are bought and sold, the amount of recycling people are paid to do falls below efficient levels due to the externalities prevalent in resource extraction, material production, and waste generation. With externalities present, market prices provide inadequate incentives for recycling. When the private market determines the price for recycled paper, for instance, the price is formed without regard for the external benefits from avoided deforestation and pollution, protected human health, and averted landfill disposal. Thomas Kinnaman (2006) lists the external costs of landfill disposal as the loss in property values near the landfill, the threat to groundwater from possible leakage, the release of carbon dioxide and methane, and the congestion and air pollution created when transporting waste to the landfill. So that potential recyclers will bear more of the

---

1 See www.econlib.org/library/Columns/y2007/Mungerrecycling.html.

2 See www.epa.gov/circulareconomy/what-circular-economy.

Figure 8.5

The Market for Recycled Paper

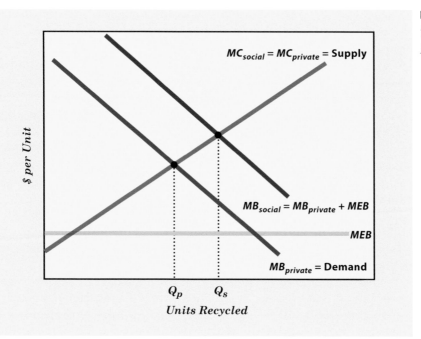

$MC_{social} = MC_{private} = \text{Supply}$

$\$ \text{ per Unit}$

$MB_{social} = MB_{private} + MEB$

MEB

$MB_{private} = \text{Demand}$

Q_p Q_s

Units Recycled

Given the marginal external benefit of recycling, the social marginal benefit exceeds the private marginal benefit. The privately optimal quantity of recycling is Q_p, which falls short of the socially efficient quantity Q_s. Policies intended to bring recycling efforts to the efficient level include legislated recycling, per-unit waste fees, moral suasion, promotions, curbside pickup, Coasian side payments, and subsidies.

burden of waste disposal that is placed on society at large, Kinnaman suggests a landfill tax set equal to the marginal external cost of solid waste disposal—about $10 per ton—minus any "host fees" already paid to the local area by the landfill operators.

Imperfect information compounds problems with externalities. Even when market participants care about the harm they cause to the environment and society, some lack the knowledge that, for example, a ton of paper made from recycled pulp saves approximately 17 trees, 4,200 kWh of electricity, 7,000 gallons of water, and 3 cubic yards of landfill space, while avoiding the emission of 60 pounds of airborne effluents. Ding et al. (2023) found that awareness of the consequences of wasting resources was a notable influence on the intention to recycle. Recycling rates in Escambia County, Florida, tripled after an ambitious education campaign. These findings suggest that market failure in recycling results from both externalities and imperfect information.

Figure 8.5 applies the social efficiency analysis introduced in Chapter 3 to the case of recycled paper. The market demand reflects the private marginal benefit derived from the market for products made of recycled paper. The market supply reflects the private marginal cost of recycling. We will assume the private and social costs of recycling are the same (that is, that there are no negative externalities), while the private and social benefits differ by the marginal external benefit of recycling as discussed earlier in this section. In the presence of these positive externalities, private incentives result in recycling at the inefficient low level Q_p.

There are several available remedies for this inefficiency. The socially efficient level, Q_s, can be attained if the marginal external benefits are internalized or the marginal cost is reduced, which subsidies or Coasian side payments could achieve. Many communities lower the marginal cost of recycling by subsidizing curbside pickup and drop-off sites. If estimated correctly, Q_s can also be reached via mandated recycling, as occurs in Japan for home appliances, and in the United States for oil and tires, among many other examples. Remedies also include per-unit waste fees, moral suasion, advertising, and the promotion of markets for products made from recycled materials.

The key economic message is that only when the *social* marginal benefits and costs are considered can the efficient decision be made. Extreme positions for or against recycling improve with focus on the criterion that social marginal cost should equal social marginal benefit. For accuracy, benefit estimates must include the value of avoiding all the problems with resource extraction, use, and disposal. The benefits from recycling also include revenues from the resale of some recycled materials, protected resources (avoided mining, deforestation, and oil drilling), and cost saving from decreased municipal solid waste production. For example, in New York City it costs $150 less to process a ton of recycling than to dispose of a ton of waste.[3] Cost estimates must include the time and energy required to collect, sort, and transport recycled materials and the expenditures on recycling equipment, containers, and labor.

Recycling can also be a teaching tool. Exposure to the recycling process teaches children about resource scarcity and some of the workings of a circular economy. The exercise of recycling engages participants in environmental stewardship that might be the precursor for broader investments in volunteerism and environmentalism. Involvement in recycling makes people more aware of the level of waste they generate and may promote conservation. And recyclers derive good feelings sometimes described as a "warm glow" from doing what they feel is right.

As technology improves and recycling trends continue, so will the efficiency of recycling. In the past, manufacturing plants have logically been located close to the source of virgin materials, which is often not so close to the source of recycled materials and in many cases is oceans away from where the recycled goods are collected. In the future,

3 See https://cbcny.org/advocacy/testimony-state-new-york-city-recycling.

Old tires can become new welcome mats and sandals.

manufacturing plants can be built closer to the source of recovered materials—New York rather than Oregon in the case of paper—improving the efficiency of recycling heavy and bulky materials that are difficult to transport. The production of goods close to the locations of their use has the added benefit of decreasing private and social transport costs between the production sites and stores. Innovation is also bringing technology that allows recyclable materials to be separated from trash, sorted, and processed with greater ease.

Recycling Policy

Pay-as-You-Throw Fixed-rate, all-you-can-eat restaurants eliminate the financial disincentive for overindulgence and thereby lead their customers to overeat. Likewise, fixed-rate trash pickup promotes excessive waste generation because the private marginal cost of filling another trash bag is just the trouble of taking it to the curb. Trash programs that charge by the bag are analogous to à la carte restaurants and provide incentives for recycling and responsible consumption. Per-unit fees for trash collection, the basis of "pay-as-you-throw" programs, are now in place in 46 U.S. states and many other countries. Some insurance companies are adopting "pay-as-you-drive" programs that provide similar incentives to drive less. In Australia, a program aimed to decrease the water consumed by flushing toilets among other uses has humorously been dubbed "pay-as-you-go."

Charting the Course to Sustainability

In recent decades, global leaders have taken important first steps toward sustainability, which an ancient Chinese proverb reminds us are the beginning of every long journey. Any meaningful change in the sustainability of human behavior would likely come as the result of long-term, concerted efforts. Several international initiatives seek to lay the foundation for critical progress on this front.

In earlier groundwork, the Oslo Ministerial Roundtable of 1995 defined the goal of sustainable production and consumption (SP&C) as

> the production and use of goods and services that respond to basic human needs and bring a better quality of life, while minimizing the use of natural resources, toxic materials, and emissions of waste and pollutants over the life cycle, so as not to jeopardize the needs of future generations.

In 2000, all 191 countries that were then members of the United Nations agreed to help reach eight Millennium Development Goals for 2015 that included, "To assure environmental sustainability." After that goal was not satisfactorily met in 2015, the United Nations established the 17 sustainable development goals for 2030, as illustrated earlier in this chapter. Meanwhile, the World Business Council for Sustainable Development (WBCSD) and the United Nations Development Program (UNDP) have promoted their own brand of sustainability, eco-efficiency, which is

> reached by the delivery of competitively priced goods and services that satisfy human needs and bring quality of life while progressively reducing ecological impacts and resource intensity throughout the life cycle to a level at least in line with the earth's estimated carrying capacity.

The WBCSD identified the following paths to eco-efficiency:

- Reduce the material intensity of goods and services

- Reduce the energy intensity of goods and services

- Reduce toxic dispersion

(continued)

- Enhance material recyclability

- Maximize sustainable use of renewable resources

- Increase material durability

- Increase the service intensity of goods and services

The UNDP and the WBCSD emphasize the profitability of eco-efficiency and clean production. They also catalyze sustainable development partnerships between industries and governments and promote sustainability at a global level. Major international efforts continue, for example, with annual conferences of the parties to the United Nations Framework Convention on Climate Change, hosted by Azerbaijan in 2024 and Brazil in 2025.

The incentives of pay-as-you-throw programs appear to matter. For example, Thomas Barry (2017) found that pay-as-you-throw programs in Massachusetts lower the average household's volume of trash by between 11 and 26 percent. In Germany, Jeurgen Morlok et al. (2017) found that charging by the weight of trash leads to an 86 percent recycling rate, whereas charging by the bag (or bin) in Germany leads to a recycling rate of around 70 percent.

Bottle Bills   *Bottle bills* are laws that require a deposit on recyclable cans and bottles. Deposits of five to ten cents apply to standard glass, plastic, and aluminum beverage containers. Consumers pay the deposit at the time of purchase and receive the same amount back upon returning the containers to any participating store for recycling. If a consumer is not sufficiently motivated by the incentive to recycle and throws a bottle on the ground, the incentive remains for passersby to return the bottle and obtain the deposit. Bottle deposits are in place in 10 U.S. states, Canada, and 23 other countries.

The incentives of these policies matter as well. A study of seven states found that, after passage of a bottle bill, there was a 70 to 83 percent reduction of beverage container litter and a 30 to 47 percent reduction in total litter. Michigan policymakers attribute a 6 to 8 percent reduction in overall waste flow to their bottle bill, and residents of New York saw aluminum can recycling increase by 64 percent and glass bottle recycling increase by 74 percent after the passage of a bottle bill. A study by Franklin Associates, Ltd., estimated that the New York bottle bill saves the energy equivalent of 2 million barrels of oil each year.[4]

---

4   See www.bottlebill.org/index.php/benefits-of-bottle-bills/bottle-bills-promote-recycling-and-reduce-waste.

On the negative side, the patchwork of states with bottle bills presents perverse incentives for those living near the border between a state with a bottle bill and a state without one. Retailers in states with bills worry that customers will purchase their beverages in neighboring states where no deposit is required. In Hawai'i, border crossing is not a problem but space is. Shopkeepers there and elsewhere point out that the storage of increasing numbers of recycled containers places a strain on space constraints. Major beverage producers oppose bottle bills because they necessitate more staffing and accounting to track the bottles and deposits. The slow adoption of bottle bills within the United States reflects this contention. At the same time, most of the states with bottle bills have had them for more than 30 years, so apparently to know one is to like one.

Broader Policies Toward Sustainability

More ambitious initiatives are needed to meet sustainability goals and the strong sustainability criterion in particular. Many of the environmental policies discussed in this text would contribute to sustainability. The three proposals featured in this section have appeared in many contexts and were synthesized by Ecological Economics professor Robert Costanza as a possible foundation for policy consensus.

Natural Capital Depletion Tax

A substantial tax on national capital depletion (NCD) could constrain natural capital exploitation to sustainable levels while serving interests on both sides of the sustainability issue. An NCD tax would lead to the conservation of natural capital as a precaution against the distinct possibility that ample substitutes are not forthcoming. It would also promote the development of substitutes to the extent that they are available. In that way, Costanza suggests that an NCD tax should appeal to weak and strong sustainability advocates alike.

As with any tax, the size of the tax would be a source of debate. The tax would fall heavily on fossil fuels. As mentioned in Chapter 7, economists have estimated the social marginal cost of gasoline, for example, although a compromise among the various estimates would be necessary. Acquiescing in a ballpark figure, Costanza et al. (2015, p. 249) writes,

> It would be helpful to have better quantitative measures of these perceived costs, just as it would be helpful to carry along an altimeter when we jump out of an airplane. But we would all prefer to have a parachute than an altimeter if we could only take one thing.

In this story, the parachute for an environment in freefall is a tax on natural capital depletion that errs on the safe side by being large enough to reliably slow the descent.

An NCD tax would fall largely on consumers of energy and energy-intensive products. Because those with low incomes spend a larger fraction of their income on energy than those with high incomes, the tax would be regressive. As a remedy, the revenues from the NCD tax could be used to decrease income taxes disproportionately for low-income individuals, making the overall tax effect neutral or progressive (placing a higher burden on the rich). As another variant, thousands of economists are among the advocates of a *carbon fee-and-dividend program* that collects a tax on fossil fuel sales and rebates an equal portion of the tax revenue back to every consumer.[5] The average consumer would receive a rebate equal to their tax payment (less their share of administrative costs). In the likely event that high-income consumers use more carbon-based energy than low-income consumers, high-income consumers will pay more than the rebate, and low-income consumers will pay less than the rebate, making the program progressive.

Precautionary Polluter Pays Principle

Environmental policy strategists struggle with uncertainty about the future. Maybe climate change will gravely disrupt agriculture. Maybe pollution will choke the food chain or cause cancer rates to skyrocket. Maybe we will not find physical capital to replace depleted natural capital. Rather than taking a stance on one side or the other, the precautionary polluter pays principle (4P) takes the bottle bill concept and makes a safety net out of it.

The idea of the **precautionary polluter pays principle** is that those who take risks with natural resources would post a bond large enough to cover the best estimate of the worst-case scenario for future environmental damages. For example, those building a landfill would have to post an assurance bond sufficient to pay for cleanup costs in the event of groundwater contamination. This causes the resource users to internalize the risk burden associated with their behavior. If subsequent damages are not forthcoming, or it can be demonstrated that the worst-case scenario is not as bad as originally thought, the users receive the portion of the bond (plus interest) corresponding with the absolved risk. Activities that cause no harm end up with no loss from the deposit. If damage does occur, the deposited funds are used to repair and compensate for those damages.

SUSTAINABILITY

---

5 See https://clcouncil.org/economists-statement/.

Ecological Tariffs

On their own, NCD tax and 4P policies do not motivate sustainability outside their jurisdictions, and present problems with unfair competition from firms in countries with lower environmental standards. If Country A adopts an NCD tax and Country B does not, firms in Country A will have higher costs and thus difficulty competing with imports from Country B. Ecological tariffs could prevent the short-run competitive advantages of low standards and provide incentives for improved environmental policies among exporting countries. In effect, the imposition of tariffs can bring the cost of unsustainable production up to the cost of sustainable production.

Countries with comparable NCD taxes and 4P policies could engage in free trade on fair terms. Exports from countries without such policies would be hit with ecological tariffs equal to the estimated savings from not producing under the designated policies for sustainability. Having leveled the playing field, the tariff revenues could then be invested into the protection of natural resources to help remedy the underlying environmental neglect.

Whether or not these policies are implemented as stated, they represent prototypes for available policy directions. They are starting points and options to build on, and they exemplify the ways in which economic incentives can be managed to achieve sustainability targets.

Summary

With the goal of providing intergenerational equity in opportunity, sustainability can be seen as a constraint on growth in deference to the future. Sustained intergenerational welfare would require the maintenance of capital levels, arguably including physical capital, natural capital, human capital, and social capital. This goal can be achieved by harvesting renewable resources only at the rate of replacement, exploiting depletable resources only as fast as renewable substitutes are found, and creating waste no faster than it can be assimilated into the environment.

The appropriate criterion for sustainability is under debate. Advocates of weak sustainability feel that natural and physical capital are substitutes and that the maintenance of the total capital stock would provide nondecreasing welfare. They believe that short-run investments of natural capital will be rewarded with sustainable long-run payoffs from physical capital. Supporters of strong sustainability argue that natural and physical capital are complements. They feel that neoclassical economists place too much faith in the unrealized substitutability of natural and physical capital and too little emphasis on pollution and other externalities. In their view, welfare can only be maintained by a nondecreasing stock of natural capital.

Policies that promote sustainable practices include pay-as-you-throw trash programs and bottle bills, both of which have increased recycling rates. The efficiency of recycling particular materials depends on fluctuating markets for end products, technology, and the cost of disposal alternatives. Those conducting efficiency analyses must be careful to consider all the social costs and benefits of recycling. Costs include the value of time spent sorting materials and the expense of collection and processing operations. Benefits include the avoided externalities from energy use, pollution, and resource use associated with waste disposal and the processing of virgin materials.

More ambitious policy options could drive decision-makers to internalize the costs of their actions. A natural capital depletion tax would steer manufacturers toward conservation. A precautionary polluter pays principle would require deposits as security against potential environmental tragedies. And ecological tariffs would prevent unfair price competition from producers whose practices are unsustainable.

The debates over reasons and methods for sustainability rage on, but the concept has become a fixture within and beyond the fields of environmental and natural resource economics. Armed with an awareness of the relevant issues and approaches, you are now equipped to contribute to discussions that will help determine your future.

• • • • • • • • • • • • • • • •

Problems for Review

1. Suppose each period's net marginal benefit curve for oil is a straight line with a vertical intercept of $100 and a horizontal intercept of 50 barrels. Assume the discount rate is zero.

    a) What is the dynamically efficient allocation in each period if the total quantity of oil is 60 barrels?

    b) What is the dynamically efficient allocation in each period if the total quantity of oil is 120 barrels? *Illustrate your answers with graphs similar to Figure 8.4.*

2. Suppose Period 1's net marginal benefit curve for platinum is a straight line with a vertical intercept of $26 million and a horizontal intercept of 70,000 tons.

    Period 2's net marginal benefit from platinum is $26 million *for every unit from the first to the last.*

    a) *Assume the discount rate is zero. Draw a two-period model for this situation and indicate the dynamically efficient allocation of platinum in each period.*

    b) *Assume instead that the discount rate is 100 percent. On the graph for part a, add a line showing the present value in Period 1 of the net benefit in Period 2 if the discount rate is 100 percent and indicate the dynamically efficient allocation of platinum in each period. (The present value formula in Chapter 5 will be useful here.)*

3. This year, Nearland has a GDP of $200 billion, savings of $100 billion, $70 billion in depreciation of natural capital, and $30 billion in depreciation of physical capital. Farland has a GDP of $500 billion, savings of $150 billion, $100 billion in depreciation of natural capital, and $70 billion in depreciation of physical capital. Calculate the Pearce-Atkinson index of weak sustainability, Z, for each of these countries and indicate whether they satisfy the weak sustainability criterion.

4. Choose one policy from the chapter that you could implement locally. What do you think would be the biggest challenge to implementation and how would you overcome it?

5. Suppose Eastland imposes a safe minimum standard that 20 percent of the original forests must be left standing to assure adequate biodiversity, oxygen creation, and carbon sequestration. According to this chapter
   a) *How can one argue this policy neglects cost-benefit analysis?*
   b) *How can one argue this policy satisfies cost-benefit analysis?*

6. In a circular economy, products are designed for reuse and recycling for as long as possible.
   a) *Explain how one product you use (other than the example in the reading) could be redesigned to help create a circular economy.*

   b) *How would you suggest that consumers be incentivized to adopt your redesigned product?*

7. The world population is growing by about 70 million people per year. The growth in population and the depreciation of physical capital violate the assumptions of which "rule" as described in the chapter? Draw a modification of the total capital line in Figure 8.1 that would satisfy a desire for constant per-capita welfare given increasing population growth.

8. Take a moment to reflect on the sustainability of your activities in the past week. Consider your acquisitions and your modes of travel.
   a) *What was your least-sustainable activity over the last week?*
   b) *What costs and benefits did you consider when deciding to carry out that activity?*
   c) *Was your decision socially efficient?*
   d) *Do you think that physical capital could replace the natural capital you used most?*

9. Some people claim that fervent recycling is the result of a "religion" of conservation that makes people feel good despite a lack of concrete reason. Suppose that: (1) the curves in Figure 8.5 exclude the value of such good feelings; (2) the good-feelings benefit is the same for each unit; and (3) the per-unit benefit exceeds the marginal external benefit.

Recreate Figure 8.5. Add a private $MB_{combined}$ curve with the height of the $MB_{private}$ curve plus the constant per-unit good-feelings benefit. Add the new MB_{social} curve with the height of all the benefits to society. Would the quantity of recycling at the intersection of the private $MB_{combined}$ curve and the MC curve be excessive from the standpoint of social efficiency? How can you tell?

10. Among the "broader policies toward sustainability" discussed in this chapter
   a) Which policy prevents the competitive advantages of low environmental standards?
   b) Which policy has a variant that can help low-income households in particular?
   c) Which policy provides protection against the worst-case scenario of environmental damages?

11. *This problem requires the use of algebra.* Consider the scenario given in problem 1 with a total quantity of 60 barrels of oil. Note that in each period, MB = 100–2Q. What is the dynamically efficient allocation if the discount rate is 10 percent per period?

12. *For those who have studied budget constraint/indifference curve diagrams and want a challenge.* Draw a graph with trash pickup services on the horizontal axis and "other goods/services" on the vertical axis. Draw two budget constraints, one reflecting a fixed fee for unlimited trash pickup and one reflecting a pay-as-you-throw program. Use indifference curves to explain the likely effects of these two policies on the amount of trash individuals generate.

websurfer's challenge

1. Find a website that describes cutting-edge sustainable living. This might involve cohousing, green architecture, freeganism, or sustainable architecture.

2. Find a policy proposal relating to sustainable development put forth by an ecological economist and evaluate its message.

3. Find a website that speaks against some aspect of sustainable development and evaluate its message.

4. Find a website that discusses recycling policies or levels in your area and summarize your findings.

Key Terms

Ecological sustainability

Ecosystem resilience

Environmental sustainability

Hartwick's rule

Human sustainability

Precautionary polluter pays principle

Safe minimum standards

Social sustainability

Strong sustainability

Sustainable

Sustainable development

Survivability

Weak sustainability

Internet Resources

Bottle Bill Resource Guide:
www.bottlebill.org

Fostering Sustainable Behavior:
www.cbsm.com

International Institute for
Sustainable Development:
www.iisd.org

List of Sustainable Development
Organizations:

*www.greendreamer.com/journal/
environmental-organizations-nonprofits-for-
a-sustainable-future*

Towards Sustainability blog:
www.towards-sustainability.com

United Nations 2030 Agenda for
Sustainable Development:
sdgs.un.org/2030agenda

Further Reading

Anderson, David A. *Treading Lightly: The Joy of Conservation,
Moderation, and Simple Living.* Danville, KY: Pensive Press, 2009.
A collection of information and ideas about sustainable living.

Barry, Thomas W., IV. "When Trash Costs Money: Analyzing the
Impact of Pay-As-You-Throw Programs in Massachusetts." *Journal of
Environmental and Resource Economics at Colby 4, no. 1* (2017): Article
3. An empirical study of the effects on household behavior of new
per-unit charges for trash disposal.

Castle, E. N., and R. P. Berrens. "Endangered Species, Economic
Analysis, and the Safe Minimum Standard." *Northwest Environmental
Journal 9* (1993): 108–130. Explains the distinction between cost-bene-
fit analysis and the safe minimum standard.

Costanza, Robert, John H. Cumberland, Herman Daly, Robert Goodland, Richard B. Norgaard, Ida Kubiszewski, and Carol Franco. *An Introduction to Ecological Economics*. New York: CRC Press, 2015. Not a stand-alone textbook but a thoughtful introduction to the field of ecological economics.

Daly, Herman E. "Sustainable Growth: An Impossibility Theorem." *Development 40*, no. *1* (1997): 121–125. Argues that the constraints of natural capital would prevent any sustained level of growth.

Daly, Herman E., and John B. Cobb, Jr. *For the Common Good: Redirecting the Economy Toward Community, the Environment, and a Sustainable Future*. Boston, MA: Beacon Press, 1989. Introduces the index of sustainable economic welfare and presents comparisons between it and the gross national product.

Ding, Lili, Zhimeng Guo, and Yuemei Xue. "Dump or Recycle? Consumer's Environmental Awareness and Express Package Disposal Based on an Evolutionary Game Model." *Environment, Development and Sustainability 25*, no. *7* (2023): 6963–6986. https://doi.org/10.1007/s10668-022-02343-1. Finds that consumer awareness of environmental issues removes resistance to recycling despite the external nature of most of the benefits.

Hartwick, J. M. "Intergenerational Equity and the Investing of Rents from Exhaustible Resources." *American Economic Review 67* (1977): 972–974. An important note that demonstrates the Hartwick rule.

Hicks, John R. *Value and Capital*. Oxford: Oxford University Press, 1975. An overview of economic theory by a Nobel Laureate.

Kinnaman, Thomas C. "Policy Watch: Examining the Justification for Residential Recycling." *Journal of Economic Perspectives 20*, no. *4* (2006): 219–232. An overview of recycling issues including cost, benefits, and landfill fees.

Morlok, Jeurgen, Harald Schoenberger, David Styles, Jose-Luis Galves-Martos, and Barbara Zeschmar-Lahl. "The Impact of Pay-As-You-Throw Schemes on Municipal Solid Waste Management: The Exemplar Case of the County of Aschaffenburg, Germany." *Resources 6*, no. *8* (2017): 1–16.

Pearce, David, and Giles Atkinson. "Measuring Sustainable Development." In *Handbook of Environmental Economics*, edited by Daniel W. Bromley. Cambridge, MA: Blackwell, 1995, 166–181. A useful overview of weak and strong sustainability and international trade issues.

Solow, Robert. "On the Intergenerational Allocation of Natural Resources." *Scandinavian Journal of Economics 88* (1986): 141–149. A classic discussion of sustainability across generations.

Tiller, Kelly H., Paul M. Jakus, and William M. Park. "Household Willingness to Pay for Drop-off Recycling." *Journal of Agricultural and Resource Economics 22*, no. *2* (1997): 310–320. A CVM study of the value of recycling opportunities in rural/suburban areas.

Vitousek, Peter M., Paul R. Ehrlich, Anne H. Ehrlich, and Pamela A. Matson. "Human Appropriation of the Products of Photosynthesis." *Bioscience 36*, no. *6* (1986): 368–373. Estimates a specific range for the proportion of solar energy that is appropriated by humans.

CHAPTER 8

Permaculture, a form of sustainable living, gets its name from contractions of permanent and culture as well as permanent and agriculture. Crystal Waters Permaculture Village is part of the Global Ecovillage Network. Learn more at https://crystalwaters.org.au/.

Population, Poverty, and Economic Growth

After exploring wilderness areas around the world and producing more than 115 films about the planet and its creatures, Jacques-Yves Cousteau called population growth "the primary source of environmental damage." For millennia, the world population was nearly constant. Over the past century, growth has become exponential, doubling every 30 to 40 years. Some of today's elderly have watched the world population quadruple.

Population growth rates are generally high in the early phases of economic development. A rising population means more mouths to feed, more homes to build, and more goods to manufacture. Later phases of development can be a double-edged sword if not carefully implemented. A rising standard of living reduces poverty and population growth, but haphazard economic growth can assault the environment and natural resources.

Simon Kuznets (1955) found that as income per capita grew, income inequality first increased and then decreased. More recently, economists studying pollution have observed the same increasing and then decreasing pattern in response to income growth. Unfortunately, like population growth, growing income levels worsen the mounting international problem of waste generation, which appears to buck the Kuznets up-and-down scenario and just keep increasing.

Thomas Malthus (1798) feared that a lack of self-discipline would put population growth on a collision course with limited resources in the nineteenth century. He was wrong about the timing, but the logic of his argument—that a growing population must at some point outstrip the planet's resources—has kept the

DOI: 10.4324/9781003428732-9

concern alive. This chapter introduces the economics of population growth, the effects of income growth, and the determinants of economic growth's impact on the environment.

Population Growth and Resource Scarcity

Thomas Malthus

Even in the eighteenth century, the concept that the human population might outgrow its resources was not novel. In the preface to his 1798 *Essay on the Principle of Population*, Malthus himself wrote, "It is an obvious truth, which has been taken notice of by many writers, that population must always be kept down to the level of the means of subsistence." Nonetheless, with his eloquence and sway, Thomas Malthus has become the most famous of population theorists. The crux of Malthus's argument was that the insatiable human appetites for sex and food would lead to a disaster of resource inadequacy. He wrote, "towards the extinction of the passion between the sexes, no observable progress has hitherto been made."[1] Malthus argued that unchecked human reproduction would increase populations *geometrically*, causing them to double every so many years, while food supplies could only increase *arithmetically* by a constant amount per period due to diminishing marginal returns from land.

Malthus came to this conclusion after noting that in the United States, where resources were not a binding constraint, the population reportedly doubled every 25 years. As for food, Malthus reasoned that nowhere on Earth could food supplies double every 25 years, and at best they could grow by the amount of the 1798 food supply every 25 years. Starting with a food supply that would feed the 7 million people in his homeland of England, he felt the best-case scenario was adequate food for 14 million people in 25 years, 21 million people in 50 years, 28 million people in 75 years, and so on. If the population were unchecked and doubled over each 25-year period, there would be 14, 28, and 56 million people in 25, 50, and 75 years, respectively. Figure 9.1 illustrates the dreary conclusion that the food supply would fall short of consumptive needs after 25 years.

The **population growth rate** is the birth rate minus the death rate plus immigration minus emigration. Malthus warned that population growth rates that exceeded growth rates for food supplies would lead society into what is now called a *Malthusian population trap*, in which the inadequacy of food supplies limited the size of the population. In the absence of deliberate restraints on population growth, natural checks

---

1 See Malthus (1798, p. 50).

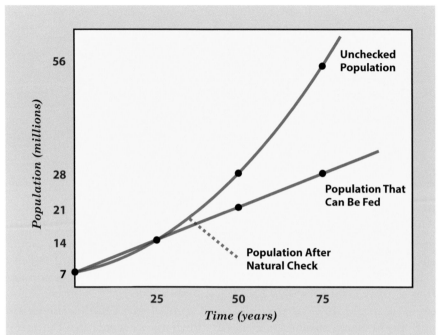

Figure 9.1

Malthusian
Population Trap

Malthus reasoned that with the population doubling every 25 years and the means of subsistence increasing by a constant quantity each period, the population would inevitably exceed the available supplies of food and other resources. Rather than being able to continue unchecked as on the solid red line, the natural checks of famine, disease, and war would force the population back below the level for which there is food, as indicated by the dotted red line.

would be inevitable. Malthus first described the checks on population as the misery of war, famine, and disease, along with havoc, hatred, and vice. In the second edition of his essay, he added the check of moral restraint, explained as "delaying the gratification of passion from a sense of duty."

Malthus's theory of stagnating per-capita output is consistent with the human experience over most of history. However, the **agricultural revolution** of farming advances in the late eighteenth and nineteenth centuries, such as crop rotation and plows drawn by domesticated animals, resulted in rapid increases in food production. Then, the **green revolution** of chemical and technological advances in the twentieth century brought hybrid grains, chemical fertilizers, pesticides, mechanized harvesters, and irrigation equipment. These changes allowed food production to grow geometrically over most of the twentieth century and allayed the Malthusian fear of famine in many regions. England now has a population of 67 million and ample food to serve them.

The next section describes decreases in the birth rates of developed countries, which have dealt a further blow to Malthusian predictions.

Although increases in food production and decreases in population growth have dispelled past forecasts of doom, uncertainty remains about the future. Global climate change, drought, and environmental toxicity could affect food supplies, while another doubling of the world population is likely before stability becomes a reality. The risks of environmental catastrophe and the possibilities of technological advancement are ever-present but difficult to quantify. Perhaps we'll colonize the moon and obtain needed resources from outer space, but prudent policymakers will respond with caution to gloomy and glowing predictions alike.

The Economics of Population Growth

Since the time of Malthus, unforeseen advances in birth control have served as a partial substitute for moral restraint from sexual behavior. In many areas, population growth has been further curbed by improved opportunities for education, employment, and health care. These changes reflect new cultural attitudes toward women and their rights, as well as advancements in transportation, communication, and education systems.

Children have costs and benefits as do goods and services, placing the decision to have children under the purview of microeconomic analysis. It costs money to feed, clothe, and education children, but the benefits can be great. Children become workers in the home, earners in the workforce, and caregivers as their parents grow old. As women gain access to better educational opportunities and higher wages, the opportunity cost of their time spent raising children increases. Career women tend to put off marriage and children until later in life. As incomes, sanitation, and health care improve, it is no longer necessary to conceive as many children to ensure that some will survive. As care for the elderly improves and machinery replaces labor-intensive jobs, the value of children for old-age security and labor decreases.

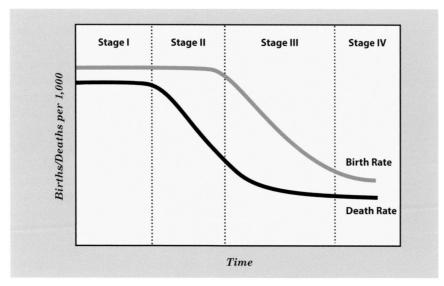

Figure 9.2

*Stages of
Demographic
Transition*

With improvements in income, health care, and sanitation, death rates tend to drop before birth rates, bringing a period of rapid population growth before birth rates drop as well.

The theory of demographic transition defines four stages of population growth for developing nations determined by the relative frequency of births and deaths. Figure 9.2 illustrates the following stages:

1. **High birth rates and high death rates.**
   Early in the development process, a scarcity of educational opportunities, jobs, and family planning resources keeps birth rates high. Poverty, unsanitary conditions, and health-service limitations cause high and fluctuating death rates. Relatively low rates of population growth are the result.

2. **High birth rates and declining death rates.**
   As development progresses, improvements in public health services, better nutrition, and safer living conditions increase life expectancy and decrease death rates. These changes are not immediately accompanied by a change in birth rates.

3. **Low death rates and declining birth rates.**
   Birth rates eventually fall as well, influenced by changing cultural attitudes and the availability of education, employment, and birth control.

4. **Low birth rates and low death rates.**
   After both birth rates and death rates have fallen, nations experience relatively low population growth rates once again.

Western Europe developed roughly according to this theory, reaching Stage III at the beginning of the twentieth century, and it is now

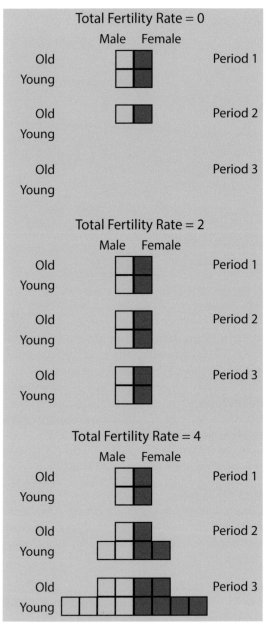

Figure 9.3
*Growth and
Total Fertility*

The total fertility rate matters. Beginning with two males and two females living two periods each and reproducing only when young, the population in Period 3 is 0, 4, or 12, depending on whether the total fertility rate is 0, 2, or 4, respectively.

experiencing nearly identical birth rates and death rates. The **total fertility rate** (TFR) is the average number of children each woman has in her lifetime. The United States has a TFR of 1.8 and an annual population growth rate of 0.5 percent. The populations of many less developed nations are growing more rapidly, as in Sudan, where the TFR is 4.54 and the growth rate is 2.7 percent. Population growth places stress on regional food supplies. The UN Food and Agriculture Organization reports that between 720 and 811 million people cannot obtain adequate nutrition,[2] and 3.1 million children under five die each year due to malnutrition.[3]

Figure 9.3 illustrates the influence of the total fertility rate in a simplified situation in which people are either young or old. Between one period and the next, the young reproduce, the young become old, and the old die. Each green square represents a male, each purple square represents a female, and half of each age group is assumed to be female. In Period 1 there is one male and one female in each age group.

2   See www.fao.org/state-of-food-security-nutrition.

3   See www.theworldcounts.com/challenges/people-and-poverty/hunger-and-obesity/how-many-people-die-from-hunger-each-year.

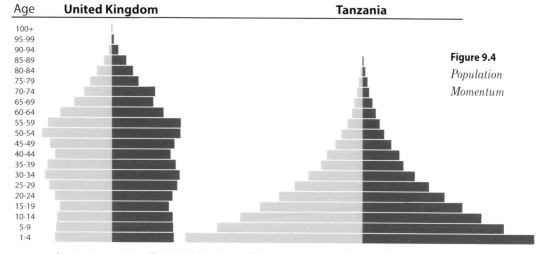

Age | United Kingdom | Tanzania

Figure 9.4

Population Momentum

Age structure matters. The United Kingdom and Tanzania each have about 67 million citizens. However, because the United Kingdom's population is concentrated at non-childbearing ages and most of Tanzania's population is of childbearing age, the growth rate is 0.34 percent in the United Kingdom and 2.96 percent in Tanzania.

The top scenario of Figure 9.3, a total fertility rate of zero, means there is no reproduction. Between Period 1 and Period 2, those who were old in Period 1 die, and those who were young in Period 1 become old. There are no new young people because there were no births. In Period 3, the old from Period 2 are dead, and there is no one left. Strict celibacy and the resulting TFR of 0 led to the demise of the Shaker religion that began in England and spread to America in the nineteenth century. The regional TFR of 0.76 in Heilongjiang, China, is reportedly the lowest in the world.

A total fertility rate of 2, as in the middle scenario, allows the young generation to exactly replace itself before turning old. In each new period there are two new young people to replace the two people who grew old. In reality, a TFR of 2 will lead to a slight decrease in population over time because, unlike in this model, some people die early in life.

With a total fertility rate of 4, each young pair is replaced by two young pairs in the next period. The population triples from 4 to 12 between Period 1 and Period 3. The population would be 24 in Period 4 and 48 in Period 5. In 2023, there were 26 countries with TFRs of 4 or more, including Niger (6.73), Angola (5.76), Mali (5.45), and Benin (5.39).

The mathematics of population growth are complicated by the age and gender structure of the population. For a given population size and total fertility rate, a population can have zero, negative, or positive population growth, depending on its age structure. A population that is disproportionately beyond childbearing age or of a single gender carries less reproductive momentum than younger, more gender-balanced populations. The average age in many developed nations is increasing, while shorter life expectancies and rapid growth contribute to younger populations in many developing countries. Figure 9.4 illustrates the age structures in the United Kingdom and Tanzania. Although each country has roughly 67 million citizens, the large number of Tanzanians of childbearing age give the country tremendous population momentum. The population growth rate is 2.96 percent in Tanzania and only 0.34 percent in the United Kingdom. When a high proportion of the population is young, even when the total fertility rate falls below the replacement rate of just over two children per woman, the population can continue to grow for several decades as larger, younger generations replace themselves while the older generations that pass away are relatively small.

In terms of policy approaches, some efforts to reduce population growth have addressed determinants of the demand for children with opportunities for education, employment, and care for the elderly. For example, Grameen banks in Bangladesh make loans to help women start small businesses. The subsequent employment increases the opportunity cost of bearing children, and loan recipients receive family planning information.

The cost of avoiding unwanted pregnancies is sometimes decreased with subsidized birth control. In India, population control efforts have included expanded energy availability so that young adults might stay up watching television rather than engaging in sex. Other policies target the marginal cost of children: Between 1979 and 2015, China had a one-child-per-couple policy that imposed fines of as much as half of a couple's annual salary for excessive births. Some couples were punished

Figure 9.5
World
Population
Growth

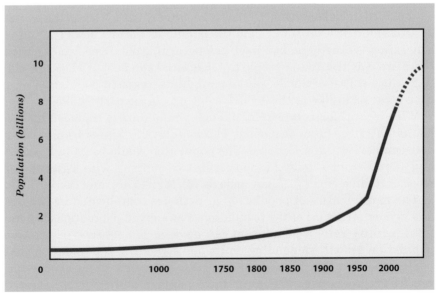

Exponential growth in the world population has produced a growth diagram in the shape of a J. This is expected to level off to form an S at a population of around 10 billion people.

with the loss of land grants, loans, supplies, jobs, and Communist Party membership, while doctors received bonuses for sterilizing patients. The goal was to limit China's population to 1.2 billion in 2000. The population reached 1.3 billion in 2000, but the total fertility rate decreased from 5.01 in 1970 to 1.8 in 2000. This change came at the expense of alleged human rights violations and large numbers of abortions.[4]

Figure 9.5 illustrates the creeping world population growth until around 1900 followed by exponential growth since that time. The *J*-shaped population expansion—some have called it an explosion—is expected to level off to form an *S*-curve at a world population of around 10 billion.

Population, Poverty, and Other Determinants of Waste

The Growing Problem of Municipal Solid Waste

The demand for goods grows with populations and incomes, resulting in increased production and consumption. With consumption comes waste. Residents of Europe, the United States, and Japan—a relatively wealthy 16 percent of the world population—are responsible for about 80 percent of natural resource consumption on an annual basis. The annual purchasing habits of the average U.S. citizen require the use of 25 tons of raw materials, and the 4.2 percent of the world population living in the United States are responsible for an estimated 12 percent of the world's waste production.[5]

After decades of municipal solid waste dumping nearby, "Glass Beach" in Hawai'i is made up almost entirely of crushed bits of glass and pottery that have washed up on shore.

4   See, for example, www.npr.org/2016/02/01/465124337/how-chinas-one-child-policy-led-to-forced-abortions-30-million-bachelors.

5   See https://environmentamerica.org/articles/how-much-trash-does-america-really-produce/.

Municipal solid waste (MSW) comes from households, small industrial enterprises, and municipalities. In the U.S. it is made up of food waste (24 percent); plastics (18 percent); paper and cardboard (12 percent); rubber, leather, textiles (11 percent); metals (10 percent); wood (8 percent); yard trimmings (7 percent); glass (5 percent); and miscellaneous other materials (5 percent). U.S. residents generate about 4.9 pounds of municipal solid waste per person per day, more than any other Organization for Economic Cooperation and Development (OECD) country except Austria. The residents of Romania, Japan, and Türkiye, for example, produce less than half as much waste per capita.[6]

reality check

Trains of Trash and Barges of Garbage

There are many places where trash will fit, but few places where trash is welcome. As growing populations generate more waste than existing landfills can hold, urban garbage goes on some amazing adventures. Trash in Amsterdam often ends up in canals and makes its way to the North Sea. As a clever solution, a social enterprise called The Great Bubble Barrier uses air compressors powered by renewable energy to form a bubble curtain that traps plastics and pushes trash into a catchment system.

Trash and sewage traveling by train from New York to a landfill in Alabama recently raised a stink when train cars filled with 10 million pounds of human feces were stranded in Parrish, Alabama. The foul loads sat in Parrish for 2 months, halted by an injunction obtained by a neighboring community to prevent the material from traveling through their town. Eventually, the waste was transferred to trucks and sent on the final leg of its journey.

The most famous vacation for homeless trash was taken on a garbage barge. On March 22, 1987, Captain Duffy St. Pierre departed New York with 3,186 tons of trash aboard the *Mobro 4000*. New York's landfills were full, and the intent was to ferry the trash to Morehead City, North Carolina, where it could be turned into methane fuel. North Carolina officials turned the trash away after hearing it might contain medical

(continued)

6 See https://data.oecd.org/waste/municipal-waste.htm.

waste. The *Mobro* was then rejected by Louisiana and thwarted by the Mexican Navy in the Yucatán Channel. Belize and the Bahamas were among the six states and three countries that refused the trash. The *Mobro* returned to New York, where the cargo was incinerated. The 430 tons of ash were laid to rest where the journey began, in Islip, New York, after traveling for 3 months and 6,000 miles.

The longest refuse cruise lasted 16 years. A barge called the *Khian Sea* carrying 15,500 tons of ash from incinerated trash had no place to go. Starting in Pennsylvania, the ash was hauled to Bermuda, Honduras, and Chile but found no takers. A load of 2,500 tons of the ash was dumped on a beach in Haiti, while the remaining 13,000 tons went "deep sea diving." Haiti rejected the dumped ash and sent it to Florida. Florida paid $614,000 to send the ash by rail to Hagerstown, Maryland. From there it was trucked to the Mountain View landfill in Franklin County, Pennsylvania.

In support of this type of practice, in 1991, Lawrence Summers, then World Bank chief economist and later U.S. Secretary of the Treasury, wrote in an internal memo:

> The measurements of the costs of health-impairing pollution depends on the forgone earnings from increased morbidity and mortality. From this point of view a given amount of health-impairing pollution should be done in the country with the lowest cost, which will be the country with the lowest wages. I think the economic logic behind dumping a load of toxic waste in the lowest-wage country is impeccable and we should face up to that.
>
> (See https://en.wikipedia.org/wiki/Summers_memo.)

Summers apologized for the memo the next day. He seems to have forgotten that there are many types of costs beyond forgone earnings; environmental costs are among them. It follows from better economic logic that countries with low wages often have large populations, and thus large labor supplies that suppress wage rates. To dump toxic waste in those places would be to cause inefficiently large losses in human life.

As growing urban populations produce mounting volumes of municipal solid waste, we can send it on vacation, but what goes around tends to come around. Ditto for memos.

Figure 9.6

Trends in MSW Generation

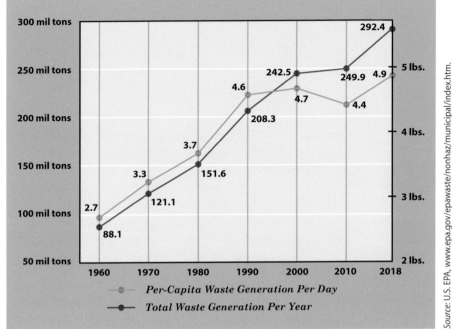

Per-Capita Waste Generation Per Day

Total Waste Generation Per Year

Source: U.S. EPA, www.epa.gov/epawaste/nonhaz/municipal/index.htm.

Figure 9.6 indicates the growth in MSW generation in the United States. The waste per person dropped between 2000 and 2010, although population growth caused the total amount of waste to rise even then. Societal attitudes influence waste volume. Per-capita waste generation increased by only 12 percent during the 1970s, a decade that spawned the EPA and Earth Day. Waste increased by 24 percent during the "me" decade of the 1980s. An increase of 2 percent in the 1990s and a decrease of 6 percent and 2000s coincided with renewed environmentalism spurred by famous garbage fiascos (see the Reality Check), scientific acknowledgment of global climate change, and increasingly visible effects of pollution. And a 17-percentage-point reversal (from a 6 percent decrease to an 11 percent increase) between 2010 and 2018 accompanied a period of political polarization and a diminished role for the EPA.

The previous section explained that development can help stabilize population growth, and the section that follows describes evidence that development can lead to higher standards for environmental quality. In those respects, there are reasons for developing countries to emulate the performance of developed countries, but in terms of waste generation, the opposite is true. The growing scarcity of landfill sites will require policymakers in developed countries to bring their levels of waste generation closer to the levels of their poorer neighbors.

Belgium, Britain, Germany, Luxemburg, and the Netherlands reportedly have less than 10 years of landfill capacity left.[7] The same is true for many U.S. states. The number of U.S. landfills has declined from 7,924 in 1988 to less than 2,000 today, although the average size of landfills has risen. As existing dumpsites are filled and urban sprawl pushes new sites further from urban centers, the direct and external costs of transporting and land-filling MSW increase. Beyond transportation issues, potential external costs include groundwater contamination at landfills and toxic ash emissions from the 12 percent of U.S. MSW that is incinerated.[8]

Demographic Trends and the Determinants of Waste[9]

The demographic determinants of waste become relevant as the characteristics of the world population change, and as we have opportunities to slow or accelerate these changes. Income, education, agrarian populations, population density, age structure, and ethnic makeup are all shifting. The global median income increased by over 150 percent over the past 30 years. Despite the tragic setbacks of the COVID-19 pandemic, education levels are rising worldwide. The percentage of adults in OECD countries with any college degree increased from 30 percent in 2010 to more than 40 percent in 2023.

The effects of income and education on MSW generation are theoretically indeterminate. Consumption increases with income, but so do expenditures on pollution-abating and waste-reducing products such as solar panels and durable consumer goods. Income affords opportunities to spend money on conservation. Recycling bins, composting systems, and products made from recycled materials generally cost more than their less ecological alternatives.

On the other hand, Blagoeva et al. (2023) find that, in most cases, individuals in the European Union with higher incomes generate more municipal solid waste. Given the higher opportunity cost of their time, Lackman (1976) reasons that high-income groups will prefer to spend money on disposable items rather than spending time on repairable or returnable goods. Subsequent empirical research by Anderson (2005), Zia et al. (2017), and others supports these findings.

Education may teach alternatives to resource exploitation and convey the repercussions of waste. Some secondary school curricula require

---

7 See https://essutility.co.uk/landfill-waste-not-want-not/.

8 See http://theconversation.com/garbage-in-garbage-out-incinerating-trash-is-not-an-effective-way-to-protect-the-climate-or-reduce-waste-84182.

9 Data on trends were obtained from the U.S. Census Bureau (www.census.gov), the Population Reference Bureau (www.ameristat.org), and Zippia.com. Additional findings on the determinants of waste are from Anderson (2005).

The effect of income on resource use depends on how the income is used. High incomes encourage consumption, but also the purchase of environmentally friendly goods, such as this bench made from recycled plastic.

students to construct solar water heaters and learn conservation techniques. College-level courses such as Environmental Economics teach about climate change and resource conservation. And many college campuses bring students into contact with recycling bins and environmentalism. For such reasons, college-level education is found to have a negative effect on waste-generation levels.

Attitudes and culture also matter. Modern Eskimo and Aleutian cultures carry remnants of historical reverence for the animals that sustained their communities.[10] A common goal is that no piece of flesh or bone should go to waste. These groups are found to produce less waste, and their representation in the U.S. population is growing.

Age may play an important role in resource management. Today's seniors, who lived through the imposed frugality of the Great Depression (or were closer to the generation that did), generate less waste than younger adults or children. Over the next 25 years, the under-20 and 20-to-64 populations are projected to decrease by 1 to 2 percent, and the over-65 population is projected to increase by 40 to 56 percent. It remains to be seen whether future generations of elderly citizens will be as frugal as those with closer proximity to the Depression era.

10 For example, whaling ship captain Burton "Atqann" Rexford wrote: "The bowhead [whale] is our brother. Our elders tell us that the whales present themselves to us so that we may continue to live. If we dishonor our brother or disturb his home, he will not come to us anymore." See www.boem.gov/about-boem/public-engagement/native-whalers-view.

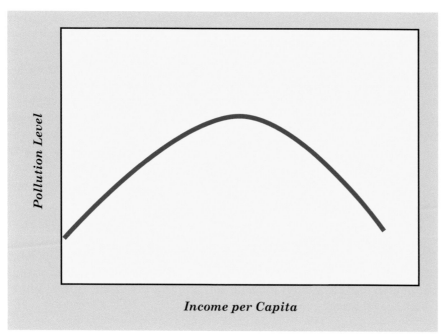

Figure 9.7

*Environmental
Kuznets Curve*

Research reveals an inverse-U-shaped relationship between some pollution levels and income per capita. This suggests that with the right approach, economic growth need not be accompanied by environmental degradation.

Economic Growth and the Environment

Economic growth can help the poor by providing more jobs and higher incomes. Simon Kuznets (1955) reported an inverted-U-shaped relationship between income and income inequality, suggesting that growth influences the distribution of income as well as its level. The effects of very high and very low incomes on various types of pollution suggest a similar relationship between income and environmental impacts.

There is considerable evidence of an inverted-U-shaped relationship between per-capita income and various measures of environmental damage.[11] Figure 9.7 illustrates this relationship with what is known as the **environmental Kuznets curve** (EKC). Among the problems exhibiting the EKC relationship in studies are emissions of sulfur dioxide, particulates, nitrogen oxides, carbon monoxide, and lead; urban air pollution; deforestation; and river pollution by fecal contamination, oxygen depletion, and heavy metals. For some other substances, including municipal solid waste and carbon dioxide, the existing studies find a positive relationship between income and substance levels.

11 See Leal and Marques (2022) for a review of the literature.

It is conceptually clear that higher incomes could lead to higher environmental standards and an increased ability to pay for pollution abatement. Findings in support of the environmental Kuznets curve are evidence that there are ways for economic growth to tread lightly on the environment. Findings to the contrary remind us that pollution abatement requires more deliberate efforts than are taking place. Chapter 5 described many ways in which economic growth could occur in concert with environmental goals. This relationship between economic growth and pollution rests in part on decisions by the population about how to exercise their affluence and shape their technology.

Paul Ehrlich and John Holdren modeled the relationship between environmental impact and growth with the equation:

$$\text{environmental impact} = \text{population} \times \text{affluence} \times \text{technology}^{12}.$$

Using the first letters in impact, population, affluence, and technology, this is abbreviated as $I = P \times A \times T$ or *IPAT*. Defining *affluence* as output per person and *technology* as pollution per unit of output, this equation forms the identity:

$$\text{environmental impact} = \text{population} \times \frac{\text{output}}{\text{person}} \times \frac{\text{pollution}}{\text{output}}$$

$$= \text{population} \times \frac{\text{pollution}}{\text{person}}.$$

This equation suggests that we could cut the environmental "footprint" of human behavior by the same proportion, say 10 percent, by cutting the population *or* the output per person *or* the pollution per unit of output by 10 percent, or with some combination of the three reductions. Efficient policies should target the one out of these three factors that can be decreased at the lowest cost.

Ehrlich and Holdren used the *IPAT* relationship to show that technological progress that reduces the amount of pollution per person will offer no improvement if accompanied by a proportional increase in the population. Some scholars feel that market pressures will avert environmental problems by increasing the prices of goods that become scarce and that the market mechanism will prevent a simultaneous escalation of population, output, and pollution. Ehrlich's stance is: "If I'm right, we will save the world. If I'm wrong, people will still be better fed, better housed, and happier thanks to our efforts."[13]

12 This is a modern adaptation of the original equation that appeared in Ehrlich and Holdren (1971). Dietz and Rosa (1994) provide a history of the concept and an overview of empirical research on the relationship.

13 See Ehrlich (1971, p. 179).

Summary

Over the next half century, the population of the 48 least-developed countries will nearly triple. There is a net gain of more than two people per second in the world, one person every 18 seconds in the United States, and one person every 20 seconds in Brazil. The population in many developed countries is becoming older, wealthier, more diverse, and better educated. Some of these trends are favorable to the environment and some are not. Education, age, and ethnic diversity can bring environmental sensitivity that is otherwise difficult to convey. Increases in income have a positive influence on the generation of municipal solid waste. Evidence of an environmental Kuznets curve, however, shows that growth in income per capita can have a moderating effect on some types of pollution. Accompanied by the right attitude, affluence can elevate environmental standards and the willingness to pay for conservation measures.

Rapid population growth in developing countries, like rampant materialism in developed countries, creates environmental challenges. Without adequate policy responses, these and related trends will hasten resource depletion and expand municipal solid waste production. Even with stable per-capita waste and pollution levels, increasing populations result in growing environmental burdens. Concurrent decreases in the number of available landfills place added pressure on existing wilderness areas and environmental sinks. The resource and disposal needs of growing populations also lead to heated disputes, the resolution of which is the topic of Chapter 15.

• • • • • • • • • • • • • • • • •

Problems for Review

1. Explain how each of the following would affect population growth.

   a) *An economy that was primarily agricultural becomes primarily industrial, and child labor is prohibited in the industrial sector*

   b) *Employment rates and wages increase for women*

   c) *Infant mortality rates decrease*

   d) *The country adopts a system of support for the elderly*

2. The number of years it takes a population to double is found by dividing 69.3 by the population growth rate. For example, if the growth rate is 5 percent, the doubling time is 69.3/5 = 13.9 years. Calculate the doubling time for the following countries:

   a) *Angola: growth rate 3.5 percent*

   b) *Denmark: growth rate 0.2 percent*

   c) *Columbia: growth rate 1.0 percent*

   d) *Cayman Islands: growth rate 1.6 percent*

3. If you presided over an international organization and wanted to increase the population doubling time in Bangladesh, what specific policy would you suggest? Explain your choice.

4. Panama's annual birth rate has fallen from 32 to 18 per 1,000 since 1975, and its death rate is steady at 5 per 1,000. In which stage of demographic transition is Panama? Belgium is in Stage IV of demographic transition. Describe the characteristics of that stage.

5. Suppose that fruit flies on Kiwi Island live for 3 days. On their first day they are considered young. After their first day, the young reproduce. The total fertility rate is 2 and half of all offspring are female. On the second day the flies are middle-aged and cannot reproduce. On the third day they are considered old, and they cannot reproduce. All fruit flies live to old age. Indicate the number of fruit flies on Kiwi Island on days 2, 3, and 4 under each of the following scenarios:

   a) On day 1 there are equal numbers of males and females in each age category, including six old flies, four middle-aged flies, and two young flies.

   b) On day 1 there are two males and two females in each age category.

   c) On day 1 there are 100 young male flies and 100 middle-aged female flies.

   d) On day 1 there are equal numbers of males and females in each category, and there are six young flies, four middle-aged flies, and two old flies.

6. Starting with the situation in Part (d) of Problem 5, on what day would the population stabilize, in that it would be the same as the day before, in each of the following situations:

   a) With no change from the scenario in Problem 5.

   b) Fruit flies live for 4 days rather than 3 days but can still only reproduce after their first day.

   c) One young male and one young female fruit fly die each day before reproducing.

7. Income and education levels are on the rise in many parts of the world.

   a) How do you think your own post-college boost in income (after you land your first "real" job) will affect your consumption behavior?

   b) How do you think your decision to attend college will affect the timing of any children you might have relative to if you ended your schooling after high school?

   c) Do you think most people would provide similar answers? What does this suggest about policy and preparation for related demographic changes?

8. In the context of the *IPAT* equation, explain the meaning of
    a) *affluence*
    b) *technology*

9. Singapore is intent on lowering the environmental impact of its economy. The annual population growth rate in Singapore is 2.5 percent, and per-capita output is increasing by 3 percent per year. To reduce the environmental impact, the country is subsidizing new technology for recycling, solar energy, and waste minimization. If the subsidies reduce the pollution per unit of output by 6 percent, will that be sufficient to reduce Singapore's environmental impact? *Hint: To increase a value by, say, 5 percent, multiply by 1.05. To decrease a value by 5 percent, multiply by (1−0.05).*

10. Suppose the relationship shown by the Kuznets curve applied to oil usage.
    a) *Would oil usage be highest in a country with low income, moderate income, or high income?*
    b) *What could explain a Kuznets-type relationship between high income and oil usage?*

websurfer's challenge

1. Find one website with modern arguments against population growth and one that argues that population growth is not a problem. Evaluate the primary arguments on each side.

2. Find a website that indicates the growth rate for your state or region. Calculate the population doubling time for that area using the formula provided in Problem 2. Discuss whether this rate of growth is problematic in regard to the environment.

Key Terms

Agricultural revolution
Environmental Kuznets curve
Green revolution

Municipal solid waste
Population growth rate
Total fertility rate

Internet Resources

EPA Office of Solid Waste:
www.epa.gov/osw/

Population Reference Bureau:
www.prb.org

Population Connection:
www.populationconnection.org

United Nations Population
Fund:
www.unfpa.org

U.S. Census Bureau Population
Clocks:
www.census.gov/popclock/

Further Reading

Anderson, David A. "The Determinants of Municipal Solid Waste."
Journal of Applied Economics and Policy 24, no. *2* (2005): 23–29.
An empirical study of the determinants of municipal solid waste
levels.

Andreoni, James, and Arik Levinson. "The Simple Analytics of the
Environmental Kuznets Curve." *Journal of Public Economics 80*, no.
2 (2001): 269–286. Presents a model in support of the environmental
Kuznets curve on the basis of increasing returns to pollution abatement
technology.

Blagoeva, Nadezhdac, Vanya Georgieva, and Delyana Dimova.
"Relationship between GDP and Municipal Waste: Regional Dispari-
ties and Implication for Waste Management Policies." *Sustainability
15* (2023): 15193. Finds a positive relationship between per-capita
GDP and MSW generation in most of the European Union. The case of
Bulgaria receives detailed analysis.

Dietz, Thomas, and Eugene A. Rosa. "Rethinking the Environmental Impacts of Population, Affluence and Technology." *Human Ecology
Review 1*, no. *2* (1994): 277–300. An update on the *IPAT* relationship,
including an overview of past developments.

Ehrlich, Paul R. *The Population Bomb*. Cutchogue, NY: Buccaneer
Books, 1971. A provocative discussion of population dynamics and the
environment.

Ehrlich, Paul R., and John P. Holdren. "Impact of Population
Growth." *Science 171* (1971): 1212–1217. The initial presentation of the
IPAT concept, and influential arguments for greater attention to popu-
lation growth issues.

Galor, Oded, and David N. Weil. "Population, Technology, and Growth: From Malthusian Stagnation to the Demographic Transition and Beyond." *American Economic Review 90*, no. *4* (2000): 806–828. Explores the dynamics of population and growth at three levels of economic development.

Harbaugh, William T., Arik Levinson, and David Molloy Wilson. "Reexamining the Empirical Evidence for an Environmental Kuznets Curve." *Review of Economics and Statistics 84*, no. *3* (2002): 541–551. Questions the inverted-U-shaped income—pollution relationship after looking at data from different locations and years.

Hilton, Hank, and Arik Levinson. "Factoring the Environmental Kuznets Curve: Evidence from Automobile Lead Emissions." In *Controlling Automobile Pollution*, edited by W. Harrington and V. McConnell. London: Routledge, 2018, 41–56. https://doi.org/10.4324/9781351161084. Reports empirical evidence of the Kuznets relationship between income and automobile emissions.

Kuznets, Simon. "Economic Growth and Income Inequality." *American Economic Review 45*, no. *1* (1955): 1–28. Finds an inverted-U-shaped relationship between income and income inequality, much like the possible relationship between income and pollution that was later named after Kuznets.

Lackman, Conway L. "A Household Consumption Model of Solid Waste." *Journal of Economic Theory 13* (1976): 478–483. Suggests that wealthier people choose disposable items due to a relatively high opportunity cost of time.

Leal Patricia H., and Antonio C. Marques. "The Evolution of the Environmental Kuznets Curve Hypothesis Assessment: A Literature Review Under a Critical Analysis Perspective." *Heliyon 8*, no. *1* (2022): e11521. https://doi.org/10.1016/j.heliyon.2022.e11521. A review of more than 200 articles from the last 25 years of research on the environmental Kuznets curve.

Malthus, Thomas R. *An Essay on the Principle of Population*. London: Printed for J. Johnson, 1798. www.ac.wwu.edu/~stephan/malthus/malthus.0.html. The famous progenitor of many modern discussions of population growth and resource scarcity.

Zia, A., S. A. Batool, M. N. Chauhdry, and S. Munir. "Influence of Income Level and Seasons on Quantity and Composition of Municipal Solid Waste: A Case Study of the Capital City of Pakistan." *Sustainability 9* (2017): 1568. An examination of the determinants of MSW levels in Asia.

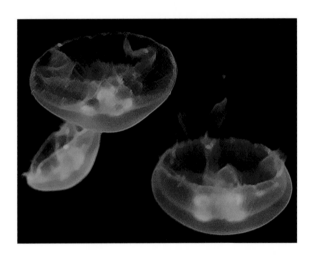

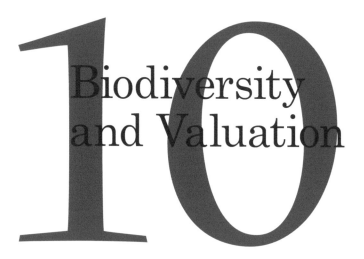

Biodiversity and Valuation

The virtues of diversity are among the basic tenets of economics. We derive more utility from relatively balanced amounts of two goods, such as sunshine and rain or food and clothing, than from just one or the other. This lesson applies to living things as well. Biological diversity, or biodiversity, refers to the variety of eco-systems, species, and genetic differences within species. From an ecocentric standpoint, biodiversity is critical to life. Decomposers like bacteria eat high-level carnivores or omnivores like humans, which eat lower-level carnivores like wild boars,[1] which eat herbi-vores like rats, which eat primary producers like corn plants. With this complex interdependence among wildlife, the loss of a family or species can threaten the sustenance of the food chain.

The domino effect of losses resulting from extinction does not just extend up the food chain. Take the example that frogs eat insects, snakes eat frogs, birds eat snakes, and so on. The loss of the Suriname redtail boa (Boa c. constrictor) might threaten not only the laughing falcon (Herpetotheres cachinnans) and other species above it in the food chain, but also the dark rice-field mosquito (Psorophora columbiae) and other members of the Insecta class because snakes keep frog populations in check and thus help insect populations to survive. Table 10.1 summarizes the classification system for plants and animals.

From an anthropocentric viewpoint, biodiversity is important not only for the sustenance of the food chain atop which we sit, but for medicinal cures, fibers, petroleum substitutes, agents for the

---

1 For wild boar recipes, see www.dartagnan.com/wild-boar-recipes/.

DOI: 10.4324/9781003428732-10

restoration of soil and water, natural beauty, tourism, pride, and education. Aspirin, penicillin, steroids, digitalis, and morphine are among the many medicines that have already come from nature. Tourism, the purpose of which is often to enjoy biodiversity, is among the largest industries in the world.

Economists have developed primarily anthropocentric techniques for estimating the value of biodiversity. These techniques allow for informed, although imperfect, decisions about the advisability of economic development at the expense of wildlife. This chapter identifies the many facets of biodiversity's

Development threatens the habitat of the Masai giraffe (Giraffa camelopardalis tippelskirchi) in Tanzania.

value, reviews common valuation techniques, and reports findings from the application of these methods.

Biodiversity Loss

Like Shakespeare, we have some windows into nature's infinite secrecy. Fossil remains provide information about the number of plant and animal species that existed at approximate points in time going back hundreds of millions of years. Estimates of the numbers of living species in various locations and classifications come from biological inventories. For example, in remote Ecuadorian rain forests, Terry Irwin of the Smithsonian Institution sprays a fog of biodegradable insecticide into the jungle, sending insects onto collection sheets below for cataloging and enumeration. Most estimates place the total number of living species at between 7 million and 100 million. About 1.75 million species have been classified with Latin names.

Biodiversity is naturally in a mild state of flux, with some species becoming extinct and new species emerging. Table 10.2 details five mass extinctions over the past 500 million years. The causes of these extinctions are under debate but may have included cosmic impacts, glaciations, volcanic eruptions, and associated climate change. Erupting

Scientific Names, From the Most to the Least Inclusive	The Human	The Tickseed Sunflower
Kingdom	Animalia	Plantae
Phylum	Chordata	Spermatophyta
Class	Mammalia	Angiospermae
Order	Primates	Asterales
Family	Hominidae	Asteraceae
Genus	Homo	Bidens
Species	Homo sapiens	Bidens aristosa

Table 10.1
The Hierarchical System of Scientific Names

Extinction	Approximate Number of Years Ago	Families Lost
Ordovician	438 million	25 percent
Devonian	360 million	19 percent
Permian	248 million	54 percent
Triassic	208 million	23 percent
Cretaceous	65 million	17 percent
Quaternary	0 (ongoing)	To be determined

Table 10.2
The Great Extinctions

volcanoes send large doses of sulfates into the atmosphere and eject ash clouds that could lower global temperatures. Dust clouds from meteorite impacts have a similar effect. Glaciers lower temperatures, and their accumulation of ice causes sea levels to fall, threatening marine life. Glacial melting has the opposite effect of raising sea levels and threatening land species.

The ongoing quaternary extinction is the direct result of human behavior. Ecosystems around the world are undergoing changes in composition, structure, and function caused by human activity. Like the other causes of mass extinctions, we too have sent sulfates into the atmosphere, contributed to global climate change, and directly impinged on biodiversity. Deforestation, agricultural practices, urban sprawl, overhunting and fishing, and pollution take their toll. Scientists estimate that present extinction rates of between 200 and 100,000 species per year are 1,000 to 10,000 times higher than the prehuman rates of extinction.[2] Among the hardest hit so far are large mammals, birds, beetles, and amphibians.

Funding limitations and opportunity costs force the prioritization of conservation efforts and constrain policies that would protect

2 See https://wwf.panda.org/discover/our_focus/biodiversity/biodiversity/ and http://advances.sciencemag.org/content/1/5/e1400253.full.

Figure 10.1

Global

Biodiversity

Hotspots

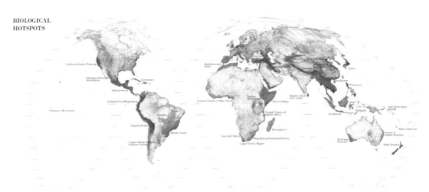

BIOLOGICAL
HOTSPOTS

Source: © 2017 Richard J. Weller, Claire Hoch, and Chieh Huang, Atlas for the End of the World, http://atlas-for-the-end-of-the-world.com.

endangered species. The selection of species to preserve is influenced by their value, as discussed later in this chapter. Environmental policies can also target specific regions that provide habitat for large numbers of species. Figure 10.1 indicates such biodiversity "hotspots." According to environmental analyst Norman Myers and his colleagues, "25 hotspots contain the sole remaining habitats of 44% of the Earth's plant species and 35% of its vertebrate species, and these habitats face a high risk of elimination."[3] The "hottest of the hotspots" are Madagascar, the Philippines, Sundaland, Brazil's Atlantic Forest, and the Caribbean. Behavior in North America and elsewhere affects the biodiversity in these regions. The United States produces a quarter of the human-made greenhouse gases that threaten to overheat sensitive species in the hotspots, and many of the hotspots are sources of lumber, seafood, agricultural goods, and manufactured products for the developed countries.

The "biodiversity hotspots" marked in red and orange support the richest biology on Earth and the largest number of species living nowhere else.

As policymakers prioritize areas for preservation, land values must be considered alongside biological data to establish cost-effective strategies. In areas of similar size, Brazil's Atlantic Forest contains about 20,000 plant species and 1,361 vertebrate species, while central Chile contains 3,429 plant species and 335 vertebrate species.[4] However, if land costs ten times as much in the Atlantic Forest as in central Chile, more species can be saved with a given expenditure by purchasing land for conservation in central Chile and in similar locations. With the money it would take to protect the Atlantic Forest, ten areas similar to central Chile could be protected for the benefit of some 34,290 plant species and 3,350 vertebrate species.

3 See Myers et al. (2000, p. 853).
4 See Myers et al. (2000, p. 854).

If all species were valued equally, an efficient policy would shift dollars to the location where the largest fraction of a species could be saved per dollar[5] until the savings per dollar was equalized across locations. The logic is that if more benefit is achieved per dollar in one location than in another, more dollars should be spent in that location to take advantage of its greater "bang per buck." With diminishing marginal returns, the marginal benefit per dollar spent in the funded location will eventually fall in line with the marginal benefit of spending elsewhere.[6] When species have differing values, efficiency is achieved in the same way, except that the marginal benefit is measured in terms of the *value* of the particular species saved, rather than simply the *number* of species saved. Methods of distinguishing between species of greater or lesser value are discussed next.

Models of Biodiversity Loss

Humans are at the helm during this sixth surge of extinctions. We have the prerogative to make difficult choices about rates and locations of growth and pollution. We control the acceleration of biodiversity loss with decisions about whether our policies

- *Fragment wilderness areas with roads*

- *Drain wetlands*

- *Introduce non-native species that become pests to wildlife*[7]

- *Clear forests*

- *Convert grasslands for agriculture or development*

- *Alter the global climate*

- *Extract natural resources*

- *Emit air, water, light, and noise pollution*

- *Otherwise encroach on natural habitat.*

---

5 For example, if a species could be saved for $1 million, we could say that one one-millionth of a species is saved per dollar.

6 See Chapter 2 for a discussion of diminishing marginal returns and an in-depth explanation of the efficiency of equalizing the marginal benefit per dollar spent on inputs or outputs.

7 For example, the zebra mussel (*Dreissena polymorpha*) is native to the Caspian Sea region of Asia. It was found in Lake Erie in 1988, probably carried there by transoceanic ships. These mussels now live in all of the Great Lakes, the Mississippi River, and inland lakes as well. Beyond clogging water intake pipes for industries, zebra mussels filter plankton out of the water, removing an important food source for larval fish and breaking a link in the food chain. See www.usgs.gov/faqs/what-are-zebra-mussels-and-why-should-we-care-about-them.

This global exercise in the allocation of scarce resources among the competing ends of agriculture, development, production, recreation, and wilderness areas helps to determine the variety of species that coexist on the planet. Economic analysis can assist with the daunting decisions of (1) which species to preserve and (2) what values to place on living things. This section describes two useful approaches.

Cost-Benefit Applications

Reining in the quaternary extinction would mean substantial changes for humans. The housing, transportation, diet, and material possessions common in developed countries all come at a high environmental price. In some cases, biodiversity loss results from a conscious decision to substitute a set of material assets (such as a subdivision) for a set of natural assets (such as a forest and its inhabitants). How much biodiversity loss is too much? Cost-benefit analysis provides one answer to this question. It should be clear that the costs and benefits referred to in this type of economic analysis do not necessarily involve the transfer of money. Of course, there are monetary payments for the costs of maintaining natural areas, and money is received for user fees, ecotourism, and medicinal cures, but many of the costs and benefits are non-monetary. Humans value the mere existence of plant and animal species and the ability to see them and preserve them for future generations. Economists are able to estimate those values, and for the sake of cost-benefit analysis, it is common to convert all of the costs and benefits into monetary values. So non-monetary values are not neglected in well-designed economic analysis, they are simply measured in terms of a standard monetary unit—dollars in our case—for the purpose of comparison.

Biodiversity can be quantified in terms of genetic variation, types of ecosystems, or the number of families or species. This chapter refers to the number of species preserved. Similar conclusions would apply to investigations using most alternative measures of biodiversity as well. Other things being equal, it is prudent to begin with those efforts that preserve species at the lowest cost. Property for a new farm may cost only a few hundred dollars more per acre in an area with no endangered species than in an environmentally sensitive area, providing a relatively inexpensive opportunity for preservation. On the other hand, it might cost several million dollars more to build a multi-story parking structure as a substitute for paving two or three acres of forest near a production facility.

The preservation of different species also necessitates different levels of expenditure. Many species of rodents, grasses, and cockroaches may well outlive humans even if we spend nothing to preserve them. On the other hand, in recent history, the preservation of species of whales, woodpeckers, and snakes has held commercial interests in fishing,

Industrial pollution threatens the weedy sea dragon (Phyllopteryx taeniolatus), found only in Tasmania and along the southern coastline of Australia.

logging, and transportation, respectively, at bay. As more and more species are preserved, increasingly expensive steps must be taken in the name of biodiversity.

While the costs of preserving additional species increase with the number preserved, the corresponding benefits are likely to decrease. The species with the top priority for preservation, presumably *Homo sapiens*, is valued quite highly. Additional species provide large benefits, but the more species that exist, the less of a contribution each additional species makes to available diversity. This is especially true within particular taxonomic groups. The first species within the *Porifera* (sea sponge) phylum[8] contributes to diversity in a way that the ten-thousandth species does not.

One could line up all species in ascending order of their marginal cost of preservation, starting, say, with the brown-banded cockroach (*Supella longipalpa*) and the brown rat (*Rattus norvegicus*), and ending with *Homo sapiens* for whom most expenditures are made. It is also possible to line up species in descending order of their marginal benefit, starting with *Homo sapiens* and ending, perhaps, with the ten-thousandth species of sponge, the brown-banded cockroach, and the brown rat. As these examples illustrate, there is no reason to think the species with the largest marginal benefit of preservation is also the species with the smallest marginal cost of preservation. Some species are arguably more important than others, which can justify the preservation of species that are not necessarily the least expensive to preserve. Thus, it would be overly simplistic to construct a graph with increasing marginal costs and decreasing marginal benefits for the same arrangement of species along the horizontal axis to determine the efficient degree of preservation. Instead, the net marginal benefit (marginal benefit minus marginal cost) could be estimated for each species. For efficiency, those

8   Sponges are a phylum with many families and about 10,000 species.

species with a positive net marginal benefit should be saved, and priority should be given to those species with the largest net marginal benefit. The prioritization of species for preservation according to their net marginal benefit is easier said than done. The next section describes a specific approach for doing so.

The Noah's Ark Model

Metrick and Wietzman view the decision of which species to preserve as the "Noah's Ark Problem."[9] Like Noah in the story of Noah's Ark, we have some control over what species can survive on the "ark" we call Earth. There is a cost associated with efforts to preserve species, exemplified by the expense of conservation projects and the forgone profits that could be earned if the species' habitats were not preserved. The value of a species in the Noah's Ark model is measured in terms of direct benefits (commercial, recreational, and emotional) and contributions to diversity (distinctiveness in comparison to the closest relative). The task of prioritizing species to board the ark given limits of space, time, and money involves a fourth variable: the increased likelihood of survival attributable to protection. Some species will persist regardless of intervention; others are vulnerable but can survive with assistance.

For notation, D represents the value of a species' distinctiveness, B is the value of its direct benefits, S is the percentage increase in survivability resulting from protection, and C is the cost of preservation efforts. The societal gain per dollar of expenditure on preservation is thus

$$\frac{(D+B)S}{C}.$$

According to this model, social welfare is maximized when assistance is prioritized for species that provide the greatest additional benefit per dollar spent on protection. A species is more likely to be granted space on the ark if it is relatively distinct, beneficial, likely to survive if given protection, and inexpensive to protect. Conservation efforts should continue until the marginal benefit from additional effort is less than or equal to the marginal cost. To the extent that the values of D, B, S, and C can be estimated, the Noah's Ark model can guide the prioritization of species for protection. Even if the numbers cannot be closely estimated, the model is useful in highlighting the variables and formula policymakers should consider to the extent possible. The Reality Check that follows demonstrates the applicability of this model.

9 See Weitzman (1998) and Metrick and Weitzman (1998).

Are We Loading the Right Species Onto the Ark?

Andrew Metrick and Martin L. Weitzman collected data on the four criteria of the Noah's Ark model—direct benefits, distinctiveness, survivability, and cost—and examined whether actual rankings correspond with the logic of the model. For measures of society's rankings of species, they looked to the nomination process for protection under the Endangered Species Act. This included counts of positive comments made about species, the decision whether or not to include species on the protected list, and public expenditures to protect species.

Direct benefits were measured in terms of species' size and taxonomic class. As a measure of distinctiveness, Metrick and Weitzman determined whether a species was the sole representative of its genus and whether it was a subspecies. For survivability, they used a 1 to 5 ranking of endangerment created by the Nature Conservancy. For cost, they used a variable indicating whether or not recovery of the species conflicts with public or private development plans.

The findings indicate that humans place a high priority on large, cuddly "charismatic megafauna" such as bears and cats. More surprisingly, they suggest that society spends more money on less endangered species than on more endangered species, and expenditures do not increase significantly for more unique species. Further, society is more likely to spend money on an animal whose preservation conflicts with development plans than on those that could be saved at a lower cost. In other words, according to this study, our current strategies for environmental protection do not coincide with what most economists would recommend in terms of maximizing social welfare.

Valuing Costs and Benefits

Cost-benefit analysis is central to the determination of appropriate levels of biodiversity. Policies that affect biodiversity are available at every expense level, and decision-makers must decide which efforts are worthwhile. This warrants a further breakdown of the types of values and their measures. To miss, or discount, costs or benefits might lead society to neglect worthwhile species or load too many onto the ark. Values can come from the direct use of a species, the option to use a species,

Metrick and Weitzman found that humans place the highest priority on saving large, cuddly looking "charismatic megafauna" such as the kangaroo (Macropus canguru).

the ability to preserve a species for other humans to enjoy, or simply the knowledge that certain animals and plants are able to survive. In some cases, economists can estimate these values on the basis of market prices for purchases of the species or associated goods. As described next, "contingent valuation surveys" and "hedonic pricing models" provide estimates for the many species and other environmental assets not sold in markets. Many of the same techniques can be applied to any asset, natural or not, be it a mountain, a river, clean air, or the Louvre museum.

Types of Value

Environmental assets provide value to users and nonusers alike. **Use value** comes from the firsthand enjoyment of resources and their by-products. Users of forests, for example, benefit from hiking trails and natural beauty or from harvesting the forests for timber. Businesses related to ecotourism, safaris, and hunting use plant and animal species to attract their customers, and farmers sell plant and animal products in markets.

There are also people who never enter a forest or see a particular animal species but who nonetheless place a *nonuse* or **passive-use value** on them. Passive-use value can be divided into the **option value** people place on the option to use a resource in the future and the **existence value** that is unrelated to any possibility of ever using the resource or its by-products. Existence value comes from the **bequest value** of knowing that preservation allows others to use a resource and the **sympathy value** of knowing that a resource is alive and well. The presence of bequest value is evident in the efforts of recyclers, conservationists, and environmentalists and are epitomized by the work of Johnny Appleseed, who planted trees to benefit another generation.[10] Sympathy value drives the establishment of nature preserves with no human access and protests against the cutting of trees in remote areas. Perhaps symbolic of sympathy value, a young woman named Julia Butterfly Hill lived in a 1,000-year-old Redwood tree deep in the woods for 2 years to protect it from loggers.[11] In policymaking, our motives for preservation are typically a combination of the many sources of value.

Value is often measured in terms of people's stated or implied willingness to pay or willingness to accept payment for a resource. An individual's **willingness to pay** (WTP) is the largest amount of money she or he would be willing to pay in exchange for the resource. An individual's **willingness to accept** (WTA) is the smallest amount of money she or he would accept to forego the resource. Note that WTA is determined from the standpoint of already having the resource, and WTP is determined from the standpoint of lacking the resource. Because positions of greater wealth (having versus not having the resource) lead individuals to place higher valuations on goods they desire, the WTA is generally higher than the WTP.

Figure 10.2

Categorizing Types of Value
When performing cost-benefit analyses of environmental assets, it is important to consider every type of value these assets provide.

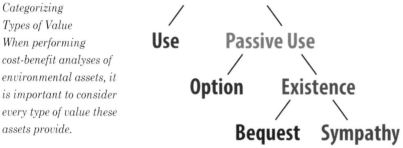

VALUE FROM ENVIRONMENTAL ASSETS

Use Passive Use

Option Existence

Bequest Sympathy

---

10 Johnny Appleseed, whose real name was John Chapman (1774–1845), traveled through what are now the states of Ohio, Indiana, and Illinois, planting apple seeds and tending hundreds of square miles of apple seedling nurseries.

11 Read more about Ms. Hill's protective action at www.circleoflifefoundation.org.

Measures of Value

Direct valuation methods estimate monetary values either on the basis of prices paid in markets for the environmental assets to be valued or using "contingent valuation" survey techniques that ask respondents what they would pay for assets in hypothetical scenarios. **Indirect valuation methods** use observable behavior to infer the monetary value of assets that are not sold in markets. For example, the **avoidance expenditure method** measures the value of an environmental asset as the expenditure avoided by its presence. Consider the $94 billion spent on bottled water each year in the United States. The presence of clean water would allow Americans to avoid that expenditure, so clean water would be worth at least $94 billion per year in the United States. Other indirect methods of placing values on environmental assets look at expenditures on complementary goods such as travel or try to tease the importance of the assets out of property values and wages, as explained next.

Market Prices Some forms of biodiversity are traded in the marketplace; many others have associated market products. Exotic orchids, birds, fish, and reptiles, among many plants and animals, are bought and sold like televisions. The market prices of health products, food, and clothing attributable to biodiversity indicate the minimum valuation current users place on the sources of those products and services. We can learn about the use value of national parks by considering the travel costs paid to see them, and the prices people pay for medicines, dyes, and fruits taken from them.

Recall from Chapter 2 that *consumer surplus* is the difference between what consumers would be willing to pay for a purchase and the price they actually do pay. Whenever a consumer surplus exists, the price people pay is below the value of the good to consumers, which makes expenditures on the whole a conservative estimate of the value of the good. As a result, it is important to consider consumer surplus when estimating values on the basis of prices.

In a perfectly competitive market with complete information and no externalities, economic theory predicts that price will equal marginal cost. Conveniently, in such a market, a product's price would approximate both the cost and the benefit associated with the last unit sold.

When there is no market for a species, preferences are sometimes revealed by expenditures on trips to see it, spending to defend it, or participation rates in efforts to preserve it. First proposed by economist Harold Hotelling in a letter to the National Parks Service in 1947, the **travel cost method** (TCM) involves using consumers' willingness to pay for travel to experience environmental assets as an indirect indication of the use value of those assets. Since consumer surplus plays a key role in this type of analysis, this discussion begins with a review of the relationship between consumer surplus, demand curves, and marginal value.

Consumers typically receive diminishing marginal utility from their purchases and will buy more of a good or service until its price surpasses the value of the marginal utility it provides. This behavior is illustrated in Figure 10.3. Suppose ecotourism trips to Madagascar cost $1,300. George values the first trip at $1,355 because it allows him to visit a new location and see lemurs in their natural habitat for the first time. The second trip is worth $1,325 to him because it's a great place, but having been there before, he finds some of the novelty has worn off. Tired of traveling and having seen it twice before, George values the third trip at $1,265. George will take the first and second trip but not the third because the first two are worth more to him than their cost and the third is not.

The right panel of Figure 10.3 shows that, in a market with many consumers, the observed market price can provide a close approximation of the value of the last unit purchased. In the market for ecotourism trips, the price of $1,300 is roughly equivalent to the marginal value of the twentieth (last) unit sold. In small markets, and for units prior to the last unit sold in large markets, how closely the price approximates the marginal value of the units sold depends on the amount of consumer surplus received. Consumer surplus separates the price and the marginal value. In the left panel of Figure 10.3 we see that George gained $1,355 – $1,300 = $55 worth of consumer surplus from the first trip to Madagascar and $1,325 – $1,300 = $25 from the second trip. That

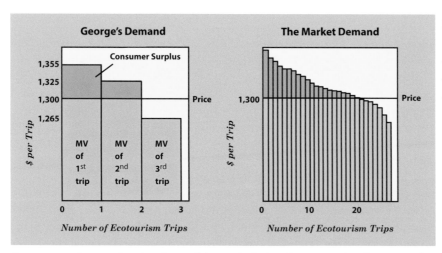

Figure 10.3
The Relationship Between Prices and Values

Consumers purchase goods and services until the marginal value of one more unit no longer exceeds the price. George will take two ecotourism trips but not a third because a third is worth $1,265 to him and would cost $1,300. In a market with many consumers, it is likely that the price closely resembles the marginal value of the last unit, as illustrated in the right panel.

Figure 10.4

*The
Relationship
Between
Demand
Curves and
Consumer
Surplus*

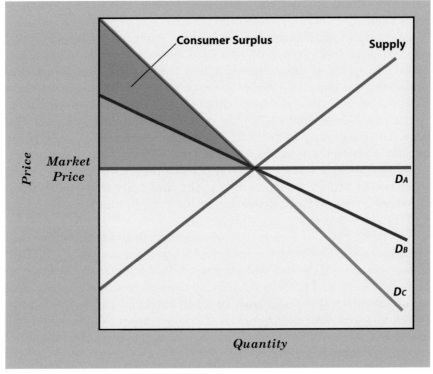

If the market demand curve resembles D_A, the market price is equal to the value of the product to each consumer in the market. If the demand curve resembles D_B, the market price indicates the value of the last unit, but earlier units have larger values to consumers. The difference between the value to consumers and the actual payment is consumer surplus. If the slope of the demand curve is even steeper, like that of D_C, the price is an even more conservative estimate of the product's value.

sizable consumer surplus makes the $1,300 price a less accurate indicator of the marginal value of the trips.

Figure 10.4 demonstrates how, for a given equilibrium price and quantity, consumer surplus increases with the steepness of a straight-line demand curve. When the market demand curve is horizontal as shown by D_A, there is no consumer surplus, and the market price indicates the value to consumers of each unit of the product. For instance, if the demand curve for travel to Madagascar were horizontal and not downward sloping as in Figure 10.3, the total expenditure on travel to go there, $1,300 × 20 = $26,000, would indicate the total value received by the travelers. If the demand curve resembles D_B, the market price indicates the value of the last unit purchased, but consumers place values on the other units that exceed the price, so they receive consumer surplus that separates the value from the travel cost. If the demand curve is even steeper, like D_C, consumers receive a correspondingly high consumer surplus, and the price is an even more conservative estimate of the product's value.

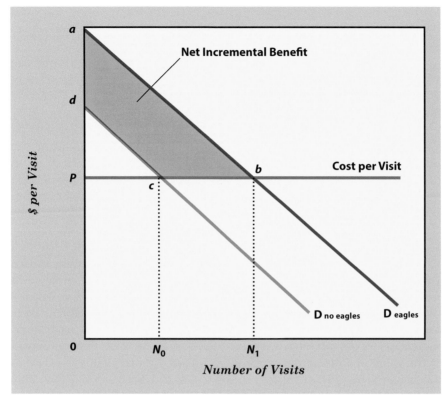

Figure 10.5
Valuing Incremental Changes with the Travel Cost Method

Without eagles, this wilderness area provides a total benefit of dcN_00 to visitors, for which they pay PcN_00. The introduction of eagles shifts the demand curve out and increases the total benefit to abN_10. After subtracting the cost of travel, time, and entry fees, the net incremental benefit is measured by the shaded increase in consumer surplus, $abcd$.

The travel cost method has evolved to include means of capturing values that are part of consumer surplus. Travel expenditures, visitation rates, and demographic data can be used to estimate demand curves.[12] The area under the demand curve and out to the actual number of visits made represents the per-period use value of biodiversity and other amenities at a site.

Figure 10.5 illustrates demand curves for visits to a wilderness area before and after the introduction of bald eagles (*Haliaeetus leucocephalus*). Before the introduction, the total benefit to visitors is area dcN_00, which they enjoy at the cost of PcN_00, including travel costs, time, and entry fees. The incremental value of an environmental change can be measured by changes in the demand curve. If the introduction of bald

12 The traditional TCM derives demand curves by considering visitation levels among many visitors located various distances from a site and therefore incurring differing travel costs to get there. The hedonic TCM looks at users facing various costs and choices among sites with varying attributes. For technical information on the derivation of recreational demand curves, see Van Kooten and Bulte (2000, pp. 113–120).

eagles would increase demand, as indicated in Figure 10.5, the total benefit would become abN_10, and the net benefit would be the orange area $abcd$.

The TCM receives broad use for ecotourism destinations around the world. For example, Wubalem et al. (2023) used the TCM to estimate a $68.5 million annual recreational value for Lake Tana in Ethiopia. Sohrabi et al. (2009) used the TCM to evaluate the value of a forest in Iran. They found the recreational value of the forest to be 25 times its value for timber production, which informed relevant policy decisions about whether to preserve the site for recreational use. Grilli et al. (2017) used the TCM to estimate that tourists are willing to pay $1,009 per day to fish for salmon in Irish rivers. The results revealed opportunities for local communities to earn revenue from tourism after commercial fishing was banned to avoid unsustainable salmon harvests.

There are several limitations to the travel cost method. Although it is appealing to analyze actual travel costs rather than values drawn from hypothetical scenarios, some economists argue that travel costs themselves can be subjective and unobservable.[13] The TCM generates estimates of use values but not option or existence values. The TCM is appropriate for estimating values to recreational users but not to commercial users and others who reside near a resource. And TCM estimates of the value of a destination should be adjusted for the value of extraneous benefits, such as the beauty of the trip to the destination, which may be embedded into travelers' willingness to pay for the whole excursion.

Contingent Valuation　To estimate the values humans place on clean water, clean air, or particular animal species, one approach is simply to ask them. The **contingent valuation method** (CVM) involves asking people to state their willingness to pay, contingent on a hypothetical scenario. CVM surveys can collect data on wildlife like worms and mosses for which there are few associated expenditures that would reveal their value. This method can assist in the estimation of both use and passive-use values.

CVM studies have employed four primary question formats. *Open-ended questions* simply ask participants for their maximum willingness to pay for the environmental improvement being studied. *Dichotomous choice methods* provide a single value that can either be accepted or rejected. The sample question about river preservation in the next paragraph is an example of the dichotomous choice method. *Payment cards* with several values printed on them are sometimes shown to participants, who are asked if any of those values are close to their maximum willingness to pay. In *bidding games*, participants receive values sequentially, either in ascending order until a value is rejected or in descending order until a value is accepted. Some empirical

13 See Randall (1994).

research suggests that the question format does not have a statistically significant effect on the estimated willingness to pay, although the payment card method seems to improve response rates.[14]

Several other pitfalls can bias CVM results if not handled adequately. **Hypothetical bias** results from respondents who do not take the hypothetical situation seriously or provide unrealistic responses because they do not actually have to pay the amounts of money they assign to resources. Saying you would pay $100 to save the manatee is easier than reaching into your pocket and handing over the money. For this reason, survey designers word questions to encourage a mindset similar to that which would exist if the money discussed were real. The following example comes from a CVM study of rivers in the "Four Corners" region of the western United States that provide habitat for nine species of threatened or endangered fish:[15]

Suppose a proposal to establish a Four Corners Region Threatened and Endangered Fish Trust Fund was on the ballot in the next nationwide election. How would you vote on this proposal? Remember, by law, the funds could only be used to improve habitat for fish. If the Four Corners Region Threatened and Endangered Fish Trust Fund was the only issue on the next ballot and it would cost your household $_____ every year, would you vote in favor of it? (Please circle one.) YES/NO

The researchers filled in the blank beforehand with a randomly selected amount between $1 and $350. They concluded that a typical household was willing to pay about $268 ($515 in 2024 dollars) to preserve the critical habitat of the endangered fish species.

If it is apparent from the survey that a study will determine whether to save, say, the manatee, respondents might exaggerate their WTP if they want to save the species or understate their WTP if they oppose the preservation. This is called **strategic bias**. Because survey results are suspect if the intent is transparent and vulnerable to strategic bias, studies often conceal or disguise their intent.

If you were asked for your WTP to save a particular tree, might your value be influenced by the benefits of the surrounding forest? When many benefits are associated with a particular asset or action, **embedding effects** lump the value of benefits not being evaluated in with the value of benefits intended for evaluation. As another example, if people are asked what they would pay for a decrease in benzene emissions achieved with a decrease in paper production, their stated values might incorporate the value of reductions in other pollutants associated with paper production. To avoid embedding effects, CVM studies must

BIODIVERSITY AND VALUATION

---

14 See Reaves et al. (1999).

15 See Ekstrand and Loomis (1998). A similar study by Ojea and Loureiro (2010) found a median household WTP of $27 (in 2024 dollars) to protect a fish called the hake and the Norwegian lobster in south-western Europe.

carefully clarify the change to be valued and/or include follow-up questions to discern what benefits respondents have associated with their stated willingness to pay.

Respondents are also subject to the following types of bias:

Information bias: They have insufficient information about the asset being valued.

Interviewer bias: They are influenced by the person who is conducting the survey.

Payment vehicle bias: They are influenced by the type of payment mentioned in the survey, such as taxes or donations.

Sampling bias: Those selected as respondents do not represent the larger population.

Self-selection bias: Those with strong opinions may be more likely to respond.

Starting point bias: Respondents are influenced by the values listed in the survey.

These and related sources of bias make the survey design of utmost importance and necessitate careful scrutiny of survey-based conclusions.

Despite these caveats, CVM studies are popular for their versatility in valuing the reintroduction or preservation of environmental assets. In a summary of past CVM estimates, Richardson and Loomis (2009) reported average per-person, per-year WTP values of $20 for turkeys, $26 for bighorn sheep, $29 for sea turtles, $55 for dolphins, and $59 for bald eagles. As another example, Trujillo et al. (2016) used the CVM method to estimate that coral reefs in a marine protected area in the Caribbean Sea are worth $16.3 million (all figures are in 2024 dollars).

Hedonic Pricing   The **hedonic pricing** approach evaluates differences in the prices of goods or services caused by (in this context) environmental assets or liabilities. If workers are willing to accept a lower wage for planting trees in a forest than for planting seeds on a farm, other things being equal, the difference in wages reflects the workers' valuation of being in the forest. Likewise, if the price of hotel rooms is higher in areas with more biodiversity than in areas with less, other things being equal, this difference reflects a use value for biodiversity. *Regression analysis* allows economists to study the hedonic (quality adjusted) effects of chosen variables like pollution or biodiversity levels on prices while holding the effects of other measurable price determinants constant. Hedonic methods have also become dominant in economic research into the value of our own species, as discussed in the upcoming Reality Check.

A downside of hedonic pricing methods, as with other price-based valuations, is that there are often passive-use values that do not enter into the prices of products associated with biodiversity. For example, the effect of the Cockscomb Basin Wildlife Sanctuary on room rates at the nearby Jaguar Reef Lodge in Belize does not indicate the existence or option values of the sanctuary's sensitive biodiversity to citizens of, say, Sweden. Another drawback is that it typically takes large datasets to establish statistically significant relationships between prices and environmental variables while holding other price determinants constant. While demographic and employment variables are readily available, environmental variables are harder to come by in large collections of data. For example, at the time of this writing, the EPA is 5 years behind in posting some of its data.

Making Use of the Numbers

Armed with estimates of the value of animals, we are now better prepared to evaluate policy proposals. The estimated value of the human animal suggests drinking water regulations that saved lives for $500,000 each were a bargain, but the particular uranium and asbestos regulations that saved lives for $30 million to $100 million exceeded our own willingness to pay for a statistical (unidentified) life.

As another example, animals becoming roadkill along highways is a serious problem that can be solved with substantial expenditures on infrastructure. The decision was made to spend about $6 million on barriers to deter wildlife access to a road through Florida's Paynes Prairie State Preserve, where an estimated 100,000 animals were killed annually. Knowing the value of wildlife, the number of animals lost each year, the cost of the solution, and an appropriate discount rate, straightforward mathematics can inform difficult decisions about preservation efforts.[16]

The present discounted value of a *perpetuity* (an unending series of benefits or costs), such as the lives saved by a wildlife barrier, equals the value per period divided by the discount rate per period. Suppose 100,000 turkeys would be saved by a wildlife barrier each year and the annual discount rate is 0.03 as discussed in Chapter 5. The present discounted value of the wildlife barrier would be:

$$\frac{100,000 \times \$20}{0.03} = \$66,666,667$$

In this case, a $6 million wildlife barrier would be well worth the cost.

---

16 See www.scientificamerican.com/article/roadkill-literally-drives-some-species-to-extinction/.

reality check

Findings: The Hedonic Value of Homo Sapiens

Economists often seek values for natural assets to inform policy decisions about what expenditure level is warranted to preserve those assets. The same is true for *Homo sapiens*, whom myriad environmental policies are designed to save. The EPA recommends a value of $10.05 million for each unidentified human life that would be saved by a policy under consideration. This is considered an average value to be applied regardless of the age, income, or demographic characteristics of the people being protected. Some arsenic emission standards save lives for less than $100,000 each. Some asbestos regulations save lives for over $100 million each.[1] As with owls and woodpeckers, we must decide where to draw the line in terms of expenditures per life saved.

Over and above the value we place on other animals, one might think we consider ourselves priceless. This is not the case. If human life were invaluable, every risky activity would involve an infinite expected value of loss (the risk of death times the infinite value of life), and every policy that might save a life would warrant whatever expenses it entails. The fact that people leave the safety of their homes, drive in cars, and walk across streets demonstrates that we place finite values on human life.

Hedonic pricing studies have estimated the values of species including *Homo sapiens*. Viscusi and Masterman (2017) analyzed the data from a large collection of past studies and estimated the value of a *statistical human life* at $12.1 million (2024 dollars). A **statistical life** is not the life of a particular individual, but the value of an unidentified life expected to be lost due to a small chance of losing many different people's lives, such as a 1-in-10,000 chance of death faced by 10,000 people. Economists can measure the value of a small risk of death by examining the trade-offs between wages and risks of occupational fatalities, controlling for other determinants of wages. If the average worker will accept an extra $1,000 per year in exchange for a 1-in-10,000 annual risk of death, that implicitly places a $1,000 value on 1/10,000th of a life. One can extrapolate to find an entire statistical life to be worth

(continued)

---

1 See Tengs et al. (1995).

If there are a variety of options regarding the length (or height or material quality) of the wildlife barrier, the fact that the total benefit would exceed the total cost for a particular length does not imply that building a barrier of that length would be efficient. As explained in Chapter 2, there are often a number of policies under which the total benefit would exceed the total cost, but only one policy that would achieve efficiency by equating the marginal benefit and the marginal cost, and thereby maximizing the difference between the total benefit and the total cost. Anyone considering a wildlife barrier or any other means of protecting biodiversity will receive the largest net benefit by adding protection until the marginal benefit of another unit would no longer exceed the marginal cost.

Summary

Economic analysis can inform the choice and design of policy instruments to achieve biodiversity goals. Although none of the valuation methods is infallible, given the wide array of policy options and their varying costs, it is often better to have some idea of the value of wildlife than no idea. Cost-benefit analysis is a useful tool for allocating natural resources when costs and benefits are carefully considered; otherwise, it can be misleading. There are several categories of value that should not be overlooked. *Use value* is

Do you have a sympathy value for this caterpillar species?

derived from the actual use of a resource or its by-products. *Passive-use value* is not associated with the hands-on use of a resource but is derived from its mere existence and the ability of others to benefit from it. Passive-use value includes *option value* associated with the ability to use a resource in the future and *existence value* from knowledge that a resource exists. Existence value includes *sympathy value* from awareness that the resource has survived and *bequest value* from being able to pass the benefits of the resource on to future generations.

One's *willingness to accept* is the smallest amount of money one would willingly accept to forego a resource. One's *willingness to pay* is the largest amount of money one would be willing to pay in exchange for a resource.

The tools of economics lend themselves to the valuation and management of biodiversity, at least from an anthropocentric viewpoint. Prices indicate the marginal value users receive from marketed natural resources. Travel costs paid to reach an ecotourism destination indicate the lower bound for the use value of that destination. Economists use *hedonic pricing* methods to evaluate the effect of biodiversity on market prices, such as room rates at hotels, home prices, and wages. With that information, economists can extrapolate to estimate the value of biodiversity itself. The *contingent valuation method* uses sophisticated surveys to summon revealing responses about the value of resources to society.

By the means explained in this chapter, economists gain critical estimates of values for comparison with preservation costs as society addresses the prudence of production and development projects. The appropriate management of biodiversity must rest on carefully considered values that are direct and indirect, present and future, financial and emotional, and well-informed.

•   •   •   •   •   •   •   •   •   •   •   •   •   •   •   •   •   •   •

Problems for Review

1. Provide your own specific example (not found in this textbook) of each of the following:

   a) An option value

   b) A bequest value

   c) A sympathy value

2. Using the names of specific living creatures such as fleas, elephants, and storks, write out a food chain that starts with plankton and ends with humans. Why is an understanding of the food chain important to the economic analysis of plant and animal species?

3. What is the most you would be willing to contribute as a one-time payment for the return of the passenger pigeon (*Ectopistes migratorius*)? What is the least you would accept in a one-time payment for the loss of the

white-crowned pigeon (*Columba leucocephala*)? Explain why these values are different, or why they are the same.

4. Suppose tourists spend $8 billion annually to visit wilderness areas in Utah.

   a) *This information could be used to estimate the value of Utah's wilderness areas using which method?*

   b) *Identify two types of value that this method fails to capture.*

5. Suppose the Bland Hotel is identical to the nearby Bellevue Hotel except that the Bland Hotel overlooks a parking lot, and the Bellevue Hotel overlooks a forest. Rooms at the Bland Hotel cost $70 per night, and rooms at the Bellevue Hotel cost $90 a night. Counting each night in a hotel room as a *room night*, people are willing to pay for 50,000 room nights each year at the Bellevue Hotel despite the option of staying at the Bland Hotel. Assume the demand curves for each hotel have the same slope.

   a) *Developers could pave the forest next to the Bellevue Hotel and create a parking lot. After deducting the costs of development and upkeep, the new parking lot would provide $750,000 per year in benefits to society. Would the new parking lot provide a net gain to society? Explain your answer using specific numbers.*

   b) *If the demand curve for the Bellevue Hotel were steeper than the demand curve for the Bland*

Hotel, would that support the argument for or against building the new parking lot? Explain your answer with reference to consumer surplus.

6. Consider the difference in hotel room prices discussed in Question 5.

   a) *Which valuation approach is based on this type of information?*

   b) *A value of which of the types shown in Figure 10.2 is revealed by this information?*

7. Do you feel it is appropriate to make decisions about species preservation based only on anthropocentric values? Why or why not?

8. If you were Noah and you could preserve half of the species on the planet, how would you decide which species to provide with boarding passes? Be specific. Your method does not need to be one of those described in this chapter.

9. Suppose there is a 1 in 1,000 chance that a snake will bite you while you are walking to class, and if you are bitten, you will certainly die unless you are carrying antivenom.

   a) *What is the most you would be willing to pay for a dose of antivenom?*

   b) *Based on your answer to part (a), what value do you place on a statistical life?*

10. Describe a weakness of the CVM method. What makes this method attractive despite potential flaws?

Key Terms

Avoidance expenditure method

Bequest value

Contingent valuation method

Direct valuation methods

Embedding effects

Existence value

Hedonic pricing

Hypothetical bias

Indirect valuation methods

Option value

Passive-use value

Strategic bias

Statistical life

Sympathy value

Travel cost method

Use value

Willingness to accept

Willingness to pay

Internet Resources

UK Joint Nature Conservation
Committee:
https://jncc.gov.uk

Conservation International, a website
dedicated to biodiversity:
www.conservation.org

Environment Canada information on
Biodiversity:
www.canada.ca/en/environment-climate-
change/services/biodiversity.html

Ecotourism Explorer:
www.ecotourism.org

U.S. Fish and Wildlife Service:
Endangered Species:
www.fws.gov/program/endangered-species

Further Reading

Anderson, David A. "Evaluating Policies for Sustainability: The
Neglected Influence of Visual Images." *International Journal of
Environmental, Cultural, Economic and Social Sustainability 1*, no.
5 (2006): 1–8. Finds that the value humans place on environmental
assets depends critically on the level of personal exposure to those
assets.

Berrens, Robert, Philip Ganderton, and Carol Silva. "Valuing
the Protection of Minimum Instream Flows in New Mexico." *Journal of
Agricultural and Resource Economics 21*, no. *2* (1996): 294–309. A CVM
study of the value of river habitat.

Ciriacy-Wantrup, S. *Resource Conservation: Economics and
Policies*. Berkeley, CA: Division of Agricultural Sciences, University of
California, 1963. The seminal work on the safe minimum standard con-
servation strategy.

Ekstrand, Earl R., and John Loomis. "Incorporating Respondent
Uncertainty When Estimating Willingness to Pay for Protecting Critical Habitat for Threatened and Endangered Fish." *Water
Resources Research 34*, no. *11* (November 1998): 3149–3155. A CVM
study of the value of river habitat in the Four Corners region, where
water is in high demand for multiple uses.

**Grilli, Gianluca, John Curtis, Stephen Hynes, and Gavin
Landgraf.** "The Value of Tourist Angling: A Travel Cost Method

Estimation of Demand for Two Destination Salmon Rivers in Ireland."
Economic and Social Research Institute 570 (2017): 1–17. A study of
tourist expenditures to fish in Irish rivers, along with the price elasticity of demand for fishing services.

Kaiser, Jocelyn. "How Much Are Human Lives and Health Worth?."
Science 299 (2003): 1836–1837. A discussion of the application of value-
of-life estimates to policy decisions.

Metrick, Andrew, and Martin L. Weitzman. "Conflicts and Choices
in Biodiversity." *Journal of Economic Perspectives 12*, no. *3* (1998):
21–34. A discussion of the Noah's Ark method and how its implications
compare to reality.

**Myers, Norman, Russell A. Mittermeier, Cristina G. Mitter-
meier, Gustavo A. B. da Fonseca, and Jennifer Kent.** "Biodiversity
Hotspots for Conservation Priorities." *Nature 403* (2000): 853–858.
Describes the analytic methods and findings of research into hotspots
that harbor the greatest concentration of plant and vertebrate species.

**Naeem, S., L. J. Thompson, S. P. Lawler, J. H. Lawton, and R.
M. Woodfin.** "Empirical Evidence That Declining Species Diversity
May Alter the Performance of Terrestrial Ecosystems." *Philosophical
Transactions of the Royal Society (London, B.) 347* (1995): 249–262.
A fascinating examination of the relationship between species
"richness" and function.

Ojea, Elena, and Maria L. Loureiro. "Valuing the Recovery of Over-
exploited Fish Stocks in the Context of Existence and Option Values."
Marine Policy 34, no. *3* (2010): 514–521. https://doi.org/10.1016/j.
marpol.2009.10.007. A CVM study of the existence and option values of
marine life in Europe.

Randall, Alan. "A Difficulty with the Travel Cost Method." *Land
Economics 70* (1994): 88–96. Argues against the use of the TCM to
derive value estimates for cost-benefit analysis, saying that travel costs
are unobservable.

Reaves, Dixie W., Randall A. Kramer, and Thomas P. Holmes.
"Does Question Format Matter? Valuing an Endangered Species."
Environmental and Resource Economics 14, no. *3* (1999): 365–383.
Compares the results of three different formats for CVM questions.

Richardson, L., and J. Loomis. "The Total Economic Value of
Threatened, Endangered and Rare Species: An Updated Meta-

Analysis." *Ecological Economics 68*, no. *5* (2009): 1535–1548. An overview of contingent valuation studies that finds that newer studies estimate higher willingness to pay values for wildlife than older studies.

Sohrabi, Saraj B., A. Yachkaschi, D. Oladi, Teimouri S. Fard, and H. Latifi. "The Recreational Valuation of a Natural Forest Park Using the Travel Cost Method in Iran." *iForest 2* (2009): 85–92. www. sisef.it/iforest/show.php?id=497. A travel cost study of the Abbas Abad Forest with policy implications for harvest decisions.

Tengs, T. O., M. E. Adams, J. S. Pliskin, D. G. Safran, J. E. Siegel, M. C. Weinstein, and J. D. Graham. "Five-Hundred Life-Saving Interventions and Their Cost-Effectiveness." *Risk Analysis: An Official Publication of the Society for Risk Analysis 15*, no. *3* (1995): 369–390. https://doi.org/10.1111/j.1539-6924.1995.tb00330.x. Provides estimates of the cost per life saved for hundreds of government policies.

Trujillo, Juan C., Bladimir Carrillo, Carlos A. Charris, and Raul A. Velilla. "Coral Reefs Under Threat in a Caribbean Marine Protected Area: Assessing Divers' Willingness to Pay Toward Conservation." *Marine Policy 68* (2016): 146–154. Uses dichotomous choice CVM models to estimate the total benefit of coral reefs to recreational divers.

Van Kooten, G. Cornelis, and Erwin H. Bulte. *The Economics of Nature.* Malden, MA: Blackwell Publishers, 2000. A rigorous overview of natural resource management, including valuation techniques.

Viscusi, W. Kip, and Clayton J. Masterman. "Anchoring Biases in International Estimates of the Value of a Statistical Life." *Journal of Risk and Uncertainty 54*, no. *2* (2017): 103–128. A discussion of the state of the art in calculating the value of an unidentified human life.

Weitzman, Martin L. "The Noah's Ark Problem." *Econometrica 66*, no. *6* (1998): 1279–1298. Discusses the theory and practice of the Noah's Ark method.

Wubalem, Atalel, Teshale Woldeamanuel, and Zerihun Nigussie. "Economic Valuation of Lake Tana: A Recreational Use Value Estimation through the Travel Cost Method." *Sustainability 15*, no. *8* (2023): 6468. https://doi.org/10.3390/su15086468. Finds that the recreational value of a lake justifies not only the preservation of the lake for recreational purposes, but also sizable expenditures on environmental cleanups and amenities.

11

International and Global Issues

efore his plane disappeared in the Nevada desert, Steve Fossett circled the globe on a sailboat, floating under a balloon, and in an airplane without refueling. Our planet is small enough for solo circumnavigation, and many of the planet's environmental problems likewise transcend national borders. Greenhouse gases, among other uniformly distributed pollutants, float like Fossett and his balloon across the globe. On the ground are the similarly intricate international issues of deforestation, endangered species, and polluted seas, to name a few.

Consider the market for ivory. Among those favoring the legal trade of ivory are some residents of Japan, Taiwan, and China, who treasure it, and some hunters in Botswana, Namibia, and South Africa, who would like to supply it. Proponents of a complete ban on the trade of ivory include many residents of Kenya, Zambia, and Liberia, who value elephants as part of their natural heritage, and many residents of the United States, India, and European nations, who place a high existence value on charismatic mega-fauna. Without a ban, the elephant population fell steeply from 1.3 million in 1979 to below 750,000 in 1988. With the complete ban that started in 1990, affluent buyers could not obtain their treasures, and impoverished hunters lost their incomes. In 2018, the U.S. Fish and Wildlife Service lifted a ban on the import of elephant tusks by hunters (but not the routine sale of ivory), again inflaming the international controversy over endangered species policy.

While many environmental and natural resource issues can be addressed unilaterally, large and complex threats exemplified by

DOI: 10.4324/9781003428732-11

*global climate change and species endangerment cannot be con-
quered without cooperation. Global commerce may foster communication and mutual understanding among nations, though
not without posing environmental threats of its own. This chapter focuses on environmental and natural resource problems with
remedies that involve international cooperation, and discusses the
organizations, policies, and agreements that govern resource allocation and the environment.*

Globalization and the Environment

The Good, the Bad, and the Ugly

Globalization is the spread of influence and interdependence among
economies and people around the world. In terms of its implications,
globalization means different things to different people. Globalization
is both feared and revered for the exports and influences that extend
across regional boundaries. The exchange of ideas and information
can foster education and promote the understanding of other cultures.
Cooperation can lead to important medicinal cures and technological leaps. It is often mutually beneficial to allow regions to specialize
according to their comparative advantages and then trade to achieve a
desired balance of goods and services.

Why, then, are some groups fervently opposed to globalization? Critics
voice several concerns. They worry that the influence of multinational
corporations can trump that of democratically elected representatives
and create an unsettling concentration of power among those driven by
profit motives. Intensified globalization might lead to a relatively homogeneous world market, causing cultures to lose their identities. And
without ecological tariffs or similar policies as discussed in Chapter 8,
production might easily be shifted to countries with low environmental
and humanitarian standards, increasing the exploitation of human and
natural resources. This section explores the contentious paths of cooperation, homogenization, and exploitation that stem from the growing
trend of globalization.

Cooperation International cooperation is economically prudent in
several contexts. Specialization and trade provide the benefits of comparative advantage and economies of scale. Shared information, about
clean energy technology, for example, can provide a public good for those
who would otherwise have to reinvent products or processes on their
own. International cooperation to assemble financial and human capital
can contribute to the success of environmental research, as with efforts
to track and propagate endangered species. And increased legal cooperation improves the ability to enforce regional environmental policies.

Cooperation also permits the sharing of risk. Much like the purchase of health insurance, with sick individuals receiving funds from the collective pool of premiums, international organizations can pool funds and disburse them for relief from environmental disasters, including oil spills, droughts, storms, and floods. When Hurricane Otis slammed Mexico in 2023, UNICEF[1] provided clean water and sanitation equipment, and the Red Cross assisted with food, shelter, and medical care.

Cooperative educational efforts include the work of the United Nations Educational, Scientific, and Cultural Organization (UNESCO). That organization sponsors educational programs on climate change, sustainability, biodiversity, and conservation, among other topics, across nearly 200 member nations.[2] Each year, the World Bank helps millions of people gain access to improved water supplies.[3] Cooperative efforts to address five other types of environmental problems are discussed later in the chapter.

Homogenization The sharing of ideas leads to homogenization, for better or worse. Signs of change include lost languages and dialects and new eating habits in the 120 countries now served by McDonald's. Fast-food restaurants themselves are of concern to some environmentalists,[4] who eschew excessive packaging, the clearing of forest for cattle and their feed, and the inefficiency of eating high on the food chain.[5] To critics of globalization, the fast-food

Spices on display in India. Homogenization can result in the loss of cultural diversity. McDonald's, KFC, and Domino's already compete with traditional Indian fare.

franchises are symbolic of a more general dissemination of unsustainable practices and lifestyles. These environmental issues are aside from the inevitable conclusion that homogeneous cultures would be dull.

There is concern that seductive corporate advertising and contagious materialistic values could create a world of people with insatiable appetites for consumption. If the world population consumed at the rate of those in the United Arab Emirates, the United States, Kuwait, Denmark, and Australia, for example, it would take four to five planets

---

1 UNICEF is the United Nations Children's Fund, with offices in 126 countries. The acronym comes from the organization's original name, the United Nations International Children's Emergency Fund.

2 See www.unesco.org.

3 See www.worldbank.org.

4 See www.mcspotlight.org/issues/environment/index.html.

5 It takes more than 10 times as much acreage and water to produce a pound of meat as it does to produce a pound of wheat. For related discussions, see Ehrlich and Ehrlich (1979).

like Earth to supply the resources. Even with current global lifestyle standards, the average human is consuming 30 percent more resources than the Earth can sustain.[6] Other things being equal, one might expect a cultural give-and-take in which citizens from high-consumption countries gained ideas about how to live simply from countries with more sustainable economies. Yet with an imbalance in marketing savvy, a balance in influence is unlikely.

Of course, international trade has been going on for millennia, and its influences are nothing new. America has adopted plant and animal species, fashions, and cuisine from its trading partners since Asian and European visitors first arrived. Newer, however, are the marketing mechanisms by which firms persuade other cultures to sidestep tradition and adopt their products. If large companies selling sportswear, meals, electronics, and chemicals, among others, are becoming increasingly adroit at changing the customs of international customers, the assertions of globalization critics are noteworthy—there may indeed be new problems stemming from the old practice of global trade.

Exploitation   In theory, global trade could permit a substantial spreading of wealth. That has not yet been the case in practice. The share of global income received by the poorest 20 percent of people remains at or below 1 percent. It is increasingly convenient for profit-maximizing firms to shop across countries for manufacturing locations where labor or environment standards are low. When the goal is production at the lowest possible cost, both workers and the environment can suffer. For example, Keho (2023) finds that globalization worsens the environmental quality in developing countries.

Many modern trade policies include provisions to protect the environment. For example, the United States-Mexico-Canada Agreement (USMCA) includes explicit commitments to protect endangered species, combat illegal wildlife trafficking, and reduce marine litter. Even so, the difficulties of funding, monitoring, and enforcing environmental protection policies across international borders render many such efforts inadequate. Improved systems of ecological tariffs and advancements in monitoring technology may eventually offer a solution. At this point, however, the possibility of environmental and labor exploitation should be considered among the costs of global trade.

Organizations

The elements of successful policy initiatives are somewhat different at the international and global level than at the national or local level. Cultural and governmental influences are muted at borderlines. Even the strongest armies have difficulty enforcing policy among rogue

---

6 See www.panda.org/lpr/08/.

nations. International law is only as binding as the multinational commitments that back it up. At the same time, humans share sources of oxygen, environmental sinks, habitat for endangered species, and the repercussions of environmental degradation. Our vital interests in international oversight have spawned several organizations of considerable strength and controversy. Given their awesome responsibilities, it is of value to be familiar with them. It will add meaning to your daily perusal of environmental economics in the media, and given your interests, you may well work for or lead one of them in the future.

The United Nations The United Nations (UN) began in 1945 with 51 countries seeking to

> *Maintain international peace and security; to develop friendly relations among nations; to cooperate in solving international economic, social, cultural and humanitarian problems and in promoting respect for human rights and fundamental freedoms; and to be a centre for harmonizing the actions of nations in attaining these ends.*[7]

The organization now boasts 193 member countries, 40 agencies and organizations, and an extensive array of programs and bodies.

On the environmental front, the UN "family of organizations" includes the Global Program on Globalization, Liberalization and Sustainable Human Development; the Inter-Agency Committee on Sustainable Development; the International Seabed Authority; the UN Food and Agriculture Organization; and the UN Environment Program. The latest activities of these organizations are summarized online at www.unsystem.org. Major efforts include the sustainability goals discussed in Chapter 8 and a Beyond GDP movement to value "what counts" and not just market production as discussed in Chapter 5.

The United Nations has focused global attention on the environment with its annual UN Climate Change Conferences. These gatherings have set forth a plethora of initiatives and commissions, including the Rio Declaration on Environment and Development (Agenda 21),[8] the Kyoto Protocol for greenhouse gas emissions, the Paris Agreement on climate change, and the 2030 Agenda for Sustainable Development.[9]

The UN Global Compact is the world's largest corporate sustainability initiative.[10] Covering human rights, labor, and the environment, the Compact's principles include the development and use of environmentally friendly technologies and initiatives to promote greater environmental responsibility. It also supports the **precautionary**

7 See www.un.org/aboutun/basicfacts/unorg.htm.

8 See www.un.org/en/development/desa/population/migration/generalassembly/docs/globalcompact/A_CONF.151_26_Vol.I_Declaration.pdf.

9 See https://sustainabledevelopment.un.org/post2015/transformingourworld.

10 See www.unglobalcompact.org.

principle, which is to take precautionary measures against serious threats to humans or the environment, even if scientific research to establish the causes and effects of those threats is incomplete. These principles are representative of the UN's stance on the environment.

The World Trade Organization   In 1995, the World Trade Organization succeeded the World War II era General Agreement on Tariffs and Trade (GATT) as the global authority on rules of international trade. WTO policies are the result of negotiations among its 164 member nations. The WTO's General Agreement on Trade in Services (GATS) is the services equivalent to GATT, which covers merchandise. "Services" represents 60 percent of global production and 21 percent of global trade. These trade agreements are designed to create "a credible and reliable system of international trade rules; ensuring fair and equitable treatment of all participants (principle of non-discrimination); stimulating economic activity through guaranteed policy bindings; and promoting trade and development through progressive liberalization."[11]

The Trade-Related Aspects of Intellectual Property Rights (TRIPS)[12] agreement covers copyrights, undisclosed information, patents (as for new varieties of plants), and related intellectual property issues. Article 27 of TRIPS allows members to exclude from patentability those products or processes the "commercial exploitation" of which is necessary to protect plants, animals, or the environment.[13] The International Council of Chemical Associations supports patent protection under TRIPS. Enforceable patents encourage inventions to preserve the ozone layer, reduce energy consumption, decrease the toxicity of pesticides, grow carbon-sequestering plants, and upgrade products and processes that currently harm the environment.

In the past, the WTO has come under fire for allegedly placing commercial interests ahead of environmental protection. Neither the intent nor the result of WTO actions is easily pigeonholed. It is clear, however, that environmental economics was on the minds of the founders of the WTO. The preamble to the 1994 Marrakech Agreement establishing the WTO states that

> relations in the field of trade and economic endeavor should be conducted . . . while allowing for the optimal use of the world's resources in accordance with the objective of sustainable development, seeking both to protect and preserve the environment and to enhance the means for doing so in a manner consistent with their respective needs and concerns at different levels of economic development.[14]

11 See www.wto.org/english/tratop_e/serv_e/gatsqa_e.htm.
12 See www.wto.org/english/tratop_e/trips_e/intel2_e.htm.
13 See www.cptech.org/ip/health/cl/cl-art27.html.
14 See www.econ.iastate.edu/classes/econ355/choi/wtomara.htm.

In support of this, umbrella clauses such as Article 20 of the GATT permit countries to act in the defense of human, animal, or plant life or health, and to conserve exhaustible natural resources. The trade rules allow subsidies for environmental protection. Whether or not WTO policies have exemplified efficient environmental stewardship in the past, the tone the organization projects in the present will have considerable influence on environmental economics and natural resource management in the future.

The World Bank and the International Monetary Fund   Financial ministers from 45 governments gathered in Bretton Woods, New Hampshire, in 1944 as architects of the modern international economy. With fresh wounds from the Great Depression and World War II, they sought economic stability and revitalization. To that end, they launched the World Bank to assist with rebuilding war-torn Europe. With that task completed, the World Bank turned to less developed countries, aiming to envelop them into the global economy and reduce poverty. The World Bank is now the "world's largest financier of biodiversity." Its projects encourage the maintenance of protected areas, sustainable use of biodiversity, the eradication of invasive species, and conservation through improved management of natural resources in production.[15]

Like the World Bank, the International Monetary Fund (IMF) was conceived as part of the Bretton Woods agreement. While the World Bank tries to eliminate poverty by financing development, the IMF serves as a stabilizing force for the currencies, interest rates, and economies of its 190 member countries by overseeing exchange-rate policies, lending to countries with balance-of-payment problems, and assisting with the development of monetary and fiscal policy.[16] The IMF does not target environmental concerns directly, although it notes an overlap between the environment and its agenda to promote economic growth. As discussed in Chapter 9, development results in increased resource use, but if environmental standards are a normal good, the correspondingly higher incomes will result in improved environmental protection. The IMF sees this as a reason to "embrace policy changes to build strong economies and a stronger world financial system that will produce more rapid growth and ensure that poverty is reduced."[17]

There are concerns that the World Bank and the IMF are not allocating resources in a socially efficient manner, and that the costs of their policies outweigh the benefits. The Friends of the Earth organization, for example, claims that IMF loans and related activities promote dependency on fossil fuels in countries around the world.[18] As with criticisms of the other international organizations, these charges are difficult to

15 See www.worldbank.org.

16 See www.imf.org.

17 See www.imf.org/external/np/exr/ib/2000/041200to.htm.

18 See https://foe.org/news/imf-curb-fossil-fuels/.

assess. These organizations have come to symbolize desired order for some, and the imposition of selfish ideals for others. It is nonetheless possible to identify several important questions that should be applied to policy proposals:

- *Are the proposed policies appropriate to the cultures and environments where they are being imposed and not just where they were conceived?*

- *Have all of the social costs and benefits been considered?*

- *Are developing countries receiving a fair share of the benefits derived from their human and natural resources?*

- *Have the desired standards of sustainability been applied?*

- *Are incentives aligned with the desired behavior?*

- *Are there opportunities for monitoring and enforcement?*

- *Are we learning from past mistakes?*

Special interest groups would have their own additions to this list. Most importantly, social efficiency cannot be achieved without asking many questions.

Approaches to Specific Global Environmental Threats

Deforestation

Forests provide essential habitat, food, building supplies, and medicine. They also absorb the greenhouse gas carbon dioxide, sequestering the carbon and releasing the oxygen that we breathe. Tree products, including rubber, fruit,

To slow the melting of glaciers like this one in Alaska, we may need to slow deforestation in places like Brazil.

lumber, and tropical oils, are prominent in global trade. The benefits of climate control and breathable air are public goods, generated on continents far from many of their recipients. Trees in South America, for example, are important to the global air supply and moderate climate. With localized benefits and global costs, deforestation presents a classic externalities problem.

The size of the Earth's tropical forests has fallen from 7 billion acres to 3 billion acres since 1800. An estimated 10,000 acres are cleared each day in the Amazon rainforest alone.[19] Deforestation may cause more than 100 species to become extinct on a daily basis.[20] Forest degradation and deforestation account for an estimated 15 percent of greenhouse gas emissions—more than all global transportation.[21] Myriad efforts aim to stave off excessive deforestation.

- *The Tropical Forestry Action Plan (TFAP)[22] was launched in 1985 as a joint effort of the UN Food and Agriculture Organization (FAO), the UN Development Program (UNDP), the World Bank, and the World Resources Institute. The goal was to promote international donor coordination in the development of National Forestry Action Plans (NFAPs) and to stimulate related policy initiatives and data collection.*

- *The International Tropical Timber Agreement (ITTA)[23] is a binding commodity agreement between consumers and producers of tropical timber, each of whom hold half of the votes. The ITTA was signed in 1983 at the United Nations Conference on Tropical Timber. A 2006 renegotiation of the ITTA came into force in 2011. The objective of the agreement is to promote research, collaboration, and information sharing in the interest of sustainable timber management. The ITTA governs the work of the International Tropical Timber Organization (ITTO).*

- *The Convention on International Trade in Endangered Species of Wild Fauna and Flora (CITES) is a binding international treaty established to regulate the trade of endangered species, as discussed in the next section.*

- *The United Nations Forum on Forests (UNFF), the current torchbearer for UN oversight, was established to "promote the management, conservation, and sustainable development of all types of forests, and to strengthen long-term political commitment to this end."[24] The UNFF is the current incarnation of work energized by the 1992 United Nations Conference on Environment and Development (UNCED).*

19 See https://education.nationalgeographic.org/resource/rain-forest/.
20 See www.theguardian.com/commentisfree/2021/oct/07/the-amazon-rain-forest-is-losing-200000-acres-a-day-soon-it-will-be-too-late.
21 See www.scientificamerican.com/article/deforestation-and-global-warming/.
22 See www.ciesin.columbia.edu/docs/002-162/002-162.html.
23 See www.itto.int/council_committees/itta/.
24 See www.un.org/esa/forests/index.html.

- *The* **United Nations Reducing Emissions from Deforestation and Forest Degradation** *(REDD+) effort creates financial incentives for developing countries to reduce emissions from forested lands. The "+" represents incentives for conservation and sustainable forest management.*[25] *REDD+ addresses tensions between desires in developed countries for a stable climate and desires in developing countries to generate income by harvesting forests and turning them into sites for agriculture, housing, or industry. REDD+ policies resemble Coasian-style "bribes" to countries like Brazil for not felling trees.*

The importance of forestation issues has prompted many related initiatives. A few of the more significant efforts are mentioned here. The Montreal Process[26] and Forest Europe[27] seek consensus on criteria and indicators for sustainable forest management. The Forest Stewardship Council[28] is working on certification systems that would identify wood that comes from sustainable sources. The Intergovernmental Technical Working Group on Forest Genetic Resources[29] advises the UN Commission on Genetic Resources for Food and Agriculture on issues related to forest resources. The Center for International Forestry Research[30] has identified policy-relevant research priorities in support of sustainable forest management.

The limited progress of international efforts in curtailing deforestation may result from the initiatives being difficult to implement, monitor, and enforce. As much as incentives matter, it is difficult to supersede the incentives to clear forests for roads, pasture, and construction projects. Market-based solutions promoting ecotourism and the products of living forests (fruits and nuts) have been successful but are limited in scope.

Threatened Species

High prices motivate producers to produce more. If the price of corn increases relative to the prices of other crops, farmers plant more corn. The difference in the context of many wildlife species is that domestic propagation is difficult or it yields inferior products. Farm-raised ginseng is an inferior substitute for wild ginseng, and many wild animals resist breeding in captivity. In such cases, high prices lead to

25 See www.un-redd.org.

26 See www.montreal-process.org/.

27 See https://foresteurope.org/.

28 See https://fsc.org/en.

29 See www.fao.org/forestry/86904/en/.

30 See www.cifor.org.

excessive harvests in open-access areas, and there are insufficient incentives for an increase in cultivation or breeding. The result is that many species are brought to the brink of extinction.

Unfortunately, scarcity and temptation go hand-in-hand in the context of endangered species. Increasing scarcity and restricted trade cause leftward shifts in the supply curve, raising the price further and providing all the more temptation for poaching and illegal trade in these species. This has been true for precious materials such as ivory, traditional medicines including rhino horn and ginseng, and delicacies such as shark fin.

A rhino-foot ashtray.

Photo by the Author.
© The Field Museum, #57683.

In response to these problems, the Convention on International Trade in Endangered Species of Wild Fauna and Flora treaty was established in 1975 as the only global treaty for the protection of plant and animal species from unregulated international trade. Membership includes 184 countries and continues to grow. Among the species receiving protection under the CITES treaty are the African and Asian elephant, American ginseng, giant pandas, and several types of rhinos and tigers.

In 2000, Russia and the United States signed an agreement to conserve polar bears, enhancing the 1973 Multilateral Agreement on the Conservation of Polar Bears between the United States, Russia, Norway, Denmark (for Greenland), and Canada. The agreement allows for a sustainable harvest by Alaska and Chukotka natives but prohibits the harvest of females with cubs less than 1 year old.[31]

The CITES ban on the trade of ivory eliminated legal markets, effectively decreasing the demand as from D_0 to D_1. It also eliminated the legal supply of ivory, reducing the supply curve to the illegal-only supply, S_1. Both the decrease in demand and the decrease in supply caused the quantity of ivory bought and sold to decrease. In theory, the net effect on price is indeterminate. In reality, the price fell, meaning that the price-lowering influence of the decrease in demand dominated the price-raising influence of the decrease in supply.

The protection of a species is aided by policies that decrease the demand for products made with that species. A decrease in demand lowers the equilibrium quantity—the overall objective. The corresponding decrease in price suppresses interest in poaching and illegal-market activities. A decrease in quantity can also be achieved with a decrease in supply, although the resulting increase in price may encourage illegal

---

31 See https://webharvest.gov/peth04/20041015111613/http://international.fws.gov/pdf/pbearagmt.pdf.

Figure 11.1

A Ban on Ivory

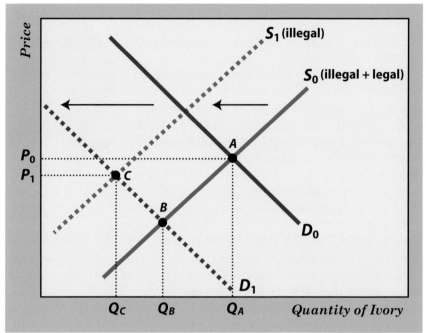

trade, as is the case with illegal drugs.[32] The CITES ban on the trade of ivory seems to have accomplished both. The ban eliminated legal markets for ivory, effectively decreasing the demand, as from D_0 to D_1 in Figure 11.1. This alone would bring the equilibrium from point A to point B, lowering the equilibrium price and the quantity. At the same time, the supply curve shifted to the left, as from S_0 to S_1 in Figure 11.1. This occurred because legal sales were eliminated and because poachers could no longer mix their ivory in with legal ivory to make it appear legitimate. A complete ban simplifies monitoring because any ivory that is discovered is clearly illegal. The end result was an equilibrium represented by point C, with a substantial decrease in the trade of ivory and therefore a decrease in the killing of elephants. Kenya, for example, lost about 3,500 elephants to poachers each year before the ban, but less than 100 in a typical year after the ban. The reported decrease in the market price of ivory, as from P_0 to P_1, suggests that the decrease in demand had a larger influence on price than the decrease in supply, as indicated in Figure 11.1.

If the dumping of stockpiled ivory on the market increased the supply from S_1 to S_2 without increasing demand or illegal-market supply, the supply from poachers would *decrease* from Q_A to Q_C. The balance of the

32 The high street price of illegal drugs motivates sellers to take great risks in selling on the black market, and encourages "pushers" to stimulate demand with free samples, loans of cash, and other aggressive marketing practices.

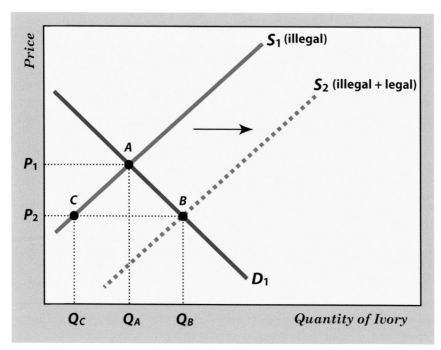

Figure 11.2

Dumping Ivory
on the Market

S_1 **(illegal)**

S_2 **(illegal + legal)**

A

P_1

C *B*

P_2

D_1

Q_C Q_A Q_B *Quantity of Ivory*

Price

equilibrium quantity, $Q_B - Q_C$, comes from the released stockpiles. If there is also an increase in demand or illegal-market supply, the amount of poaching may increase or decrease, depending on the relative sizes of the shifts.

One policy option is to dump large quantities of the animal products on the market. Just as supply restrictions increase the price, an increase in supply decreases the price, lowering the temptation for illegal activities. Indeed, in 1999 and again in 2008, large stockpiles of ivory from African countries were legally sold to China and Japan in rare exceptions to the international ban on ivory trade.

In an ideal scenario in which the legalized sale came from existing stockpiles and demand did not increase, the equilibrium quantity would increase, but the number of animals killed would actually decrease. Consider the shift from S_1 to S_2 in Figure 11.2, representing the influx of stockpiled ivory. The equilibrium moves from point A to point B. At the new equilibrium price of P_2, Q_B units of ivory are demanded, but only Q_C are supplied by the illegal market. The quantity $Q_B - Q_C$ comes from the legal stockpiles, and the quantity of ivory coming from illegal kills is *reduced* by $Q_A - Q_C$.

Carbon Leakage

Globalization also ushered in **carbon leakage**, which occurs when stricter emissions standards in one country bring about higher carbon emissions in another country with lower standards. When the cost of adopting cleaner production methods elevates prices for domestically produced goods, demand may increase for goods imported from countries with lower costs. The result is lower emissions in the importing country but higher emissions in the exporting country. Differing emissions policies can also cause firms to relocate in countries with more lenient standards, again lowering emissions in one country and raising them in another. Because carbon leakage merely shifts the location of emissions, the practice can nullify the overall benefits of emissions standards and create negative externalities for the countries with lower standards.

Even changes in international trade not specifically tied to environmental policies can shift, or *offshore*, emissions to the new production location. Both carbon leakage and the broader offshoring of emissions separate the country where products are consumed from the country where emissions occur, with important policy implications.

When consumer demand in Country A is the reason for production in Country B, carbon-reduction policies that influence consumers should be directed at Country A. For example, taxes on products with a large carbon footprint, or subsidies for products that minimize carbon emissions, belong in Country A because that is where consumption decisions are made. Policies targeting manufacturers should apply to Country B, with the caveat that the policies will fail to reduce global emissions levels if they trigger another round of carbon leakage that again relocates the production that causes the emissions.

Policymakers have developed mechanisms to address carbon leakage, some of which resemble the ecological tariffs introduced in Chapter 8. For example, the European Union's Carbon Border Adjustment Mechanism motivates cleaner production elsewhere by assuring that a price has been paid for carbon emitted in the production of certain carbon-intensive imports such as cement, iron, and fertilizer. Importers must either prove that a price has already been paid for the carbon emissions associated with those goods or pay that price by purchasing certificates that cover the carbon emissions involved.[33]

Greenhouse Gases and the Ozone Layer

Like the glass panes of a greenhouse that prevent heat from escaping, greenhouse gases in the atmosphere prevent some infrared heat energy from escaping into space, keeping the Earth about 59°F (33°C) warmer than it would be without them. Greenhouse gases that come from both natural and human-made sources include carbon dioxide, methane, nitrous oxide, water vapor, and ozone. Other greenhouse gases come

---

33 See https://taxation-customs.ec.europa.eu/green-taxation-0/carbon-border-adjustment-mechanism_en.

only from industrial processes, including hydrofluorocarbons (HFCs), perfluorocarbons (PFCs), and sulfur hexafluoride (SF_6). Substantial increases in the concentrations of greenhouse gases since industrialization[34] threaten to trap more heat in the atmosphere, causing **global climate change**.

Ozone is a form of oxygen. The oxygen we breathe is two oxygen atoms bonded together (O_2). Ozone is three oxygen atoms bonded together (O_3). *Ground-level ozone* is a harmful component of smog, produced by a reaction of sunlight, volatile organic compounds (VOCs), and nitrogen oxides (NO_x). The global concern is not so much ozone as a greenhouse gas or a ground-level pollutant, but thinning and holes in the *ozone layer* that blankets the upper atmosphere and shields the Earth from harmful ultraviolet rays. Chlorofluorocarbons (CFCs) and related industrial gases have damaged the ozone layer in many places. The satellite photo shows the seasonal ozone hole over Antarctica. Thinning of the ozone layer increases ultraviolet radiation on Earth and can affect biomass that reduces carbon dioxide, thus causing further climatic changes.

Attempts to measure ozone depletion are a good case study in international environmental cooperation. A British Antarctic survey team discovered the ozone hole in 1985, although it is now known to have existed since at least 1979. NASA studies the ozone using Total Ozone Mapping Spectrometers (TOMS). Ozone depletion measurements from TOMS are the basis for several international agreements to phase out the use of CFCs

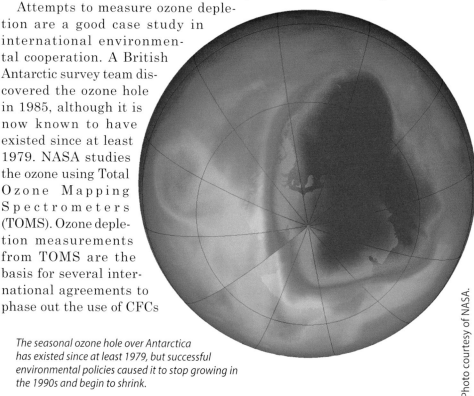

The seasonal ozone hole over Antarctica has existed since at least 1979, but successful environmental policies caused it to stop growing in the 1990s and begin to shrink.

---

34 Since the eighteenth century, for example, carbon dioxide levels have increased by 38 percent and methane levels have increased by 148 percent. See https://earthobservatory.nasa.gov/features/GlobalWarming.

and other ozone-depleting chemicals. In 1991, the former Soviet Union launched a Meteor-3 satellite carrying a TOMS instrument provided by NASA. The Japanese Advanced Earth Observations Satellite (ADEOS) carried a TOMS into orbit in 1996. As of 2023, NASA's Suomi NPP satellite monitored the ozone hole, as well as climate change, air pollution, ice cover, and vegetation across the globe.

Prior to the discovery of ozone thinning, CFCs were commonly used in refrigerators, air conditioners, and spray cans. The Montreal Protocol on Substances that Deplete the Ozone Layer, signed in 1987 and substantially amended in 1990 and 1992, is the landmark agreement on CFC reductions. The Montreal Protocol stipulated a phaseout of CFC production and use by 2000 (2005 for methyl chloroform). NASA reports that the Montreal Protocol has helped to improve the ozone layer, although the recovery will be slow because it takes up to a century for a CFC molecule to break down in the atmosphere.[35]

At the heart of international efforts toward global climate control are the annual Conferences of the Parties (COP) to the UN Framework Convention on Climate Change. In 1997, COP 3 produced the Kyoto Protocol, under which countries were assigned emissions targets for greenhouse gases depending on their environmental circumstances and economic profiles. For example, relative to 1990 levels, the European Union sought to reduce CO_2, CH_4, and N_2O emissions by 15 percent by 2010, and Japan sought a 5 percent cut in the same gases by 2012. At COP 21 in 2015, 195 countries adopted the Paris Agreement, a global deal to limit climate change with international cooperation and transparent national climate action plans. The United States withdrew from the Agreement in 2017 for reasons that remain controversial but rejoined in 2021. At COP 28 in 2023, participants showed that progress in addressing climate change was too slow, and the countries decided to accelerate all areas of climate action by 2030.[36]

Polluted Seas

The oceans covering two-thirds of the planet provide critical biodiversity, food, and carbon sequestration. Oceans are largely open-access resources and users make decisions to overfish, transport toxins, drill for oil, and dump waste without internalizing many of the costs. Problems with negative externalities are the result. For example, the oceans receive over 10 million tons of plastic each year. Accumulations of tiny bits of plastic called *microplastics* form clusters such as the massive Great Pacific Garbage Patch spanning from Japan to the West Coast of North America.

35 See https://ozonewatch.gsfc.nasa.gov/facts/history_SH.html.
36 See https://unfccc.int/cop28.

What can motivate an international agreement among countries that do not internalize the effects of their behavior? A major environmental crisis. The microplastics crisis led 175 nations to commit to developing an agreement on plastics pollution in 2024.[37] Earlier crises precipitated other agreements. The spillage of 31 million gallons of oil from the *Torrey Canyon* supertanker between England and France in 1967 prompted the Agreement for Cooperation in Dealing with Pollution of the North Sea by Oil. The agreement was signed in Bonn, Germany, by nine European countries in 1969. Under the Bonn Agreement, participants agreed to share information on contingency plans for spills,

What washes ashore—oil, trash, medical waste—is some indication of what is in the world's oceans.

alert each other to ongoing emergencies, and provide assistance to each other in the event of environmental disasters at sea.

An attempt by the Dutch ship *Stella Maris* to dump chlorinated water into the North Sea in 1971 led to the 1972 Oslo Convention for the Prevention of Marine Polluting by Dumping from Ships and Aircraft. Signed by 13 countries, the Oslo Convention established a commission to regulate and control the dumping of industrial wastes and sewage sludge into the sea and the incineration of industrial wastes at sea. These dumping and incineration activities were subsequently phased out in the northeast Atlantic during the 1990s. To regulate the land-based sources of ocean pollution, 13 countries and the European Economic Community ratified the Convention for the Prevention of Marine Pollution from Land-based Sources—the Paris Convention—in 1974.

The Paris and Oslo commissions gathered in London in 1992 to form a single entity under the Convention for the Protection of the Marine Environment of the North-East Atlantic—the OSPAR Convention. The policies under the Oslo and Paris Conventions continue under OSPAR. These include a precautionary principle described in the Reality Check and a polluter pays principle similar to that in United States Superfund legislation. Participants are also required to adopt the best available techniques (BAT) and the best environmental practices (BEP), including clean technology.

37 See www.un.org/en/climatechange/nations-agree-end-plastic-pollution.

reality _check_

The Precautionary Principle

The precautionary principle discussed earlier in this chapter entails erring on the safe side when it comes to risks to humans and the environment. The principle is a reality among international environmental agreements. Examples include the OSPAR treaty, the CITES treaty, Principle 15 of the Rio Declaration on the Environment and Development, the revised Treaty of Rome, Article 3.3 of the UN Framework Convention on Climate Change, Agenda 21 of the Rio Conference, Amendments to the Montreal Protocol on Substances that Deplete the Ozone Layer, and the UN Convention of Biological Diversity.

The Australian Intergovernmental Agreement on the Environment (IGAE) explains the precautionary principle this way:

> Where there are threats of serious or irreversible environmental damage, lack of full scientific certainty should not be used as a reason for postponing measures to prevent environmental degradation. In the application of the precautionary principle, public and private decisions should be guided by
>
> (i) careful evaluation to avoid, wherever practicable, serious or irreversible damage to the environment; and
> (ii) an assessment of the risk-weighted consequences of various options.

The precautionary principle is not without its critics. Alder (2011) suggests that some industrial chemicals prohibited by the precautionary principle would reduce more risks than they create. Arguing in favor of the principle, Gollier et al. (2000) and Steele (2006) say that by accounting for great uncertainty and risk aversion, the precautionary principle is a rational criterion for efficient policymaking. A reality check of the popularity of the precautionary principle, not to mention our speed limits, helmet laws, and insurance purchases, suggests that in some situations, humans indeed like to err on the safe side.

The Mixed Baggage of Tourism

Tourism is among the largest industries in the world. Tourism can be pro-poor, pro-development, and pro-environment, or quite the opposite, depending on the practices and policies in place. Some tourists bike or hike to their destinations and follow the maxim, "take only memories, leave only footprints";[38] others take home bulging suitcases and leave trails of waste. As with other forms of globalization and trade, tourism carries the risks of homogenization and exploitation.

During the COVID-19 era, the "staycation" trend of families staying near home for vacation improved the air and water in many cities.[39] Destination tourism requires transportation, which itself has substantial environmental impacts. The 3,105-mile round-trip from Los Angeles to Mexico City creates a carbon dioxide footprint of approximately 1,052 pounds per passenger.[40] Driving the same distance alone in an automobile that travels 25 miles per gallon would create 2,422 pounds of carbon emissions. Making the trip by train would create a carbon footprint of 1,304 pounds per person.

International tourism promotes a sharing of cultures, a transfer of wealth, and an incentive to preserve the drawing card that wildlife represents. Indeed, with tourists attracted to jungle safaris, pristine beaches, and national forests, entrepreneurs and policymakers have reason to keep wilderness areas intact. The associated airports, hotels, restaurants, gift shops, and casinos might erase the environmental gains. The effects of tourism on the local environment are assessed in several ways:

- *The **tourism carrying capacity** (TCC) is the maximum number of tourists that could visit an area simultaneously without degrading the environment or the visitor experience. For example, the level of carbon dioxide accumulation in New Zealand's Waitomo Glowworm Cave causes corrosion with more than 90 visitors per hour. The TCC is typically an approximation because it depends on the behavior of the tourists and on changing conditions at the tourist destination. Even so, an estimate can support policies that prevent limitless crowds in vulnerable areas.*

- **Limits of acceptable change** (LAC) *is a framework for determining how much human-induced change can occur in an area before the level of tourism is considered excessive. The focus is on the acceptable amount of change rather than on the number of visitors that triggers change. Because the acceptable change is determined by stakeholders, the LAC approach includes an invitation for users of the area to share their interests in tourism and environmental protection, and a consideration of plans to bring the impact of tourism within acceptable limits. The U.S. Forest Service uses this approach to balance environmental protection with recreational opportunities on public lands.*[41]

---

38 This quote is often attributed to Native American Chief Seattle, who may have said it during a 1854 speech.

39 See www.ncbi.nlm.nih.gov/pmc/articles/PMC7498239/.

40 See www.TerraPass.com.

41 See www.fs.usda.gov/detail/dbnf/home/?cid=stelprdb5346360.

- **Environmental impact assessments** *(EIAs) are analyses of the external costs and benefits of a proposed policy or development project. An EIA's focus is on environmental externalities, but the analysis often considers the effects on human health, the local economy, and society as well. The findings of an EIA are reported in an environmental impact statement (EIS). Many countries encourage consideration of the environmental consequences of tourism and related development by requiring EIAs prior to the approval of major projects such as hotel or highway construction.*

- *The* **touristic ecological footprint** *(TEF) measures the area of productive land and water needed to support tourists in a particular destination. Typical TEF calculations consider travel, lodging, sightseeing, purchases, entertainment, food, and waste. The resulting footprints can be compared across industries, activities, and locations. For example, Mancini et al. (2022) estimated the ecological footprint per tourist per day in 13 protected areas near the Mediterranean Sea. The results ranged from 0.008 hectares (1 hectare = 2.47 acres) in Lastovo to 0.026 hectares in Torre del Cerrano. This demonstrates opportunities to lower the environmental impact of tourism substantially by reallocating tourists among sites within the same region.*

The World Travel and Tourism Council, an advocacy group for the industry, reports a balanced mission of "generating profit as well as protecting natural, social and cultural environment."[42] In practice, the tourism industry is largely consumer driven, and its impact depends on the interest and awareness of those taking part. The growing trends of ecotourism, culturally and

Laws, accompanied by warnings and disincentives, are one way to reduce tourists' impact on the environment.

environmentally sensitive tourism, and voluntary payments for green energy offsets show promise for more tourists who take only memories and leave only footprints.

Government policy can also reduce the impact of tourism. The creation of new parks and hiking trails everywhere from Dominica to Greenland is allowing more tourists to appreciate nature while treading lightly. As explained in Chapter 3, taxes can reduce the quantity demanded of goods or services including tourism. The Spanish island of Mallorca is among the destinations reducing the quantity of tourists by collecting a tourist tax from each visitor. Chapter 12 describes the congestion pricing approach of increasing fees at popular times to spread out the flow of tourists. This can improve both the sustainability of tourism and the visitor experience. And laws against destructive behaviors such as tourist contact with endangered reefs and sea turtles can be effective, but only if well enforced.

Summary

From some perspectives our planet is small—one-tenth the size of Saturn and circled every 90 minutes by commercial satellites. The close proximity of human populations means no nation is isolated from environmental influences elsewhere on the planet. International environmental policies are for naught if they fail to engage multiple parties with adequate incentives and legitimate enforcement mechanisms. The United Nations, the World Trade Organization, the World Bank, and the International Monetary Fund, among other international organizations, pursue the Herculean goal of managing our global economy with appropriate sensitivity to the environment, the viability of which makes both supply and demand possible.

Policy approaches to the specific global environmental problems of deforestation, threatened species, climate change, and pollution in the seas have several common themes. In each case, international committees focus policymakers' attention, educational efforts broaden awareness, multilateral agreements garner cooperation, and economic incentives provide a motivating force. Uncertainty about problems and solutions clouds the advancement of environmental policies. The precautionary principle is a careful and popular approach: In the face of potentially irreversible environmental degradation, most organizations contend that a lack of absolute scientific certainty should not subvert safety measures.

Depending on its application, the precautionary principle may cause excessive safety. On the other hand, zeal for economic progress may result in insufficient calculations of the environmental costs of development projects. The policies and organizations you learned about in this chapter appear frequently in the media, highlighting the human struggle

An upcoming practice problem challenges you to use the economic tools explained in this chapter to reveal that burning ivory is generally not the best policy approach to the protection of elephants.

with global constraints on clean air and water, landfill space, environmental sinks, energy sources, wildlife habitat, and recreation areas. With knowledge of their successes and failures, current approaches can serve as stepping-stones to better solutions in the future.

• • • • • • • • • • • • • • • • • •

Problems for Review

1. International environmental policies would not be needed if each country enacted the policies that best served the larger world. Why is it unlikely that country-level policies on sea pollution are socially efficient, even if they are optimal for the countries that enact them?

2. Globalization permits firms to shop across countries to find low environmental standards. Explain how this practice can cause

   carbon leakage and identify one remedy for this practice.

3. The Hard Rock 100 is a running race on trails through environmentally sensitive areas of Colorado. The race is limited to 140 runners.
   a) If 140 is an estimate of the maximum number of runners that could traverse the trails without causing environmental degradation, what measure

of the effects of tourism was applied?

b) If Colorado required the race organizers to obtain an EIA, what would that mean and what four general categories of effects might be included?

c) Identify a measurement of the amount of productive land and water needed to support tourists in Colorado.

4. Explain the difference between the tourism carrying capacity and the limits of acceptable change.

5. Kenya burned 2,500 elephant tusks in a symbolic gesture against poaching. Assume the demand for ivory was unaffected by the burn. The burned tusks would otherwise have been sold legally, so without the burn, the supply would have been made up of legal and illegal ivory. With the burn, the supply was made up of only illegal ivory.

a) Using a graph like the ones in this chapter, illustrate the effect of this burn on the quantity of ivory supplied illegally by poachers.

b) How would your answer change if worldwide media coverage of the event dissuaded potential consumers of ivory?

6. Consider the scenario represented in Figure 11.2, with S_1 representing the supply of illegal ivory, which we'll assume remains constant. Let the shift from S_1 to S_2 represent an influx of legal ivory due to a partial lifting of the ivory ban.

Suppose that the relaxation of the ban increases ivory demand. Draw the largest possible increase in demand that would *not* result in an increase in the quantity of illegal ivory sold relative to the initial level, Q_A.

7. Consider once more the ivory market in which all ivory is sold at the market price. Starting from an equilibrium at point A in Figure 11.2, suppose a partial lifting of the ivory ban lowers the marginal cost of selling illegal ivory, but ivory demand is unchanged. Draw a graph that shows (as two separate shifts): (1) an increase in the supply of illegal ivory resulting from lower costs and (2) an influx of legal ivory *such that at the new equilibrium, the quantity of illegal ivory sold remains at Q_A.*

8. What do you see as the greatest *specific* cost and the greatest *specific* benefit of globalization?

9. As the director of the World Bank, what one question would you add to those listed in the section on the World Bank and the IMF as a consideration prior to the adoption of a policy?

10. Giant pandas live in China.

a) Which type of value introduced in the previous chapter makes the threat of giant panda extinction an international problem?

b) How might international cooperation lead to a solution?

1. Find one website that speaks in favor of globalization and one website that speaks against it. Briefly summarize their arguments.

2. Find a website calling for the addition of a species to the CITES preservation list. Do you agree that the species in question deserves special protection? Why or why not?

3. Find a website that describes an act of international environmental cooperation that is not discussed in this chapter. Briefly describe the project.

Key Terms

Carbon leakage
Environmental impact assessments
Global climate change
Limits of acceptable change

Precautionary principle
Tourism carrying capacity
Touristic ecological footprint

Internet Resources

EPA climate change site:
www.epa.gov/climate-change

International Monetary Fund:
www.imf.org

United Nations:
www.un.org

U.S. Forest Service Application of limits of acceptable change.
www.fs.usda.gov/detail/dbnf/home/?cid=stelprdb5346360

World Bank:
www.worldbank.org

World Trade Organization:
www.wto.org

Further Reading

Alder, Jonathan H. "The Problems with Precaution: A Principle Without Principle." *American Enterprise Institute* (2011). https://www.aei.org/articles/the-problems-with-precaution-a-principle-without-principle/. This article argues against the precautionary principle as a sound basis of policy.

Carney Almroth, Bethanie, and Håkan Eggert. "Marine Plastic Pollution: Sources, Impacts, and Policy Issues." *Review of Environmental Economics and Policy 13*, no. *2* (2019): 317–326. https://doi.org/10.1093/reep/rez012. An introduction to the problem of marine plastics, the materials involved, and the impacts of these plastics on marine and human life and health.

DeSombre, Elizabeth R. *Global Environmental Institutions*. New York: Routledge, 2017. An overview of global institutions with the goal of environmental protection.

Dickson, Barnabas. "The Precautionary Principle in CITES: A Critical Assessment." *Natural Resources Journal 39*, no. *2* (1999): 211–228. Relates the prevention of environmental damage to risk aversion and prescribes standards for the application of the precautionary principle.

Ehrlich, Paul, and Anne Ehrlich. *Population Resources Environment: Issues in Human Ecology*. Washington, DC: Population Reference Bureau, Inc., 1979. A discussion of global demands on food, resources, and the environment.

Gollier, Christian, Bruno Jullien, and Nicolas Treich. "Scientific Progress and Irreversibility: An Economic Interpretation of the 'Precautionary Principle'." *Journal of Public Economics 75*, no. *2* (2000): 229–253. States that the impact of conservation policies on humans must be considered when applying the precautionary principle.

Keho, Yaya. "Does Globalization Cause Environmental Degradation in Developing Economies? Evidence from Cote d'Ivoire Using Ecological Footprint." *International Journal of Energy Economics and Policy 13*, no. *4* (2023): 455–466. https://doi.org/10.32479/ijeep.14325. Finds that economic, social, and political globalization decrease the environmental quality in developing nations.

Komsary, K. C., W. P. Tarigan, and T. Wiyana. "Limits of Acceptable Change as Tool for Tourism Development Sustainability in Pangandaran West Java." *IOP Conference Series: Earth and Environmental Science 126*, no. *1* (2018): 012129. https://doi.org/10.1088/1755-1315/126/1/012129. A study of the limits of acceptable change in a model destination for sustainable tourism in Indonesia.

Li, Peng, and Guihua Yang. "Ecological Footprint Study on Tourism Itinerary Products in Shangri-La, Yunnan Province, China." *Acta Ecologica Sinica 27*, no. *7* (2007): 2954–2963. A study of the environmental impact of seven components of tourism, including travel, food, lodging, waste disposal, and entertainment.

Long, Cheng, Song Lu, Jie Chang, Jiaheng Zhu, and Luqiao Chen. "Tourism Environmental Carrying Capacity Review, Hotspot, Issue, and Prospect." *International Journal of Environmental Research and Public Health 19* (2022): 16663. https://doi.org/10.3390/ijerph192416663. A review of 297 articles on tourism environmental carrying capacity and frictions between economic growth and tourism.

Mancini, Maria Serena, Debora Barioni, Carla Danelutti, Antonios Barnias, Valentina Bračanov, Guido Capanna Piscè, Gilles Chappaz, Bruna Đuković, Daniele Guarneri, Marianne Lang, Isabel Martín, Sílvia Matamoros Reverté, Irene Morell, Artenisa Peçulaj, Mosor Prvan, Mauro Randone, Jeremy Sampson, Luca Santarossa, Fabrizio Santini, Jula Selmani, Capucine Ser, Iacopo Sinibaldi, Mirjan Topi, Vittorio Treglia, Simona Zirletta, and Alessandro Galli. "Ecological Footprint and Tourism: Development and Sustainability Monitoring of Ecotourism Packages in Mediterranean Protected Areas." *Journal of Outdoor Recreation and Tourism, Sustainably Managing Outdoor Recreation and Nature-Based Tourism as Social-Ecological Systems 38* (June 2022): 100513. https://doi.org/10.1016/j.jort.2022.100513. Applies ecological footprint accounting to 13 ecotourism destinations and discusses information relevant to local stakeholders seeking low-impact tourism.

Sato, Misato, and Josh Burke. *What Is Carbon Leakage? Clarifying Misconceptions for a Better Mitigation Effort.* London: School of Economics, Grantham Research Institute on Climate Change and the Environment, 2021. www.lse.ac.uk/granthaminstitute/news/what-is-carbon-leakage-clarifying-misconceptions-for-a-better-mitigation-effort/. This article explains two interpretations of carbon leakage and outlines multiple solutions to the problem.

Steele, Katie. "The Precautionary Principle: A New Approach to Public Decision-Making?" *Law, Probability and Risk 5*, no. *1* (2006): 19–31. Argues that the precautionary principle is a valid element of rational decision-making.

Zekan, Bozana, Christian Weismayer, Ulrich Gunter, Bernd Schuh, and Sabine Sedlacek. "Regional Sustainability and Tourism Carrying Capacities." *Journal of Cleaner Production 339* (March 2022): 130624. https://doi.org/10.1016/j.jclepro.2022.130624. Discusses a destination-specific method for assessing tourism carrying capacity that addresses regional sustainability and tourism development.

VEHICLE
NOISE
ORDINANCE
ENFORCED
SEC. 8.24.100
LINCOLN MUNI. CODE
FINE $100

Perspectives on Environmental Policy

*T*he pigs escaped from my grandfather's farm while my parents held their wedding reception in the farmhouse. There was no question among the revelers that the wayward swine posed a problem, but as they slogged through rain and mud, the men in tuxedos had differing perspectives on how best to remedy the situation. Some pushed pigs, some pulled, some prodded them with sticks. Others pronounced that the pigs would return to the pen if left alone. With a combination of these approaches, the pigs were eventually placed under control, whether by the most efficient means or not. Environmental policy is much the same: There is broad agreement that environmental losses are a problem, but whether it is best to push decision-makers with command-and-control policies, pull them with incentives, or prod them with the sticks of litigation and punishment is another issue. Some feel that, if left alone, the market will attain efficiency on its own.

Policy itself is pulled and pushed by the political machine, replete with myriad constituents and special interests, and limited by incomplete information and egocentric intent. Out of this melee come efforts to remedy market failure, discover environmental perils, and serve present and future constituencies—both human and otherwise—that have little or no voice in the matters. There is much to be sorted out in the realm of environmental policy.

Mancur Olson (2009) notes that the individual benefit received from many activities is small or negative when few people participate and large when they are well subscribed. For example, if only a few people drive electric cars, the recharging stations, repair

315

DOI: 10.4324/9781003428732-12

A chicken and egg problem: It's hard to get many charging stations when there aren't many electric cars, and it's hard to get many electric cars when there aren't many charging stations.

shops, and development expenditures that make electric cars convenient will not arise. Likewise, the potential benefits of recycling, clean fuels, and organic products aren't realized until a critical mass of customers supports the research and infrastructure expenditures that make them attractive. Another role of policy is to help promote participation levels that benefit all users.

This chapter discusses the two major categories of environmental policy and the optimal deterrence of environmental wrongdoing. The market-based policies introduced in Chapter 7 set a target for emissions and create incentives for polluters to achieve that target in ways the polluters choose. Command-and-control policies typically set both a target and a path by which that target shall be met. Examples of paths include the use of particular fuels, fishing equipment, or pollution-scrubbing devices. From an efficiency standpoint, we will see that market-based incentives are sometimes superior due to their flexibility. In other cases, flexibility is less important than the ability to apply and enforce common standards.

These are among many available types of policies. Support for science and technology initiatives can lead to lower-impact products and machines, including cleaner fuels and better smokestack scrubbers. The dissemination of information about environmental issues can help individuals and firms make more appropriate decisions regarding the environment. Programs to foster cooperation and trade can include environmental standards among the rules of fair play. And more effective enforcement policies can bolster the success of pollution-abatement programs.

Command-and-Control Policies

Command-and-control policies typically require certain behaviors or forbid others. The inflexibility of such stipulations can cause inefficiencies when prescribed solutions aren't the best fit for particular polluters. When given the option, for example, municipalities can achieve water quality standards by purchasing infrared lights that sterilize the output of their water treatment plants, rebuilding sewer systems to better manage flood-waters, or adding employees to monitor and reduce the illegal dumping of waste into nearby water supplies. Different solutions will be cheaper and more productive in different locations, and a rigid mandate that every municipality must purchase infrared sterilization equipment might not be the most efficient solution for every community. Regional differences in the costs of capital and labor, along with facility-specific differences in problem severity and equipment adaptability, make flexibility desirable for the attainment of technical efficiency.

Command-and-control policies apply three types of standards: **Ambient standards** designate targets for pollution concentrations at specified locations of measurement. In accordance with the Clean Air Act, the EPA has established national ambient air quality standards (NAAQS) for six criteria air pollutants as discussed in Chapter 6. The Clean Water Act requires every U.S. state to set ambient water quality standards with goals including swimmable and fishable waterways. Emissions and technology standards are then set to achieve these ambient standards.

Technology standards stipulate the use of particular methods or types of pollution-control equipment. Examples include pollution-abating catalytic converters in cars,[1] scrubbers in coal-fired power plants, and minimum heights for chimneys. Under the Clean Air Act, major new or modified sources of pollution in *nonattainment areas*—areas not meeting ambient air quality standards—are required to adopt the technology needed to emit at the *lowest achievable emission rate* (LAER). Major new or modified sources in areas meeting the standards are required to employ the *best available control technology* (BACT).[2] And existing polluters in nonattainment areas must use *reasonably available control technology* (RACT). The exact requirements under these categories are defined for specific processes and types of equipment and made available by the EPA's RACT/BACT/LAER Clearinghouse.

Emission standards require firms to reduce emissions by a designated percentage or to a set amount. An example is the National Pollutant Discharge Elimination System (NPDES) permit program authorized by the Clean Water Act. NPDES regulates point sources of

---

1 To learn how catalytic converters work, see www.howstuffworks.com/catalytic-converter.htm.

2 The European Union and others employ a similar concept, BATNEC, an acronym for best available technology not entailing excessive costs.

pollution such as industrial and municipal discharge pipes. Effluent discharge permits are granted based on the available abatement technology and the wasteload capacity of the receiving water that would ensure the maintenance of ambient water quality standards. When technology-based standards are sufficient to meet the needs of the receiving body of water, NPDES permit writers stipulate a minimum level of treatment for discharge while allowing polluters to choose their specific treatment method.

In the United States, the EPA oversees the research and review process for NPDES and other environmental standards. Proposed rules are printed in a government-wide collection of documents called the *Federal Register*. After interested members of the public have the opportunity to comment on a proposal, the EPA makes revisions as appropriate and then publishes the final rule, again in the daily *Federal Register* and in the annual *Code of Federal Regulations*. The EPA is also charged with monitoring and enforcing these standards.

At the intersection of business and government, debates swirl around whether to apply standards and which standards to apply. Technology standards are the most rigid in regard to polluters' options for compliance but can make sense when there is one clear path to solving an environmental problem, and may be justified when the provision of greater flexibility would prohibit effective monitoring. All these standards have the drawback that once standards are met, there is little incentive for further emissions reductions. The next section discusses alternative approaches that offer improved incentives and flexibility.

Market-Based Incentives

As an alternative to traditional command-and-control approaches, policymakers use the incentives of market-based policy instruments to pursue the same efficiency goals from a different angle. The market's

New Zealand isn't pulling the wool over anyone's eyes when it comes to climate change. It has five times as many sheep as people, and each sheep produces about 30 liters of methane per day. But the country plans to meet targets for greenhouse gases by taxing livestock gas emissions.

own engine of self-interest is turned against market failure. The market-based incentive might be a subsidy or tax credit, a deposit/refund program exemplified by bottle bills, one of the relatively new and innovative emissions trading programs, or a Pigouvian pollution tax.[3] For example, citing the problematic methane emissions from its 6 million cattle and 25 million sheep, the government of New Zealand plans to institute a tax on greenhouse gas emissions from livestock starting in 2025. Incentive-based solutions allow individuals and firms with differing circumstances to address environmental problems in different ways. Earlier chapters have introduced market-based solutions; this section provides more specific policy examples and elaborates on the use of tradable emissions rights.

Market Approaches to Automobile Externalities Around the World

Rush hour congestion occurs because too many drivers want to be on the same road at the same time. Some of those people *really* need to be driving then; others *kind of* need to be driving then. Those on the margin have a private marginal benefit that barely exceeds the private marginal cost of being in traffic, and if they also had to pay the marginal external cost their own car imposes on others, they would instead choose an alternative time or mode of transportation. The idea of **congestion pricing** or **peak-load pricing** is to spread out resource use by charging higher prices during peak consumption periods. This encourages those with relatively flexible schedules or good alternatives to make adjustments. Movie theaters do it with matinee prices. Hotels do it with low-season discounts. A growing number of cities have brought congestion pricing to their roads.

To drive in central London during peak driving times costs an extra $18.57 per day. If that fee approximates the marginal external cost of such driving, drivers will use busy roads during peak periods only if it is efficient, in that their marginal benefit exceeds the social marginal cost. The controversial program has enjoyed measured success, with a 20 percent reduction in traffic (20,000 fewer vehicles per day), a 37 percent increase in traffic speed, and a 14 percent increase in bus riding within its first 8 years.[4] Congestion pricing systems are now in place in cities in Australia, Canada, France, Germany, Italy, South Korea, Singapore, Sweden, and Norway. A road in Washington, D.C., has tolls that change every 6 minutes depending on the traffic volume, and New York City adopted congestion pricing in 2024.

---

3   Chapter 3 defines a Pigou tax as a tax that equals the marginal external cost of the behavior being taxed. Upon paying such a tax, the decision maker internalizes the full cost of his or her behavior.

4   See www.vtpi.org/london.pdf.

Just as cash deposits encourage the return of beverage containers for refunds in several countries, deposit programs in Greece, Norway, and Sweden encourage the return of car bodies for recycling. Deposit-return programs succeed in keeping materials out of the waste stream and placing them into new production. Among broad-ranging uses, deposits are applied to car batteries in Denmark, the Netherlands, and the United States, industrial products in South Korea, disposable cameras in Japan, and computers in Australia.

Tradable Emissions Rights: A Two-Firm Pollution Model

During the transition away from leaded gasoline in the 1980s and 1990s, countries from Mexico to Thailand used higher taxes on leaded gasoline to successfully promote unleaded varieties. Today's transition to electric and hybrid cars and to alternative fuels is being encouraged with similar tax incentives in Japan, the United States, and elsewhere. Many countries, including Canada, China, the United States, parts of Brazil, and countries across Europe, subsidize the purchase of new plug-in hybrid or electric vehicles.

Firms value the right to emit pollution because it allows them to produce goods and earn profits. An efficient level and distribution of pollution rights can be achieved by restricting emissions levels, dividing the rights to the permitted emissions among firms, and letting the firms buy and sell those rights. Figure 12.1 illustrates the marginal value of emissions for each of two hypothetical firms. This model works like the one in Figure 8.4, except there are two firms instead of two periods. Firm 1 emissions increase from left to right; Firm 2 emissions increase from right to left. The curves indicate that Firm 2 receives less value from any given amount of emissions than Firm 1, perhaps because Firm 2 is newer and operates with state-of-the-art emissions-control equipment.

The total value of emissions for a firm is indicated by the area under the firm's marginal value curve over the range of permitted emissions. If each firm is restricted to the same emissions standard of half of the total allowable emissions, the total value of emissions to Firm 1 is the sum of areas A and B and the total value of emissions to Firm 2 is the sum of areas D, E, and F.

The efficient allocation, labeled *Efficient* on the horizontal axis in Figure 12.1, equates the firms' marginal values from emissions. With the efficient allocation, the total value to Firm 1 is $A + B + C + D$ and the total value to Firm 2 is $E + F$. Relative to the equal division of emissions rights, Firm 2 loses area D, but Firm 1 gains areas C and D, for a net societal gain of area C.

The efficiency gain in this example is attractive but achieving it with government controls would require a great deal of information. It is difficult to establish the optimal total amount of pollution, although meaningful targets are available. The next challenge is to determine the values placed on pollution by the various firms. If asked to report the benefits received from the ability to pollute, each firm would have an incentive to exaggerate its marginal value of emissions in order to increase what appears to be its efficient share of the emissions allotment. It is for this challenge that tradable emissions permits become problem solvers.

Given a market for emissions permits, each firm faces a per-unit opportunity cost of emissions equal to the market price of emissions permits. Firms that do not have enough permits have an incentive to find ways to reduce their emissions to minimize the need to purchase more permits. Even if a firm has the permits needed to produce at maximum capacity, it has an incentive to keep finding ways to pollute less as long as emissions reductions cost less than the market price of permits, so that it can sell the excess permits to someone else.

Consider the situation in Figure 12.1 in which each firm begins with the right to emit half of the allowed pollution. The firm that can reduce its emissions at the lowest cost (Firm 2, due to its modern emission-control equipment) will end up selling permits to the firm for whom cleanup is relatively expensive and polluting is relatively valuable (Firm 1).

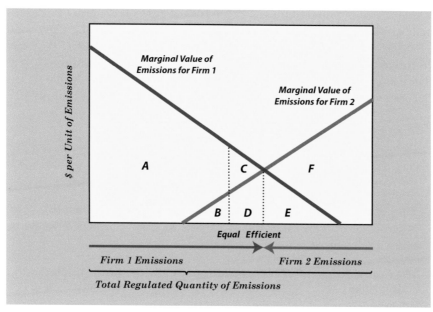

Figure 12.1
The Efficient Allocation of Pollution Rights

These sales of permits will continue until the efficient allocation of pollution is achieved at the intersection of the marginal value curves. After that point, the marginal value of emissions is higher for Firm 2 than for Firm 1. That is, the least Firm 2 would accept for a permit exceeds the most that Firm 1 would be willing to pay for another permit, so trading will cease. During this process, each firm makes decisions based on its true marginal value of emissions—a value critical to the determination of efficient levels of emissions as indicated in Figure 12.1 but difficult to ascertain by any other means.

The initial distribution of permits affects the distribution of funds between the firms. The more permits a firm receives in the beginning, the more it can sell or the fewer it must buy. But this distribution has no influence on the efficiency of the final outcome. Whether permits are all given to Firm 1, all to Firm 2, or there is some intermediate allotment, trades between the two firms are in each firm's best interest until the marginal value to each firm is the same and the efficient distribution is achieved. The same is true when there are many firms: Regardless of the initial allotment of pollution rights, trading will occur until the firms' marginal values of pollution are equal, thus marking the efficient distribution of pollution rights.

This diagram illustrates the marginal value of emissions for each of the two firms. Firm 1's emissions increase from left to right, and Firm 2's emissions increase from right to left. As indicated by the lower, green marginal value curve, Firm 2 receives less value from any given amount of emissions than Firm 1. If each firm is permitted to emit the same volume of emissions, the total value to Firm 1 is the sum of areas A and B, and the total value to Firm 2 is the sum of areas D, E, and F. The efficient allocation is achieved where the firms' marginal values are equal. With the efficient allocation, the value to Firm 1 is $A + B + C + D$, and the value to Firm 2 is $E + F$. The net gain from switching to the efficient allocation is area C.

There are three primary benefits from emissions trading programs:

1 **They allow flexibility and creativity**.
   Permissible emissions levels can be achieved via conservation, new technology, alternative energy sources, or the purchase of permits—whichever is the most feasible and affordable to a particular pollution source.

2 **They allocate emissions rights to those who value them the most**.
   Newer, cleaner firms can sell their permits to firms for whom it would be much more difficult to reduce emissions.

3 **They provide incentives for reductions below the level stipulated by regulations**.
   When emissions can be cut for less than the going price for permits, firms will do so and sell their extra permits for a profit.

Emissions trading programs also face challenges. Policies enforcing emissions targets are only as good as the choice of those targets. Environmental agencies such as the EPA face the contentious task of selecting the quantity of permits to provide. As discussed in Chapter 2, cost-benefit analysis is an available but politically volatile tool for setting standards. Politics, along with morals, emotions, and imperfect information become influential ingredients in real-world policymaking.[5] As these programs scale up, emissions trading at the international level might be less successful than trading within a nation due to monitoring difficulties. And it is important not to let the development of new emissions trading systems distract attention from more fundamental efforts to reduce consumption, conserve resources, and moderate behavior.

Tradable Emissions Rights in Practice

Emissions trading has worldwide applications. Depending on the program, pollution rights in the form of allowances, permits, or credits can be purchased at auction or obtained via one of the following mechanisms:

Banking *occurs when a firm achieves emissions levels below an established standard and is allowed to save or "bank" pollution permits for future sale or use.*

Bubble provisions *effectively allow trading within a firm or area by aggregating all sources of a specific pollutant within a "bubble" and looking only at the total amount of pollution released from the bubble. Increased emissions from one part of the bubble are acceptable if accompanied by a sufficient decrease in emissions elsewhere in the bubble. This provides the flexibility for some operations to emit particularly large amounts of pollution if others are particularly clean.[6]*

Offsets *allow increased emissions in an area that has not met environmental quality standards if they are offset by even larger reductions elsewhere in the same facility or in another facility. Offsets differ from area bubbles in that (1) offsets apply only to environmental nonattainment areas, and (2) the required reduction is larger than the increase. The 2:1 offset sanctions discussed in Chapter 6 are an example of this.*

Netting *allows existing firms to expand their operations and avoid the relatively strict emissions standards for new facilities provided that*

---

5 For perspectives on ethical and emotional arguments, see Chapter 16, Ott and Sachs (2000), and https://helpsavenature.com/what-is-environmental-ethics.

6 See www.epa.gov/archive/epa/aboutepa/bubble-policy-added-epas-cleanup-strategy.html.

their net increase in emissions falls below a set level. Netting differs from bubble provisions and offsets in that it always occurs within a single firm and allows for a net increase in pollution.

An emissions trading scheme called **cap-and-trade** is among those that set a limit on releases of particular pollutants and allocate allowances for portions of that quantity among existing firms that can then trade them on the open market. A cap-and-trade program began in the United States in 1995 to address the acid deposition problem by controlling SO_2 at the national level and NO_x in some regions. The trading of pollution allowances is similar to that of stocks and commodities. Sulfur dioxide allowances are available from allowance brokers, environmental groups, and an annual EPA auction conducted by the Chicago Board of Trade.[7] Each allowance is for one ton of SO_2 emissions. In the 2023 SO_2 auction, the market-clearing price was $0.04, and 125,000 allowances were sold.

Articles 6, 12, and 17 of the 1997 Kyoto Protocol on Climate Change set forth greenhouse gas emissions trading on a global scale.[8] To comply with emissions restrictions under the Kyoto Protocol, countries can reduce emissions in their own country, earn emissions credits by assisting with "clean development" projects in developing countries, earn credits based on land-use changes such as reforestation, or purchase credits. Unit transfers and acquisitions are tracked by the United Nations Climate Change Secretariat in Bonn, Germany.

As a component of the Kyoto Protocol trading plan, the European Union Emission Trading System commenced in 2005 as the largest permit trading scheme of its type in the world. The program covers CO_2, NO_2, and perfluorocarbons and targets a 62 percent reduction in greenhouse gas emissions relative to 1990 levels by 2030. In 2023, the European Union added Emissions Trading System 2, which covers emissions from buildings, road transportation, and small industries not addressed by the existing trading program.[9]

Innovative uses of emissions trading have proliferated. In Australia, where salt pollution by salt and coal mines threatens to make water unsuitable for drinking and irrigation, interstate salinity credit trading has led to substantial cuts in salinity levels. To limit automobile congestion and pollution, Singapore has a permit system for cars: Those wishing to drive can bid on "certificates of entitlement" at twice-monthly auctions. The U.S. Environmental Protection Agency oversees a credit trading program for phosphorus, nitrogen, and sediment releases into waterways as part of the National Pollutant Discharge Elimination

---

7 See www.epa.gov/power-sector/allowance-markets.

8 See https://unfccc.int/process/the-kyoto-protocol/mechanisms/emissions-trading.

9 See https://climate.ec.europa.eu/eu-action/eu-emissions-trading-system-eu-ets/ets-2-buildings-road-transport-and-additional-sectors_en.

System.[10] The United Kingdom's emissions trading scheme currently applies to energy-intensive industries, power generation, and aviation, and will extend to domestic maritime transportation in 2026 and waste incineration in 2028.[11]

Market Incentives and the Endangered Species Act

There are 2,367 species listed as *endangered* (in danger of extinction) or *threatened* (likely to become endangered in the foreseeable future) under the Endangered Species Act (ESA) of 1973. The United States Secretary of the Department of the Interior approves domestic plants, wildlife, and inland fish for the list and creates recovery plans for each species without regard to the cost. The Secretary of Commerce does the same for ocean-going fish and marine animals. When a species is listed, the Secretary must designate areas as protected "critical habitat" for the species. Private landowners are not compensated when the use of their land is limited by critical habitat designation.

The inflexibility of command-and-control policies is less of an issue when it comes to protecting endangered species because there are not a lot of alternative ways to save these plants and creatures. Endangered species are often those that do not adapt well to substitute habitats and do not reproduce readily in captivity. Saving these species may come down to protecting their existing habitat.

This being the case, the influence of incentives remains critically important. Given the threat of land-use restrictions without compensation, the incentive for landowners is to preemptively destroy endangered species to avoid the burden of compliance. According to U.S. Representative Richard Pombo (R-California) and supported by economic research by Lueck and Michael (2003), landowners "frequently act to eliminate habitat, for fear of losing use of their property to federal government regulations" (see www.thecgo.org/research/revision-of-the-regulations-for-prohibitions-to-threatened-wildlife-and-plants/).

Pombo and his colleague Representative Don Young (R-Alaska) proposed the Endangered Species Conservation and Management Act,

(continued)

10 See www.epa.gov/npdes.

11 See www.reuters.com/sustainability/climate-energy/uk-tighten-emissions-trading-scheme-2024-2023-07-03/.

ostensibly to remedy problems with poor incentives. The act would have compensated landowners whose property value decreased by more than 20 percent due to protective action and provided tax incentives for property owners to promote species recovery. As with many bills, however, the fine print elicited detractors. The Sierra Club said the proposed act would "gut" the ESA by permitting the destruction of endangered species habitat, delaying the listing of species, and loosening other standards for protection. The National Wildlife Federation noted that the ESA already has virtually no restrictions on property use for landowners with endangered plants and proposed its own set of incentives, including a conservation easement program that would provide tax breaks to landowners who preserve endangered species habitat. The general consensus is that even with command-and-control regulations, behavior hinges on incentives, and a policy will fail if the trail of incentives does not lead to the policy goal.

Mixed Approaches to Carbon Emissions

China and the United States are the world's first and second largest emitters of CO_2, followed by India, Russia, and Japan. Policymakers are pursuing a mix of market incentives and command-and-control regulations to limit pollution damage. The city of Beijing, China, has experimented with prohibiting owners of private cars from driving on alternating days, depending on whether the last number of their license plate is odd or even. The Chinese government also mandated a boost in renewable energy capacity that included the construction of seven giant wind farms, each with the capacity of more than 16 coal-fired power plants.

China's market-based solutions include incentive programs for commuters who find alternatives to private automobiles, subsidies for methane reduction and reforestation, and formal training programs in market-based environmentalism. The Chinese government has also hired economists to develop incentive programs including the sort of cap-and-trade systems explained earlier.

In the United States, the Department of Energy estimated that the Bipartisan Infrastructure Law (BIL) and the Inflation Reduction Act (IRA) of 2022 would reduce greenhouse gas emissions by 41 percent below the 2005 level by 2030.[12] The BIL authorized up to $108 billion for public transportation systems and environmentally friendly vehicles. The IRA committed nearly $370 billion to carbon-reducing heating and electricity

12 See www.energy.gov/policy/articles/investing-american-energy-significant-impacts-inflation-reduction-act-and.

systems, solar projects, air-monitoring technology, sustainable farming, and low-emissions transportation. Cutting-edge approaches being considered in the United States include a proposal to capture carbon dioxide emissions at the source, send them through pipelines to national forests, and inject them underground for permanent storage.[13]

India's approach to carbon includes adding forests for carbon sequestration. The Forest Conservation Rules were adopted in 2022 with interest in increasing India's forest cover from 25 percent to 33 percent and being carbon negative by 2050.[14] Japan's efforts to reach net-zero greenhouse gas emissions by 2050 include more than 200 emissions-reduction projects with partner countries supported by the Japanese government. Japan provides partners with carbon-reducing technology as part of the Joint Credit Mechanism in Article 6 of the United Nations Paris Agreement on Climate Change.[15]

Punishment and Deterrence

Deterrence via the Legal System

For 10 years, in violation of the Clean Water Act, Chemetco, Inc., allegedly discharged pollutants including zinc, lead, and cadmium into Long Lake, a tributary of the Mississippi River. Investigators say the toxic substances were released through a "secret pipe" from the company's copper smelting plant in southwestern Illinois. The case was investigated by the EPA's Criminal Investigation Division, the FBI, the U.S. Department of Transportation, the State Police, and the Illinois EPA. The U.S. District Court ordered a fine of $3,865,100 and $400,000 in restitution payments by Chemetco. Individuals accused of building or using the secret pipe received sentences of home confinement and paid fines up to $500,000. This case highlights both the risks that some people take to circumvent environmental policy and the many agencies that devote resources to monitoring and enforcement.

Environmental policy loses its influence when enforcement mechanisms are weak. Environmental abuses are difficult to police because toxic releases can happen anywhere at any time. The most vulnerable wilderness areas are often the most remote, which leads to monitoring difficulties with implications for the best approaches to enforcement. This section outlines the theory of optimal deterrence in the context of environmental policy.

---

13 See www.fs.usda.gov/news/releases/usda-forest-service-proposes-rule-facilitate-carbon-capture-and-sequestration.

14 See www.atlanticcouncil.org/blogs/new-atlanticist/why-india-could-play-a-pivotal-role-as-climate-mediator/.

15 See www.mofa.go.jp/files/100256809.pdf.

We have seen that socially optimal behavior is expected when decision-makers internalize the full costs and benefits of their contemplated actions. When illegal dumping or other misdeeds are considered, the expected value of punishment weighs into the compliance decision. The **expected punishment cost** equals

$$(\text{the probability of punishment}) \times (\text{the punishment cost if imposed}).$$

The probability of punishment is itself the product of the probabilities of apprehension, conviction given apprehension, and punishment given conviction. Thus, even if the chance of being apprehended is 80 percent, if half of those apprehended are convicted and half of those convicted pay a $10,000 fine while the rest get a warning, the probability of punishment is $0.80 \times 0.50 \times 0.50 = 0.20$, and the expected punishment cost is only $0.20 \times \$10,000 = \$2,000$.

As described in Chapter 4, risk-averse people feel a burden from the uncertain outcomes of risky behavior. Environmental crimes constitute risky behavior, and the associated risk burden (or risk enjoyment felt by risk-loving individuals) is an added component of the expected punishment cost.

Socially efficient decisions are made when marginal external costs and benefits are internalized. When a firm or household receives all the benefits from, say, illegal dumping, efficiency is achieved if the expected punishment cost equals the marginal external cost. This will cause decision-makers to internalize the marginal external cost and only deviate from policy when the marginal benefit exceeds the social marginal cost. When monitoring is difficult, as is typically the case for environmental misconduct, the low probability of punishment can be made up for with a high punishment cost. Table 12.1 indicates several combinations of punishment probabilities and costs, all of which yield an expected punishment cost of $10,000 for a risk-neutral decision maker.

If the marginal external cost of dumping is $10,000, any of these combinations will result in the efficient amount of dumping. If monitoring is costly, however, *the solution that minimizes monitoring costs is that which imposes the highest possible punishment cost.* A fine of $10 billion

Table 12.1
Punishment Alternatives

Punishment Probability	Punishment Cost	Expected Punishment Cost
1	$10,000	$10,000
1/10	$100,000	$10,000
1/100	$1,000,000	$10,000
1/10,000	$100,000,000	$10,000
1/1,000,000	$10,000,000,000	$10,000

coupled with the minimal monitoring efforts required to catch one in one million criminals would have the same deterrence effect on risk-neutral criminals as a certain fine of $10,000. The difference is that it would be prohibitively expensive to make detection certain by catching every offender. The reduction in monitoring costs is one of the reasons we see large jury awards in environmental crimes, as was the case for Chemetco. The risk burden of uncertain punishment for risk-averse criminals means that, with a one-in-one-million chance of apprehension, there is a fine less than $10 billion (depending on their degree of risk aversion) that would elicit the same behavior as a certain fine of $10,000.

In 1990, the EPA increased the penalties for hazardous waste violations under the Resource Conservation and Recovery Act (RCRA). Some fines were increased by 10 to 20 times their original size by the revised RCRA Civil Penalty Policy. Research by Sarah L. Stafford (2002, p. 294) found that within 3 years of this policy change, the number of violations per EPA inspection began a steady though less-than-hoped-for decline. Controlling for other influences, Stafford also found fewer violations in states where more citizens were members of environmental organizations. Apparently while incentives matter, so do citizens' attitudes toward the environment. Monitoring matters as well. Blundell et al. (2021) find that decreased state-level funding for monitoring and enforcement of RCRA regulations may result in increased environmental degradation.

Excessive Deterrence

The inability to monitor some types of environmental damage can lead to capricious behavior. At the other extreme, policymakers and environmentalists must be alert to the possibility of excessive precautions. One problem with combining private and public remedies is that the expected punishment, compounded by the burden of uncertainty, can exceed the marginal external cost of a particular action. There is often uncertainty about the standard for appropriate precaution—how many tests must be run before a jury would find that PondClear Corp. did enough to establish the safety of its new algae killer to fish and swimmers? With the prospect for uncapped damage awards from juries or unchecked ecoterrorism, there is no limit to the costs that might be imposed.

Excessive deterrence can also result from the combination of regulation and liability. When monitoring is effective, an overlap of legal remedies and regulations may provide production disincentives that are too strong. For instance, a combination of pollution taxes and the threat of litigation for subsequent environmental risks can bring production

Figure 12.2

Excessive

Deterrence

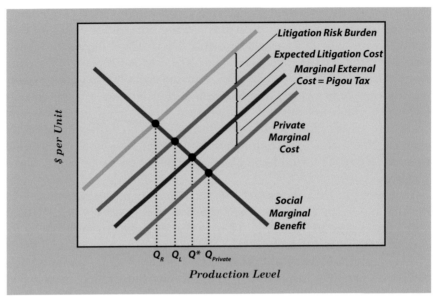

CHAPTER 12

below the efficient level. As illustrated in Figure 12.2, if Pigou taxes are set to bring about the efficient level of production, Q^*, the addition of expected litigation costs will lead to the inefficiently low production level, Q_L, and any associated risk burden[16] or ecoterrorism threat will result in production levels even further below the efficient level, such as Q_R. The moral of the story is that those who establish policy must consider the entire set of incentives so that the *combined* effect elicits the efficient level of care for the environment.

Some economists advocate regulation as an alternative to litigation. W. Kip Viscusi states

> *One cannot rely on tort liability in lieu of regulation because products liability incentives are ill-suited to the task. Not all injured parties file claims, and court awards are far below what is required to promote efficient safety incentives.*[17]

Legislation could stipulate tests and approval procedures for a new product that are *exculpatory*, meaning that their successful completion would release the producer from subsequent litigation over the issue of that product's safety. For example, suppose litigation threats make it prohibitively risky for a company to introduce an energy-saving battery for hybrid electric cars that has a small chance of leaking chemicals and triggering a multimillion-dollar jury award. The satisfaction of a specified

16 As explained in Chapter 4, a risk burden is the largest amount one would be willing to pay just to avoid
   uncertainty about the possible outcomes and know for sure that the actual cost will equal the expected cost.
17 See Viscusi (1991, p. 129) and Viscusi (2013).

set of safety standards—say a hazard warning program compliant with current regulations, the completion of 25 successful crash tests at highway speeds, and approval by a national product safety commission—could constitute an exculpatory level of care. The trick here is to set an appropriate standard for safety. This may again require cost-benefit analysis and political wrangling. However, if a panel of jurors can be relied upon *after* a claim has been filed to determine whether safety checks were adequate, a similar panel should be able to make the same determination *before* the product has been released. This would allow companies to innovate and market new products without fear of extraordinary reprisals.

With increased production comes increased pollution and related externalities. In the absence of litigation and taxes, the firm will produce the inefficiently large quantity $Q_{private}$, as indicated by the intersection of the firm's private marginal cost and marginal benefit curves. A Pigou tax equal to the marginal external cost causes the firm to internalize the externalities and produce at the socially efficient level, Q^*. If litigation over the externalities is an added risk, the marginal cost rises further, and production will occur at the inefficiently small quantity Q_L. Risk aversion on the part of the firm will result in the added cost of a risk burden and the still smaller quantity Q_R.

Activism and Vigilante Justice

Frustrated by what they perceive as insufficient policy standards, monitoring efforts, or punishments, some individuals and groups have decided to take matters into their own hands. Private eco-activism ranging from letters to politicians to throwing soup on paintings in museums[18] to violent acts of ecoterrorism add to the landscape of incentives for socially responsible behavior. In grassroots efforts, consumers have voted with their pocketbooks against everything from Coke (due to allegedly straining water resources in India)[19] to U.S. coal (due to allegedly excessive greenhouse gas emissions).[20] In a recent global survey of 24,000 consumers, 91 percent said they "wanted brands to demonstrate they are making positive choices about the planet and environment more explicitly in everything they do."[21] Some of the more successful boycott efforts have led the fast-food industry away from Styrofoam[22] and coaxed fashion retailers away from fur coats.[23] Greenpeace is

---

18 See www.npr.org/2022/11/01/1133041550/the-activist-who-threw-soup-on-a-van-gogh-explains-why-they-did-it.

19 See www.theguardian.com/world/2017/mar/01/indian-traders-boycott-coca-cola-for-straining-water-resources.

20 See www.courier-journal.com/story/tech/science/watchdog-earth/2017/11/22/climate-change-expert-boycott-drive-kentucky-clean-energy-future/888116001/.

21 See https://sustainablemedia.dentsu.com/home1/home.

22 See www.mcspotlight.org/campaigns/countries/usa/usa_toxics.html.

23 See www.scientificamerican.com/article/impact-activism-on-fur/.

Activists erected this display at Venice Beach, California, to raise public awareness about animal cruelty.

famous for its peaceful but aggressive campaigns to influence policy on nuclear threats, whaling, toxic releases, deforestation, and global climate change, among their other concerns.[24]

Like litigation and prosecution, the threat of responses from activists poses a small risk of large costs but with a vigilante twist. For example, between 1995 and 2010, the Animal Liberation Front (ALF) and the Earth Liberation Front (ELF) fought animal cruelty and urban sprawl with violent tactics.[25] The ELF allegedly burned new homes and businesses, damaged bulldozing equipment, and scrawled "meat is murder" and "if you build it we will burn it" on a restaurant and a home. Such groups have placed metal spikes in trees to make logging dangerous, destroyed labs where genetically engineered crops were developed, and released animals being kept for the harvesting of their fur.[26] While peaceful protests have influenced policy and practice in meaningful ways, violent approaches are largely ineffectual. For governments to cater to criminal actions would be to invite more crime, and so the response to ecoterrorism has been limited almost exclusively to actions against the terrorists.

Summary

There is more than one way to skin a cat and likewise to catch a pig and to control pollution. Differing perspectives complicate the formation of environmental policy, although economic theory provides guidelines.

---

24 See www.greenpeace.org/campaigns/.

25 See www.dhs.gov/publication/st-frg-overview-bombing-and-arson-attacks-environmental-and-animal-rights-extremists.

26 See www.jstor.org/stable/26297171.

There is general consensus among economists that approaches that allow flexibility and provide incentives are critical, as are allocative mechanisms that grant pollution rights to those who value them the most.

When a variety of alternative solutions are allowable and enforceable, flexibility permits firms to meet environmental standards using cost-effective methods appropriate to the age of the facility and the options available. In other situations, command-and-control regulations are appropriate, as when a single effective and enforceable solution exists. Ambient environmental quality standards can be the target for both command-and-control and market-based policies. Technical standards, emission standards, and cap-and-trade programs are all among common attempts to meet air- and water-quality standards.

Efficiency requires that the marginal cost of pollution abatement be the same across firms, which can be achieved with cap-and-trade programs that limit the total releases of specific forms of pollution and allocate tradable permits to pollution sources. The expectation is that those who derive the most benefit from polluting will buy permits from those who value the rights the least and those who can reduce their emissions at the lowest cost. Trade will continue as long as one firm can reduce emissions at a lower cost than another, resulting in an efficient outcome. Tradable emissions permits have become popular worldwide thanks to their flexibility, efficient allocation mechanism, and incentives for continued abatement.

Threats of litigation and prosecution serve as added incentives for compliance with environmental policy. The expected punishment cost of environmental misconduct is the product of the probability of punishment and the punishment cost if imposed. When other incentives for compliance are not present, efficient behavior results from an expected punishment cost equal to the marginal external cost. Monitoring costs can be reduced for a given expected punishment cost by maximizing the punishment and thereby minimizing the probability of punishment necessary to achieve the desired expected cost. When legal threats are combined with risk burdens, environmental taxes, and other pollution disincentives, the result can be an inefficiently low level of pollution and production.

Activism, and in the extreme, ecoterrorism, are private approaches to environmental justice. Violent measures generally get nowhere because a favorable response to violence would invite more violence. On the other hand, grassroots efforts demanding corporate responsibility and peaceful protests by environmental organizations have influenced firm behavior and received widespread support.

Problems for Review

1. Suppose it is your job to protect the environment using the mechanisms of banking, bubble provisions, offsets, or netting as described in the chapter.

   a) *Which of these mechanisms would be appropriate if you wanted to decrease emissions in an environmental nonattainment area?*

   b) *Which of these mechanisms could be applied if you wanted to allow a net increase in pollution?*

2. Consider an electric utility that in 2023 had enough SO_2 allowances to cover its current emissions. If the utility could have reduced its emissions at a cost of $0.02 per ton, should it have done so? Use specific information provided in the chapter to explain why or why not.

3. Suppose that, rather than regulating the amount of pollution, the government required a certain amount of pollution abatement as shown in

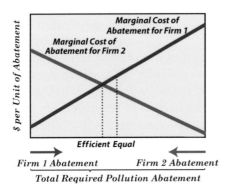

Figure 12.3 *Graph for Problem 3*

Figure 12.3. Duplicate the graph on your homework sheet. On your graph, shade with horizontal stripes the total cost of abatement if each firm must eliminate the same amount of pollution. Shade with vertical stripes the total cost of abatement if tradable permits allow the firms to reallocate abatement responsibilities as desired.

4. Draw two graphs, each with a market demand curve for pollution that has a vertical intercept of $100 and a horizontal intercept of 100 units of pollution. These demand curves can be defined by the equation $Q = 100 - P$, with Q being the quantity of pollution and P being the price per unit of pollution. Assume all pollution costs are external. Label the privately optimal quantity of pollution as Q_0 on each graph. On the first graph, draw a flat-rate Pigou tax of $25 per unit of pollution and label the resulting quantity of pollution Q^*. On the second graph, draw a vertical line at 75 units of pollution representing the total quantity of pollution allowed by an emission standard at the efficient level of pollution. Assume there is no emissions trading program available.

   a) *What level of emissions would result from the Pigou tax if that were the only cost of pollution? (Hint: the equation for demand will assist with this.)*

b) If meeting the emission standard is exculpatory for the industry, which approach is most likely to lead to the efficient level of pollution?

c) Compare the information required to establish a policy that leads to the efficient level of pollution under each approach.

5. In what situations are command-and-control regulations most appropriate? Provide a specific example of a command-and-control policy that is not from this book.

6. The Endangered Species Act stipulates that recovery plans be established for each listed species *without regard to the cost of recovery*. Using economic reasoning, explain a pro and a con of this approach.

7. Suppose the market for an herbicide resembles Figure 12.2. Draw a graph as in Figure 12.2 and label the four cost curves W, X, Y, and Z, in that order, from left to right. Unless told otherwise, assume that firms are risk neutral, there are no Pigou taxes in place, litigation is a risk, and the vertical distance between each neighboring set of cost curves is $2. The marginal external cost per unit of herbicide is also $2. For each of the following characterizations, indicate the per-unit tax or subsidy (if any) that would lead to the efficient production level if all the sellers could be characterized that way:

a) Tree huggers who feel the pain of (internalize) every environmental cost

b) Risk lovers whose enjoyment of taking chances gives them a negative risk burden equal to the expected litigation cost

c) Selfish sellers who care only about themselves

d) Above-the-laws who feel immune to litigation risks.

8. Explain how your school library could implement a congestion pricing scheme to create incentives for optimal use of the facility.

9. Describe an environmental problem that might be solved with activism and the type of activism you would recommend. Discuss the combination of incentives faced by those causing this problem and how you feel they would respond to the incentives. Be specific.

10. Suppose ChemsAreUs Corporation is contemplating the production of Agent Yellow, an update of defoliant Agent Orange. Agent Yellow has a 10 percent chance of causing $1 billion worth of unintended environmental damage and a 90 percent chance of causing no damage beyond the intended defoliation. If Agent Yellow is produced and the damage occurs under existing policies and enforcement levels, there is a 30 percent chance that ChemsAreUs will have to pay a

$3 billion fine. Use calculations of the company's expected payments and expected damage to determine whether its incentives for efficient behavior are just right, inadequate, or excessive if it produces Agent Yellow under each of the following conditions:

a) ChemsAreUs is risk neutral and faces no threat from activists.

b) ChemsAreUs is risk averse, and the mere possibility of paying the stated fine imposes a $10 million risk burden on the company.

c) ChemsAreUs is risk neutral, but in addition to the risk of paying a fine, there is a 50 percent chance that environmental activists will organize a successful boycott that causes the company to lose $40 million.

websurfer's challenge

1. Find a website that describes a command-and-control policy. Discuss whether the policy as set forth is the best way to satisfy the desired goals. Is more flexibility a possibility in this context? If you were the policymaker, explain any changes you would make to the policy and their influence on efficiency.

2. Find a website that describes an act of environmental activism. Do you feel that the activism was appropriate? Did it provide incentives for efficient behavior?

Key Terms

Ambient standards
Banking
Bubble provisions
Cap-and-trade
Congestion pricing
Emissions standards

Expected punishment cost
Netting
Offsets
Peak-load pricing
Technology standards

Internet Resources

Institute for European Environmental Policy:
https://ieep.eu/

Canadian Environmental Policy:
www.canada.ca/en/environ-ment-climate-change/services/canadian-environmental-protection-act-registry/plans-policies.html

United Kingdom Environmental Policy:
www.gov.uk/government/topics/environment

U.S. Department of Energy, Office of Environmental, Health, Safety & Security:
www.energy.gov/ehss/environmental-policy-and-assistance

EPA page on NPDES Regulations:
www.epa.gov/npdes/npdes-regulations

EPA Office of Policy:
www.epa.gov/aboutepa/about-office-policy-op

Further Reading

Arimura, Toshi H. "An Empirical Study of the SO_2 Allowance Market: Effects of PUC Regulations." *Journal of Environmental Economics and Management 44*, no. *2* (2002): 271–289. Discusses the combined influence of tradable SO_2 allowances and public utility regulations on the decision to use lower-sulfur coal.

Blundell, Wesley, Mary F. Evans, and Sarah L. Stafford. "Regulating Hazardous Wastes Under U.S. Environmental Federalism: The Role of State Resources." *Journal of Environmental Economics and Management 108* (July 2021): 102464. https://doi.org/10.1016/j.jeem.2021.102464. Examines the effects of decreased funding on monitoring and enforcement efforts for environmental regulations.

Hansjurgens, Bernd, Ralf Antes, and Marianne Keudel (eds). *Permit Trading in Different Applications*. Oxford: Routledge, 2010. A contemporary outlook on permit trading.

Lueck, D., and J. Michael. "Preemptive Habitat Destruction Under the Endangered Species Act." *Journal of Law & Economics, 46*, no. *1* (2003): 27–60. An empirical study of the influence of the ESA on timber harvests near known endangered species habitat.

Olson, Mancur. *The Logic of Collective Action*. Cambridge, MA: Harvard University Press, 2009. Develops the theory of group behavior and discusses the optimal provision of public goods.

Ott, Hermann E., and Wolfgang Sachs. "Ethical Aspects of Emissions Trading." In *Equity and Emission Trading—Ethical and Theological Dimensions*. Saskatoon: World Council of Churches, 2000. www.wupperinst.org/Publikationen/WP/WP110.pdf. A critique of international tradable emissions permit systems.

Stafford, Sarah L. "The Effect of Punishment on Firm Compliance With Hazardous Waste Regulations." *Journal of Environmental Economics and Management 44*, no. *2* (2002): 290–308. Finds that the EPA's 1991 increase in penalties led to a decrease in hazardous waste violations.

Stavins, Robert N. "What Can We Learn from the Grand Policy Experiment? Lessons from SO_2 Allowance Trading." *Journal of Economic Perspectives 12*, no. *3* (1998): 69–88. Provides a summary of the benefits and challenges of the first 3 years of allowance trading in the United States.

Viscusi, W. Kip. *Reforming Products Liability*. Cambridge, MA: Harvard University Press, 1991. Proposes solutions to the overlapping influence of liability risks and regulation.

Viscusi, W. Kip. "Regulation, Taxation, and Litigation." In *The American Illness*, edited by Frank Buckley. New Haven, CT: Yale University Press, 2013, 270–288. Discusses the interacting incentives of regulation, taxation, and litigation in the context of health risks from tobacco.

Harvested crayfish near Pensacola, Florida

13 Natural Resource Management: Renewable Resources

*P*opulations of Pacific rockfish—the 60 or so species that include *yelloweye, canary rockfish, and bocaccio—are in trouble. Trawling nets dragged across the continental shelf catch these fish and everything else in their paths. About half of the global continental shelf is now trawled and the repercussions of inefficient fishery management are widespread. Falling fish populations affect those who rely on seafood for their diet or their livelihood. The global fishing industry creates hundreds of billions of dollars' worth of commerce each year for fishers, boat and equipment manufacturers, canners, grocers, and restaurateurs. Disturbances in fish stocks also affect species above and below them in the food chain. For example, the expansion of commercial fishing operations in the North Pacific over the last half century coincided with a steep decline in populations of sea lions that eat the same types of fish caught as humans.*

*A **fishery** is an area where fish are caught—a fishing ground— or a firm in the fishing industry. **Bycatch** is fish, birds, turtles, and marine mammals accidentally caught in a fishery and discarded because they have little or no commercial value, or they do not meet regulatory requirements. Bycatch represents an estimated 38 million tons or 40 percent of the overall global catch each year. Under the Sustainable Fisheries Act (SFA) of 1996,[1] a fish species that falls below 25 percent of its natural (unfished) population must be rebuilt to 40 percent of its natural population. With the help of such policies, the percent of measurable U.S. fish stocks*

341

---

1 The SFA amended the Magnuson Fishery Conservation and Management Act of 1976, which was renamed the Magnuson-Stevens Fishery Conservation and Management Act.

DOI: 10.4324/9781003428732-13

*that are overfished has fallen in recent decades from above 40 per-
cent to below 20 percent.[2] In Europe, 40 percent of fish stocks in the
Northeast Atlantic are overfished, as are 90 percent of fish stocks
in the Mediterranean Sea.*

*Possible solutions to overfishing include the closure of fisheries,
quotas, government buyouts of fishing rights, and regulations that
require fishing equipment that minimizes bycatch. Policymakers
must decide how much fishing is ideal and how that goal should
be addressed. This chapter introduces a popular model of natural resource management using fisheries and forests as the central examples. These are both* **renewable resources** *because their
supplies can increase over a reasonable time period. The chapter
also discusses the pros and cons of available policy options and the
policy implications of uncertainty. Chapter 14 follows up with case
studies of depletable and replenishable resources.*

Fishery Management

A Biological Growth Function

Our study of fisheries begins with a discussion of growth in the number
of fish as it depends on the fish population, or **stock**, in a fishery.
Growth rates are relevant to fishing concerns because to be sustainable, fish harvests cannot exceed the growth in fish stocks. The same is
true for the harvest of any renewable resource, making the economics of
fishery management relevant to many resource-management decisions.

M.D. Schaefer (1957) modeled fishery growth as a function of the fish
stock as in Figure 13.1. The horizontal axis measures the size of the
fish stock, and the vertical axis measures the number of additional fish
spawned each period, which we will say is a year. For example, starting
with a quantity of $Stock_A$ fish, the stock will grow by $Growth_A$ fish over
the next year. For simplicity, we will assume that all fish are the same
size and that other variables that affect fish populations, including food
sources, water temperature, and pollution levels, are held constant. The
small increments of growth at low and high stock levels and the larger
growth increments at middling population levels are characteristic of a
logistic growth function. This function represents a common biological
growth pattern: There is exponential growth at low population levels

---

2   See www.fisheries.noaa.gov/national/sustainable-fisheries/status-stocks-2021.

and then declining growth as food supplies dwindle or other constraints become binding.[3]

In the fisheries example, the story can be told this way: At very low population levels, food and space are abundant but mates are scarce. As the population grows, opportunities for reproduction increase while food and space remain negligible constraints, allowing for large growth increments. As the stock approaches its **carrying capacity**—the maximum size supportable by its habitat—incremental growth declines. Fish must compete for food and places to escape predation. When the carrying capacity is reached, and at the opposite extreme where the fish stock is zero, there is no growth, and the population remains constant.

In the absence of outside influences such as fishing, the carrying capacity is a **stable equilibrium**, meaning that if the fish stock deviates slightly from this number, natural forces of growth or death will bring it back to the equilibrium. The arrow pointing to the right in Figure 13.1 indicates that a fish stock with this growth function has a natural tendency to increase until it reaches the carrying capacity. This is the case because, for each size of the fish stock between zero and the carrying capacity, the blue line is above the horizontal axis, showing us that positive growth in the fish stock occurs at that level of fish stock.

Figure 13.1

The Growth of Fish Stocks

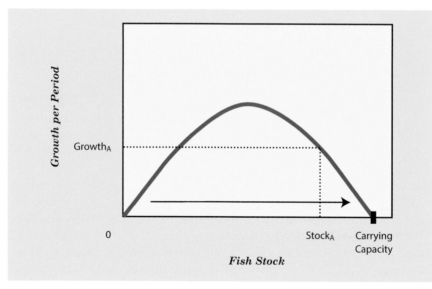

This curve shows the relationship between the size of a fish stock and the growth in that stock per period. Small fish stocks do not propagate rapidly, in part because mates are scarce. With some growth in the fish stock, food and space may still be abundant, and propagation accelerates due to the increased availability of mates. As space and food constraints become binding, growth slows until the carrying capacity of the fishery is met.

3 The equation for a symmetric logistic growth function is $y = K/(1 + \exp(a + b * x))$, where K is the carrying capacity (maximum population size), and a and b are constants that determine the shape and scale of the function.

Natural resource management policies have helped many fish stocks recover from being overfished.

That growth brings the population size further to the right until the capacity is met.

A fish stock of zero is an unstable equilibrium. With zero fish, there can be no growth in the population. If a few fish are introduced, as indicated by the rightward arrow, the population size will grow until it reaches the next equilibrium rather than trending back to zero.

Notice that in Figure 13.1, the slope is always decreasing as the stock increases, meaning that incremental increases in the fish stock lead to smaller and smaller increases in the growth per period and then to larger and larger decreases in the growth per period. In contrast, the slope of a **depensated growth function** increases initially, and then decreases.[4] For Problem 2 at the end of this chapter, you will draw a depensated growth function that begins with a positive slope, meaning that the growth per period initially increases by larger and larger amounts and then follows the pattern of the logistic growth function— increasing by smaller and smaller amounts and then decreasing by larger and larger amounts.

The graph in Figure 13.2 is a special case of a depensated growth function. A **critically depensated growth function** such as this begins with a negative and increasing slope that becomes positive and

4   See Clark (2010).

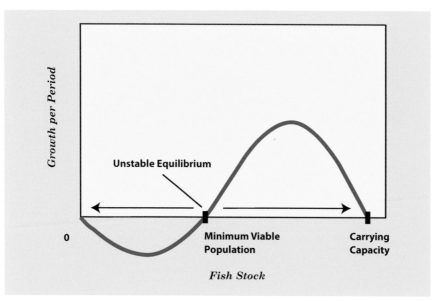

Figure 13.2
*A Critically
Depensated
Growth
Function*

A critically depensated growth function has stable equilibria at each extreme and an unstable equilibrium at the minimum viable population. A fish stock smaller than the minimum viable population cannot reproduce as fast as fish are dying and eventually collapses. Larger fish stocks continue to grow until the carrying capacity is met.

increasing and then positive and decreasing and then negative and increasing. This function may be more realistic than the simplified growth function in Figure 13.1. Rosa et al. (2022, p. 1) warn that "policies neglecting the existence of critical depensation may compromise stock rebuilding objectives and might even result in fishery collapses." Rather than suggesting that growth will occur with any positive quantity of fish, this function allows for a **minimum viable population**, below which there are too few fish to maintain or increase the population. A very small population may be unable to reproduce fast enough to replace fish that are dying. Thus, the minimum viable population is an unstable equilibrium. With slightly fewer fish, the population will fall toward the (in this case stable) equilibrium of zero, and with slightly more fish, the population will increase toward the stable equilibrium at the carrying capacity.

Reverting back to a logistic growth function for simplicity, in Figure 13.3, consider the number of fish that could be caught without reducing the fish stock. For any stock size, an annual catch equal to the annual incremental growth is sustainable. If the stock were at the carrying capacity and fishers removed a catch of size $Catch_A$, the fish stock would decline because the catch exceeds the growth rate of zero at the carrying capacity. With continued catches of size $Catch_A$, the fish stock would decline until the growth equaled the catch size, which occurs with $Stock_A$. Because $Growth_A$ equals $Catch_A$, the fishery could sustain

Figure 13.3

Growth and
Sustainable
Yield

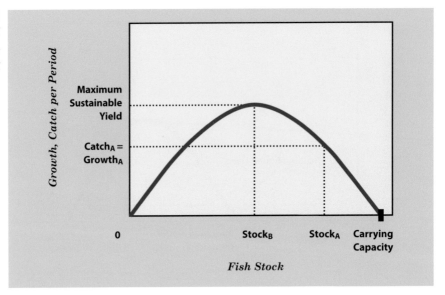

For a given stock size, a catch per period that equals the growth per period, such as Catch$_A$ when the stock level is Stock$_{A'}$ can be sustained indefinitely. If the catch per period is smaller than the growth per period, the stock will grow until the catch equals the growth per period. If the catch per period is larger than the growth per period, the stock size will fall until either the growth per period equals the catch per period or the fishery is eliminated, whichever comes first.

Catch$_A$ without further affecting the size of the fish stock. The height of the growth function indicates the largest sustainable yield at each stock size, making it a **sustainable yield function** as well. The maximum sustainable yield occurs at Stock$_B$. In the next section we will see that the socially efficient yield falls below the maximum sustainable yield.

Sustainable Yield Functions

H. Scott Gordon's model of fishing effort is useful for explaining why the maximum sustainable yield is not the socially efficient yield, and why the socially efficient yield is not chosen by unrestricted users.[5] Gordon modeled fisheries as property with shared ownership and accessibility. Many open-access resources can be modeled similarly, including game animals such as deer, wild plants such as ginseng, and even petroleum.

Gordon assumed that the number of fish caught with any given level of effort is proportional to the fish stock, which makes the yield curves showing the relationship between the fish stock and the catch per period linear. The left side of Figure 13.4 illustrates yield functions of this type. We'll measure the level of fishing effort as simply the number of fishers, although effort could alternatively be measured as the number of fishing

---

5 See Gordon (1954, p. 124).

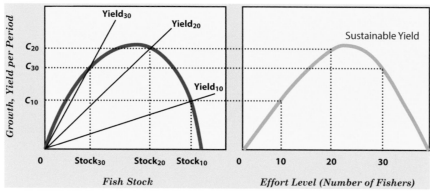

Figure 13.4

Sustainable Yield as a Function of Fish Stocks and Effort

In the graph on the left, straight lines indicate yield as a function of fish stocks for 10, 20, and 30 fishers. The curved line represents growth in the number of fish as a function of the fish stock. The same growth function indicates the number of fish that can be harvested per period without altering the fish stock: The growth function is also the sustainable yield function. The graph on the right indicates sustainable yield as a function of effort rather than as a function of fish stocks.

vessels, nets, and so on. The shape of each yield function indicates that as the fish stock increases, a given number of fishers can catch proportionately more fish. As the number of fishers increases, an even larger number of fish can be caught at each level of fish stock.

The intersection of the yield function for 10 fishers and the growth function indicates the highest sustainable yield for 10 fishers, which is a catch of C_{10} out of a total of $Stock_{10}$ fish. Note that as the level of effort rises from 10 to 20 and then 30 fishers, the sustainable yield increases and then decreases as the increasing yield functions intersect higher and then lower sections of the sustainable yield function.

The right side of Figure 13.4 maps out the relationship between the effort level and the sustainable yield. The yield function for 10 fishers reaches its sustainable yield at C_{10}, so the height of the sustainable yield function is C_{10} at an effort level of 10 fishers, and so on. The sustainable yield function increases and then decreases like the growth function, but the sustainable yield function has effort rather than fish stock on the horizontal axis.

The Choice of Effort Levels

The sustainable yield-effort relationship becomes even more interesting when it is "monetized" by multiplying the sustainable yield for each effort level by the price per fish, which Gordon assumed is constant. The result is a total revenue curve as illustrated in the top graph of Figure 13.5. For example, if 20 fishers could catch 1 million fish per year and each fish sold for 50 cents, the height of the total revenue curve above an effort level of 20 fishers would be 1 million × $0.50 = $500,000.

Figure 13.5

The Gordon Fishery Model

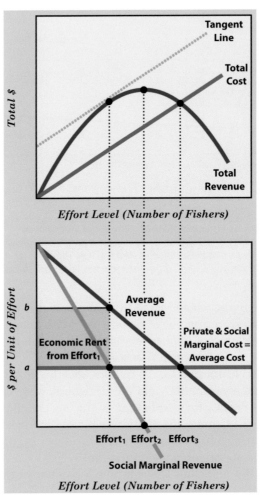

Tangent Line

Total Cost

Total Revenue

Effort Level (Number of Fishers)

Total $

$ per Unit of Effort

b

a

Average Revenue

Private & Social Marginal Cost = Average Cost

Economic Rent from Effort₁

Effort₁ Effort₂ Effort₃

Social Marginal Revenue

Effort Level (Number of Fishers)

In the top graph, the total revenue curve is found by multiplying the sustainable yield for each effort by the price per fish. The total cost comes from the collective opportunity cost of the fishers' time. $Effort_1$ corresponds with the effort level at which total revenue exceeds total cost by the greatest amount, which is at the tangency point between total revenue and the green dotted line.

Individual fishers will enter until $Effort_3$ is reached, at which point their private marginal benefit from fishing—the average revenue received by each fisher—equals the opportunity cost of their time. The societal marginal revenue curve indicates the new fisher's catch minus the losses to the other fishers due to the new fisher's entry. The socially efficient effort level is $Effort_1$, at the intersection of social marginal revenue and social marginal cost. The orange rectangle represents the maximized economic rent gained from $Effort_1$.

With effort level $Effort_2$, fishers could catch the maximum sustainable yield and thereby maximize total revenue. Let's examine the efficiency of that effort level. Gordon assumed the only cost of fishing effort is the opportunity cost of time spent fishing, which is the same for each fisher. This makes the private and social marginal cost of effort constant and equal to the average cost, as shown by the horizontal red line in the bottom graph in Figure 13.5. Each additional fisher lowers the catch of existing fishers due to added competition for the fixed number of fish. This external effect is reflected in the diminishing average revenue, which is notable because the average revenue is each fisher's reward for fishing effort. The social marginal revenue is the additional revenue an additional fisher receives for her effort minus the losses existing fishers experience due to the added fisher.

Note that the social marginal cost of each unit of effort beyond $Effort_1$ exceeds the social marginal revenue, so effort levels in excess of $Effort_1$

are inefficient from society's standpoint. **Economic rent**, the difference between what a factor of production receives and the minimum payment required to keep it in its current use, is analogous to profit in this context,[6] and it is maximized with $Effort_1$. In the top graph, $Effort_1$ corresponds with the tangency point between total revenue and the green dotted line parallel to the total cost curve, which indicates the effort level at which total revenue exceeds total cost by the greatest amount. Which effort level will identical, unrestrained fishers choose? None of the above!

Fishers will continue to join the fishing effort until, at $Effort_3$, the average revenue received by each fisher equals the opportunity cost of their time. Fishers do not stop entering when social marginal revenue equals private and social marginal cost because their *private* marginal gain from entering is the average revenue, which is higher than the social marginal revenue. By entering until average revenue equals private and social marginal cost, fishers are in fact equating their private marginal benefit and private marginal cost. A new fisher receives the same revenue as everyone else—the average revenue. If that average revenue exceeds her marginal cost, she earns economic rent.

The socially efficient effort level is $Effort_1$ at the intersection of social marginal revenue and private and social marginal cost. The orange rectangle represents the maximized economic rent gained from $Effort_1$. If there were sufficient barriers to entry or the fishers could form a successful agreement and restrict output, this rent could be divided among them. With open access, as long as economic rent exists, fishers will enter and compete for it. The open access outcome is $Effort_3$, at which point each fisher receives revenue exactly equal to her opportunity cost, and the economic rent is zero. This dissipation of rents exemplifies the tragedy of the commons introduced in Chapter 3.

The Gordon model is forward looking in the sustainability of its outcomes but static in its lack of discounting. The sustainable yield is dynamically efficient only if the discount rate is zero. Positive discount rates for future costs and benefits imply larger efficient catches today. That is, if we don't care as much about tomorrow as we care about today, we don't want to sacrifice enough today to maintain the sustainable yield for tomorrow. Colin Clark (2010) demonstrated that as the discount rate approaches infinity, the socially efficient effort level approaches the zero-rent level $Effort_3$.

Policy Responses

As the Gordon model suggests, it is challenging to achieve efficient yields at fisheries among other open-access resources. The success of policy solutions rests initially on the ability to enforce policy at all,

---

6 Economic rent is like profit but is a narrower concept, in that it is profit derived from owning or controlling a factor of production.

which requires jurisdiction. Historically, it was the maritime custom for countries including the United States to claim jurisdiction over waters and resources within 3 miles of shore. The United States, European countries, and Canada, among others, then extended fishing restrictions to 12 miles, and now a 200-mile exclusion zone is the norm.

The 1976 Magnuson Act established a 200-mile exclusive economic zone (EEZ) on U.S. coasts in which foreign ships must have a permit and can harvest only highly migratory species or fish in excess supply. The Act stipulated a maximum penalty of $100,000, 1 year in prison, and forfeiture of the unauthorized fishing vessel. Despite claims of jurisdiction, enforcement remains a problem, particularly at the distant ends of territorial waters. Some economists fear that EEZs may actually worsen overfishing problems by promoting the development of domestic fishing fleets in coastal countries, while foreign fleets continue to harvest a bit farther out in international waters.[7]

To the extent that policy is enforceable, several policy initiatives can address inefficient resource-use decisions. If the marginal cost to fishers is increased from a to b in Figure 13.5, fishers will choose the efficient effort level $Effort_1$, at which their new marginal cost equals their private marginal benefit (the average revenue). This increase in marginal cost could be accomplished by requiring fishers to fish farther from the shore or to use smaller nets, boats, or motors, among other command-and-control techniques. However, these methods of artificially increasing the cost of fishing cause a loss of the economic rent otherwise gained at the efficient level of effort. Instead, a fee equal to the difference between marginal cost and average revenue at $Effort_1$, represented by segment ab on the graph, would bring the fishers' marginal cost up to the level b that elicits optimal fishing effort. The revenues from the fee would equal the shaded economic rent otherwise obtained from the efficient effort level, and these revenues could be distributed as the governing body sees fit.

Restrictions could also limit the effort level or the size of the catch. A **total allowable catch** (TAC) is a limit on the overall amount of fish that can be caught in a fishery each year. Similar restrictions on effort are achieved by closing fisheries or limiting fishing to certain days or seasons, which is the case for lobster trapping. As with most regulations of open-access resources, effort and catch limits are difficult to monitor and enforce. The rents earned on the last fish caught leave fishers wanting more, thereby creating temptation.

A simple TAC program also elicits a race to catch fish, as individual fishers each want to harvest as much as possible as quickly as possible before the quota is reached. This competition breeds inefficient expenditures on oversized vessels and related fishing capital. A relatively new

---

7 See McKelvey et al. (2002).

And Overfish We Do

Fish have been an important source of protein since before protein sources set foot on land. Fish consumption by humans has increased dramatically over recent decades. Between 1960 and 2022, annual worldwide fish consumption per capita increased from 9 kg to 20 kg (19.8 lb to 44.0 lb). The worldwide catch was about 178 million tons per year, including more than 87 million tons of fish raised in commercial "aquaculture" farms. Growing fish harvests have strained fishery stocks. One-third of global fish stocks are fished at an unsustainable level. Under the U.S. Endangered Species Act, 2,270 marine species are considered threatened or endangered.

The steelhead trout (*Oncorhynchus mykiss*), for example, has suffered from a barrage of human activity. Beyond commercial, recreational, and tribal harvests, the U.S. Office of Protected Resources lists the following as factors in the steelhead's severe decline: logging, road construction, urban development, grazing, mining, agriculture, hydropower development, flood control, loss of large woody debris in waterways, and artificial propagation (aquaculture). The most surprising of these problems might be aquaculture. In fact, fish "farmed" in aquaculture facilities result in a considerable loss of fish in natural settings. Farmed fish are often fed wild fish for improved health and flavor, and fish that escape from farms threaten wild fish by competing for food, interbreeding, and spreading parasites and diseases.

approach involves **individual transferable quotas** (ITQs). Under this system, total allowable catches are divided into quotas for individual fishers, who may then sell their rights to shares of the TAC or buy larger shares. Being assured a particular share of the total catch, fishers need not use resources inefficiently in a rush to harvest.

The market-based ITQs approach lowers capital expenditures and allows relatively efficient fishers to purchase fishing rights from relatively inefficient fishers, leading to a lower societal cost of catching fish. New Zealand initiated the first major ITQ program in 1986. The United States operates ITQ programs for surf clams and ocean quahogs in mid-Atlantic and New England waters, for wreckfish along the south Atlantic coast, and for halibut and sablefish off Alaska. Australia, Canada, Iceland, Italy, the Netherlands, and South Africa have similar programs. Monitoring issues remain, as does controversy over the appropriate TAC.

Policy Under Uncertainty

Variations in fishery conditions make the shape of growth functions difficult to pinpoint. Even if the best available data support an estimated growth function, as in Figure 13.6, changes in the sea temperature, pollution levels, predators, or food supplies can shift the growth function up or down. If a TAC of $Catch_A$ is instituted and the annual growth is $Growth_B$, the fish stock will grow beyond the intended equilibrium level $Stock_A$. If the annual growth rate is $Growth_C$, a TAC of $Catch_A$ cannot be maintained at any stock level. The fish stock will fall until the stock is eliminated or the TAC is modified. Given a fish stock of $Stock_A$, a precautionary approach would be to set the TAC at $Growth_C$, meaning that fish populations will either remain unchanged or grow.

In a simple model with costless enforcement, *landing fees* based on the weight or value of fish caught, *catch quotas* that limit the size of the catch, and *effort quotas* that limit fishing time or gear all provide similar incentives for efficient fishery management. The complication of uncertainty makes some policies better than others. Martin Weitzman (2002) explains that when setting fishery management policy based on incomplete information, a landing fee can be superior to a catch quota. This is because less information is needed to set an optimal landing fee than to set an optimal quota. The quota belongs at the intersection of social marginal revenue and social marginal cost, whereas the optimal fee is based only on the marginal external cost of fishing. Asgeir Danielsson (2002) used mathematical models to compare the efficiency of effort quotas and catch quotas. He concluded that effort quotas are

Figure 13.6

Uncertainty and the Growth Function

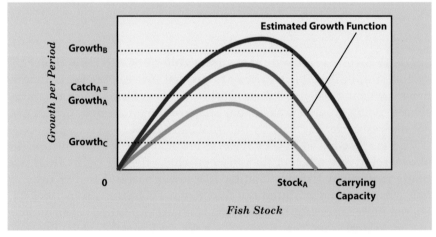

Uncertainty regarding water conditions and food supplies cloud growth function estimates and total allowable catch limits for fisheries. Overly optimistic growth estimates can lead to catch rates that reduce or eliminate fish stocks. Pessimistic estimates can lead to unexpected fishery growth.

best when the price sensitivity (or "elasticity") of demand is high and the growth rate is more variable than the catch per unit of effort. When the price elasticity of demand is low and the catch per unit of effort is more variable than the growth rate, catch quotas are more efficient than effort quotas. These studies demonstrate that policymaking with disregard for the effects of uncertainty can lead to suboptimal results.

Forest Management

After deforestation contributed to Yangtze River flooding that killed 3,656 people and caused $44 billion worth of damage, forestry policy in China took a dramatic turn toward reclamation. Since 1998, logging has been banned in 17 provinces as part of a nationwide Natural Forest Protection Plan. When such policies are enforced, as they were in China, they serve targeted areas well. Without a decrease in forest-product demand, however, logging is simply displaced—China's logging ban led to increased timber imports from Southeast Asia, where problems with flooding and soil erosion worsened. Complementary policies to decrease demand are needed to reduce deforestation on a broader scale. China addressed demand by promoting wood substitutes such as bamboo, a fast-growing member of the grass family. China encouraged the use of "plybamboo" boards and bamboo chopsticks. The government also sought to increase the supply of timber with a Grain-to-Green policy that gave grain subsidies to communities that planted trees.

China, Australia, and Canada are among the countries whose forests are largely controlled by the government. Trees on open-access lands are like fish in open-access waters and elephants in open-access forests: They present problems with monitoring and harvest-effort control. In other countries including the United States, most forest lands are privately owned and can be managed more like cattle and poultry, which changes the focus of renewable resource policy.

Forest managers must decide when to harvest a stock that is constantly growing. Like the growth function for fish in the previous section, the function for timber growth is logistic, but we'll examine the timber story using a graph with different measurements on the axes. In Figure 13.7, time (not stock) is on the horizontal axis, and the total timber volume (not incremental growth) is on the horizontal axis. In this depiction, the growth per period can be observed as the slope of the line, which makes sense because

slope = rise/run = change in volume/change in time = growth per period.

Unsurprisingly, the decision of when to harvest a tree comes down to marginal analysis. We know every activity should continue until the marginal cost equals the marginal benefit. The marginal benefit of

Figure 13.7

*Timber Volume
as a Function
of Time*

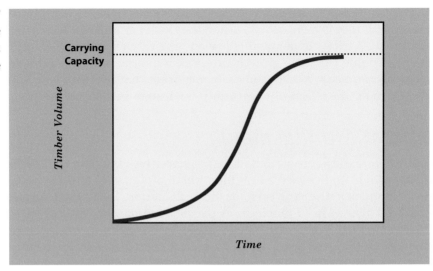

Carrying
Capacity

Timber Volume

Time

This version of the logistic growth function shows total biomass volume (in this case, timber volume) on the vertical axis and time on the horizontal axis. The starting volume for a particular section of forest can be zero; a global function would need to start at the volume of the minimum viable stand of forest. The slope of this line is the growth of timber volume per period. The growth per period increases at an increasing rate initially, then at a decreasing rate, slowing to zero at the carrying capacity of the forest.

maintaining an investment in a tree for another year is that it will grow in volume and be worth more in the next year. The marginal cost is an opportunity cost: The best return that the money invested in that tree could earn in an alternative investment. This might be the amount that could be earned by harvesting the tree and putting the revenue in the bank to earn annual interest rate R.

For the moment, assume the only benefit of waiting to harvest is the increase in timber volume and the only cost is the opportunity cost of lost interest. Later we will consider the possibility that the forest land can be repurposed after the timber is cut, but that is not always the case. Figure 13.8 illustrates a method of sustainable forest management called *selective cutting* that involves continuous use of the forest. Each period, only a fraction of the timber is cut, leaving immature trees to regenerate the forest. This method helps to preserve wildlife habitat, limit erosion, and prevent forest fires (due to forest thinning).

For our simple model we will also assume the tree under consideration can be harvested for free; no spotted owls have made a home in the tree; and no one is fond of thinking about, looking at, or hugging the tree. Such a tree should become a telephone pole or a textbook at the time when the marginal benefit of increased volume no longer exceeds the marginal cost of lost interest, which is when the tree's rate of growth equals the interest rate. The tree's growth rate is found by dividing

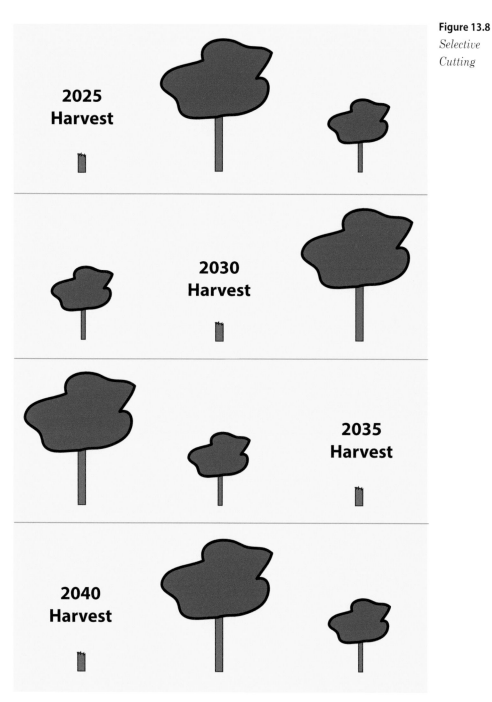

Figure 13.8
Selective Cutting

2025 Harvest

2030 Harvest

2035 Harvest

2040 Harvest

Sustainably managed eucalyptus plantations are the source of paper for many textbooks among other goods. Eucalyptus trees can be harvested after roughly 15 years. Selective cutting as shown here limits erosion and habitat destruction. In this process, only the most mature trees are harvested, leaving younger trees to regenerate the forest.

the change in volume by the total volume, so the *condition for optimal harvest is*[8]

$$R = \frac{\text{change in volume}}{\text{total volume}}.$$

Consider a tree that could sell for $500 today and assume the market price per unit of timber volume is constant. If the tree will grow in volume by 7 percent over the next year and the interest rate is 5 percent, it is better to maintain the investment in the tree for another year. The tree will be worth

$$\$500 + \left(0.07 \times \$500\right) = \$535$$

next year, whereas if the tree is sold and the money is deposited in the bank, the value next year will be

$$\$500 + \left(0.05 \times \$500\right) = \$525.$$

The decreasing slope on the right side of the growth function indicates the growth rate will eventually fall toward zero. At the point when the growth rate equals 5 percent, the money will earn as much in the bank as it will be invested in the tree for another year. In subsequent years, it will earn more in the bank.

Trees must grow for many years before harvest. The simplified harvest decision is effectively a series of comparisons between the growth rate and the rate of return on alternative investments. The optimal year for harvest is the first year in which the tree's growth rate falls to equal the annual rate of return from the next-best investment, as represented by the condition for optimal harvest.

Now let's remove some of the simplifying assumptions and see how the harvest condition changes. We will incorporate the harvest cost, the land value, and the standing value of the forest. The harvest cost is the cost of felling trees and bringing them to market. This cost should be subtracted from the timber value when comparing the returns from maintaining a tree stock with the returns from selling the timber.

---

8 This equation can be derived using calculus. Let t represent time, $V(t)$ represent the volume of the tree as a function of time, and P represent the market price for a unit of tree volume. The present discounted value (PDV) of the tree is $e^{-Rt}V(t)P$. The optimal harvest condition is found by setting the derivative of this equal to zero: $dPDV(t)/dt = e^{-Rt}V'(t)P - Re^{-Rt}V(t)P = 0$. Dividing both sides by $e^{-Rt}P$ and solving for R yields the equation given, where $V'(t)$ is the change in volume and $V(t)$ is the total volume. Note that the price drops out of the equation, as would a term for the initial cost of planting the tree, which has become a "sunk cost" that should be ignored after it is incurred.

Trees often stand on useful land that will provide a stream of net benefits from continued tree harvests or other uses. If a forest will be clear-cut rather than selectively cut, leaving the trees to grow for another year means the land cannot be sold, used as pasture for cattle, or replanted with the next generation of trees. It is appropriate to assess the value of land as the present value of the future net benefits achievable from the land. This land value is forgone for another year if the existing trees are not harvested, and the lost annual return on the land value should be included in the cost of not harvesting in a particular year.

The last consideration is the standing value of the forest itself. As China learned through tragedy along the Yangtze River, the standing value of a forest includes flood and erosion control. It also includes the ability to use the forest to sequester carbon, recreate in, look at, and promote biodiversity. If the forest provides some of the last viable habitat for the endangered black-footed ferret,[9] for example, the standing value is substantial.

To grasp the effects of these considerations, begin by rewriting the simplified condition for optimal harvest in terms of dollars. The total tree volume can be expressed in monetary terms as the timber value, which is simply the total timber volume multiplied by the market price per unit of timber. Likewise, the change in timber value is the change in volume multiplied by the market price per unit of timber. With these changes, the optimal harvest condition becomes

$$R = \frac{\text{change in timber value}}{\text{timber value}}.$$

Multiplying both sides by the timber value gives us

$$R(\text{timber value}) = \text{change in timber value}.$$

The left side is the interest to be gained by harvesting the trees now and putting the money in the bank (assumed for simplicity to be the best alternative investment) for a year. The right side is the gain from waiting another year to harvest.

Now we can add the new items. The sum of money on which interest could be earned if the trees were cut now is increased by the land value and decreased by the harvest cost. The interest rate times the resulting amount indicates the marginal cost of waiting to harvest. The marginal benefit of waiting another year is increased by the annual standing value. With these adjustments, the harvest condition becomes:

9   The black-footed ferret was thought to be extinct in the late 1970s, but a small colony was discovered in 1981. Several hundred now live in captivity, and some have been released into the wild.

$$R(\text{timber value} + \text{land value} - \text{harvest cost}) =$$
$$\text{change in timber value} + \text{annual standing value}$$

The top line is the marginal cost of waiting, and the bottom line is the marginal benefit.

The marginal cost of waiting another year to harvest is the interest that could be earned on the timber value and the land value minus the cost of harvest. The timber value, and therefore the marginal cost, is initially low, increases to a maximum as the trees grow, and eventually decreases as the trees stop growing and are subject to disease, death, and forest fires. The marginal benefit of waiting another year to harvest is the annual change in the timber value plus the annual standing value for recreation, animal habitat, scenery, carbon sequestration, and the like. This marginal benefit will increase and decrease with the growth rate. The intersection of these two lines indicates the optimal rotation period, after which time the marginal cost of waiting another year exceeds the marginal benefit.

Figure 13.9 shows the time-dependent relationship between the marginal cost and marginal benefit of waiting another year to harvest. The marginal cost of waiting moves with the timber value on which interest is forgone, starting low, increasing to a maximum as the trees grow, and then decreasing as the trees stop growing and are subject to disease, forest fires, and death.

The marginal benefit of waiting another year to harvest increases initially and then decreases, driven by the increasing and then decreasing growth rate of timber. This makes sense because the benefit of waiting another year is higher when the tree is growing more quickly. The optimal rotation period, after which the existing trees are harvested and

Figure 13.9

Optimal Forest Rotation

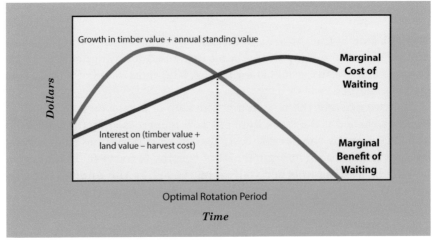

new trees are planted, is found at the intersection of these two lines. After that time, the marginal cost of waiting another year exceeds the marginal benefit.

Changing conditions alter the optimal harvest decision. An increase in the standing value of a forest, as might occur if pollution levels make carbon sequestration by trees more important, will shift the marginal-benefit-of-waiting curve upward to intersect the marginal cost curve at a longer rotation length. An increase in land value or a decrease in harvest cost will shift the marginal cost of waiting curve upward and decrease the optimal rotation length. Only if the interest rate—and therefore the marginal cost of waiting—were zero would harvest occur at or beyond the time when trees stop growing.

Forest management may involve harvest decisions for hundreds or thousands of acres of trees. In a planned forest with an optimal rotation length of t years, a steady flow of timber can be acquired by selectively cutting 1/tth of the forest each year. For example, if the optimal rotation is 20 years, 1/20th of the forest can be harvested and replanted each year. That method provides an unchanging annual harvest even though each tree will stand for t years.

Summary

Fisheries and forests naturally rejuvenate themselves, but commercial harvests often outpace natural growth rates. These and other renewable resources require careful management to attain sustainable yields. When there is open access to fisheries, economic rents are dissipated by fishers who disregard the lost yield their entrance imposes on others. The Gordon fishery model indicates the optimal effort level and the entrance fee that would bring fishers to internalize the full cost of their behavior. In addition to entrance fees, policy options include total allowable catch limits, individual transferable quotas, and limits based on seasons, locations, effort, and equipment.

The Magnuson Act established exclusive economic zones and limits on overfishing in the United States; related policies exist on every continent. Fishery policy is complicated by enforcement difficulties and uncertainty about stocks, growth rates, and yields. As fishing technology has improved and seafood has grown in popularity, the fish stocks of many major fisheries have fallen to dangerously low levels that necessitate a phase of rebuilding.

The decision of how often to harvest forests depends on a comparison of the benefit from another year's growth and the opportunity cost of the money that could otherwise be invested elsewhere. Trees should be harvested when the growth rate of timber volume falls to the level of the best alternative rate of return. Forests will be harvested more frequently given a relatively high opportunity cost of money or land, or a relatively low harvest cost or standing value. The benefits provided by standing trees include carbon sequestration, watershed protection, recreation, and wildlife habitat.

• • • • • • • • • • • • • • • • • •

Problems for Review

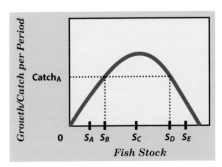

Figure 13.10 *Growth Function for Problem 1*

1. Consider the fishery growth function in Figure 13.10. Suppose $Catch_A$ fish are caught each period for as long as possible. At what level will the fish stock stop changing (because either there are no more fish or the catch equals the growth) if the beginning stock is each of the following:

   a) S_A

   b) S_B

   c) S_C

   d) S_D

   e) S_E

2. The chapter defined a depensated growth function and illustrated a critically depensated growth function.

   a) Draw a depensated growth function that is not critically depensated.

   b) Label any equilibria as stable or unstable.

   c) Explain what the shape of this function means in terms of the growth pattern of the stock.

3. Prickly pear cactus fruit is delicious fresh or in candies or jellies. Table 13.1 shows the number of prickly pears a pear picker can pick per hour in a fictional desert, depending on the number of pickers.

   a) Assuming each prickly pear sells for $1, graph the average and marginal revenue for society at each level of effort (number of pickers).

   b) Given that the opportunity cost of each picker's time is $10 per hour, label the socially optimal number of pickers and the number of pickers that will participate if the desert is an open-access resource. Explain why the pickers will dissipate the economic rent.

4. Look again at the data for Problem 3.

   a) What is the economic rent with the socially optimal number of pickers?

   b) What economic rent results from entry based on private incentives?

c) What entry fee would result in the socially optimal number of pickers?

5. Consider the situation in Problem 3 once more.

   a) Identify two policies that do not involve a fee that could result in the optimal level of effort.

   b) Explain a drawback or challenge with the administration of one of these policies.

6. Other than fish and prickly pears, based on your understanding of the model, what are two other resources the Gordon fishery model could apply to.

7. Suppose policymakers are choosing between a landing fee and a catch quota to limit overfishing. They have good estimates of the marginal external cost of fishing but lack information on the social marginal revenue and the private marginal cost. Which of those two policy options would you recommend to them? Explain your answer.

8. Consider the total growth function for timber shown in Figure 13.7. Assume the money invested in timber could instead go into an alternative investment with a constant rate of return.

   a) Draw a graph that corresponds with Figure 13.7, again with time on the horizontal axis, but with growth per year (not total volume) on the vertical axis.

b) Explain why the optimal harvest is unlikely to occur at the highest point on Figure 13.7.

c) Explain why the optimal harvest is unlikely to occur at the highest point on the graph you drew.

9. The condition for optimal harvest discussed in this chapter is for harvests on private land. Logging is sometimes allowed in national forests, where difference include (1) the harvester cannot sell or repurpose the land regardless of when trees are felled, (2) the harvester may receive little or no part of the standing value of the forest, and (3) there may be competition among harvesters to fell the best trees. Explain how each of these factors affects the marginal cost or the marginal benefit of waiting to harvest.

10. Explain how each of the following would affect the socially optimal rotation interval for trees on private land.

    a) An increase in the demand for real estate

    b) An increase in soil erosion on deforested land due to global warming

    c) Engineers perfect the harvest-cost-saving Super-Axe-Hacker (conceived by Dr. Seuss in The Lorax)

"At the current rate of decline, the last tropical rain forest tree will fall in 2045."

— THE COLUMBUS ZOO

websurfer's challenge

1. Find a website that discusses overfishing issues at a specific fishery.

2. Find a discussion of a specific policy that would affect natural resource management in your region.

3. Find a website that explains what happened to old-growth forests in the Pacific Northwest region of the United States.

Key Terms

Bycatch
Carrying capacity
Critically depensated growth function
Depensated growth function
Economic rent
Fishery
Individual transferable quotas

Minimum viable population
Renewable resources
Stable equilibrium
Stock
Sustainable yield function
Total allowable catch

Internet Resources

American Tree Farm System:
www.treefarmsystem.org

Fisheries and Oceans Canada:
www.dfo-mpo.gc.ca/index-eng.html

USDA Forest Service:
www.fs.usda.gov

Natural Resources Canada:
www.nrcan.gc.ca/forests

U.S. Fish and Wildlife Service:
www.fws.gov

National Marine Fisheries Service:
www.fisheries.noaa.gov

Further Reading

Clark, Colin W. "Profit Maximization and the Extinction of Animal Species." *Journal of Political Economy 81* (1973): 950–960. Identifies the importance of dynamic optimization elements, including discount rates and harvest costs, to the risk of species extinction.

Clark, Colin W. *Mathematical Bioeconomics: The Mathematics of Conservation.* 3rd ed. New York: Wiley, 2010. An introduction to the theory of biological conservation.

Danielsson, Asgeir. "Efficiency of Catch and Effort Quotas in the Presence of Risk." *Journal of Environmental Economics and Management 14*, no. *1* (2002): 20–33. Presents a bioeconomic model of efficient fishery management.

Garrity, Edward J. "Individual Transferable Quotas (ITQ), Rebuilding Fisheries and Short-Termism: How Biased Reasoning Impacts Management." *Systems 8*, no. *1* (2020): 7. https://doi.org/10.3390/systems8010007. Examines the human component of ITQ systems and the challenges of setting TAC levels correctly.

Gordon, H. Scott. "The Economic Theory of a Common Property Resource: The Fishery." *Journal of Political Economy 62* (1954): 124–142. A seminal article on the modeling of fishery resources.

Hartman, Richard. "The Harvesting Decision When a Standing Forest Has Value." *Economic Inquiry 14* (1976): 52–58. A classic discussion of the timing of tree harvests.

McKelvey, Robert W., Leif K. Sandal, and Stein I. Steinshamn. "Fish Wars on the High Seas: A Straddling Stock Competition Model." *International Game Theory Review 4*, no. *1* (2002): 53–69. Models the conflict between foreign distant-water fishing fleets and domestic fleets on either side of EEZ boundaries.

Poudel, Diwakar, Leif K. Sandal, and Sturla F. Kvamsdal. "Stochastically Induced Critical Depensation and Risk of Stock Collapse." *Marine Resource Economics 30*, no. *3* (2015): 297–313. https://doi.org/10.1086/680446. A study of the risks of fish stock collapse due to critical depensation in managed fisheries.

Rosa, Renato, Tiago Costa, and Rui Pedro Mota. "Incorporating Economics Into Fishery Policies: Developing Integrated Ecological-Economics Harvest Control Rules." *Ecological Economics 196*

(June 2022): 107418. https://doi.org/10.1016/j.ecolecon.2022.107418. Results from a case study indicate how important it is to the rebuilding of fish stocks that policymakers understand and acknowledge critical depensation.

Schaefer, M. D. "Some Considerations of Population Dynamics and Economics in Relation to the Management of Marine Fisheries." *Journal of the Fisheries Research Board of Canada* (now called the *Canadian Journal of Fisheries and Aquatic Sciences*) *14* (1957): 669–681. A seminal article on growth functions for fisheries.

Weitzman, Martin L. "Landing Fees vs. Harvest Quotas With Uncertain Fish Stocks." *Journal of Environmental Economics and Management 43*, no. *2* (2002): 325–338. A theoretical comparison between fishing fees and harvest quotas with and without uncertainty about fish stocks.

Most major cities are built around water, the replenishable resource that sustains life.

14 Natural Resource Management: Depletable and Replenishable Resources

*F*rom the great oceans and swamps of prehistoric times came the most popular energy sources of our time. Plants and animals store solar energy from the sun and retain some of that energy when they die. Accumulations of biomass were entombed by layers of mud and silt for hundreds of millions of years, where pressure and heat transformed them into the simpler chains of carbon and hydrogen that make up the fossil fuels we burn today. These resources provide a different type of challenge than deciding when to harvest a tree. With oil, it is no longer a question of whether to wait for the next incremental growth in the resource—any realistic discount rate makes the present value of supplies created in thousands or millions of years virtually zero.

After discussing the oil that fuels our motors, this chapter covers the water that fuels the cells in our bodies. As with motor fuel, water may be the basis for future wars. A panel of scientists told the United Nations General Assembly that, "Conflicts over water will become more common without science-based water diplomacy."[1] On a regional scale, water disputes have already raged between residents of the western United States,[2] Pakistan and India,[3] Egypt and Ethiopia,[4] and others. A worthwhile movie directed by Robert Redford, the *Milagro Beanfield War*, is based on a true story about water disputes. Dispute resolution is the topic

---

1 See www.un.org/pga/77/2023/02/07/press-release-conflicts-over-water-will-become-more-common-without-science-based-water-diplomacy-panel-tells-un-general-assembly/.

2 See https://today.usc.edu/the-water-wars-of-the-future-are-here-today/.

3 See www.express.co.uk/news/world/938329/India-Pakistan-water-conflict-dispute-indus-treaty-punjab-dams-world-war-3.

4 See www.bbc.com/news/world-africa-43170408.

DOI: 10.4324/9781003428732-14

of Chapter 15; here we'll discuss the underlying scarcity of water and ways to allocate it.

To be precise, we call fossil fuels **depletable resources** *because their time frame for renewal is beyond the scope of practical consideration. We saw in Chapter 13 that supplies of renewable resources, such as fish and forests, can increase over a reasonable period of time. Some stocks of nonliving resources, including water, can be replenished within a reasonable period of time, and we differentiate them from resources that increase via biological growth by calling those nonliving resources* **replenishable resources**. *Air and water are among the life-sustaining replenishable resources. This chapter provides case studies of oil as an example of a depletable resource and water as an example of a replenishable resource.*

Oil is a depletable resource because the millions of years needed to form new oil make renewal veritably impractical.

Oil

Oil, like coal, natural gas, metals, gems, and water, is in relatively fixed supply. We lack certainty about the supplies of these resources, although in some cases we can make reasonable approximations.[5] In 1956, M. King Hubbert predicted that U.S. oil production would resemble a bell-shaped curve, reaching what is called **Hubbert's peak** in the 1970s and then declining.[6] This prediction was accurate until around 2010, when high oil prices and new technology sent oil production climbing toward

---

5 See www.opec.org/opec_web/en/publications/340.htm.

6 For a detailed page on Hubbert, see www.hubbertpeak.com/hubbert/. For a graph of U.S. oil production in the lower 48 states, see www.hubbertpeak.com/blanchard/.

a new peak. As with Malthus's predictions about food supplies, market forces and great efforts put off the fate of previous trends.

Hubbert's theory that resource production would follow a bell-shaped pattern has guided useful projections for the availability of many other resources as well, including helium, coal, precious metals, uranium, and natural gas. New discoveries will be made. Unviable supplies will become viable with advancements in technologies and increases in resource prices. For instance, the oceans may become a source of drinking water as desalinization methods improve and water prices rise. But these resources will never grow like trees in the forest, and this knowledge should be reflected in policy decisions. For a resource with a fixed global supply, the incremental and total growth functions are flat, and the sustainable yield is zero. Economic models can factor in discount rates and increasing costs to establish the efficient distribution of these resources over time. Hubbert reminds us that stories have a beginning, a middle, and an end, and it takes deliberate action to prolong the middle and avoid the end as long as possible.

Hotelling's Rule

Harold Hotelling studied the specific relationship between the discount rate (assumed to be equal to the interest rate) and the stream of economic rent (price minus extraction cost) over time. He modeled the extraction of a homogeneous, nondurable resource with a fixed supply, such as oil. His famous conclusion, known as **Hotelling's rule**, was that in competitive equilibrium, the net price (economic rent) of a depletable resource will increase at a rate equal to the discount rate.

If the rent grows at the discount rate, then the present value of the rent stays the same over time. For example, let's say the discount rate is R, the rent in period 1 is $Rent_1$, and the rent in period 2 is $Rent_2$. According to Hotelling's rule, $Rent_2 = Rent_1(1+R)$. Using the present value formula from Chapter 5, the present value of $Rent_2$ is $Rent_2(1+R)/(1+R) = Rent_1$. Given the constant present value of rent, regardless of the timing of extraction, the present value of a depletable resource can be estimated as

$$\text{Present Value} = (\text{Price} - \text{Extraction Cost}) \times (\text{Estimated Volume}).$$

Merton Miller and Charles Upton call this **Hotelling's valuation principle**.[7]

Empirical tests of Hotelling's rule provide mixed results. The prices of depletable resources can be erratic and do not always increase consistently over time. Marginal rent can increase even when prices are

---

7 See Miller and Upton (1985).

decreasing if marginal extraction costs are falling faster than prices. Nonetheless, marginal rents themselves appear not to adhere to steady growth.[8] There are several likely reasons for this. *Known* supplies are not rigidly fixed, and some resources are not homogeneous as Hotelling assumed. Extraction industries are also capital intensive, making it difficult for them to respond quickly to anticipated price changes. And to the extent that supply is insensitive to price (inelastic), changes in demand will cause prices and rents to fluctuate relatively widely, resulting in price and rent volatility.

The Appendix for this chapter explains Hotelling's rule in greater detail and provides a model of oil allocation between two periods.

Transitions

The largest amount consumers are willing to pay for the first unit of a resource per period is called the **choke price**. This is represented graphically as the vertical intercept of the demand curve. The availability of substitutes can reduce the demand and choke price for a resource. Consider the market for gasoline, in which ethanol and hydrogen are among the potential substitutes. Remember that externalities are an important part of energy costs, and the social efficiency of choices about energy depends on whether decision-makers internalize the full costs of their behavior. Previous chapters explained how policies can help users internalize these costs. For example, ethanol and hydrogen are relatively clean-burning fuels and their use is subsidized, whereas excise taxes and per-unit extraction charges called *severance fees* often apply to oil and other fossil fuels.[9]

Gasoline is currently the fuel of choice for automobiles, but consumers can also choose biofuels or electric cars for their transportation needs. The Hyundai Nexo and the Toyota Mirai even offer a hydrogen fuel cell option. For rational, cost-minimizing consumers, the gasoline substitutes cap the willingness to pay for gasoline at the price of the lowest-cost alternative. As the cost of gasoline comes to exceed the cost of one or more substitutes, drivers will hasten their transition to alternative-fuel vehicles.

Decreased reliance on gasoline will result in a lower net marginal benefit of using gasoline in the future and thus a lower marginal opportunity cost of gasoline use today. When the opportunity cost of using something decreases, current consumption increases. The exhaustion of depletable resources thereby accelerates with the promise of affordable substitutes. In the extreme, if it were known that a renewable perfect substitute for gasoline with the same or lower marginal cost would be

---

8   For an inquiry into the empirical relevance of depletable resource theory, see Chermak and Patrick (2001).

9   For example, in Texas there is an oilfield clean-up severance tax of 0.625 cents per standard barrel of crude oil extracted.

Figure 14.1

Switch Points
for Automobile
Fuels

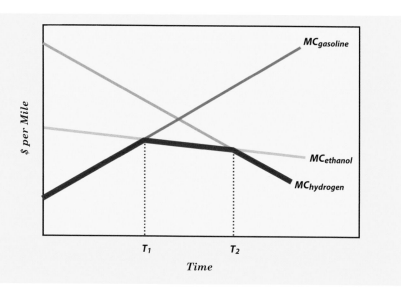

As the marginal cost of a depletable resource like gasoline increases, market forces lead to the development and use of less expensive alternatives. At T_1, ethanol becomes less expensive per mile and drivers switch to the substitute fuel. There is no spike in fuel costs because at the switch point the marginal cost of the two fuels is the same. There may, however, be costs associated with converting vehicles and changing habits that prevent an immediate or complete transition. Likewise, T_2 is the switch point between ethanol and hydrogen because that is when hydrogen becomes less expensive per mile than ethanol.

available next period, the opportunity cost of consuming gasoline this period would fall to zero, and all economically viable stocks of gasoline would be tapped.

If other resources are close substitutes, the transition from one resource to the next is straightforward. Figure 14.1 illustrates hypothetical per-mile marginal cost curves (including extraction costs and opportunity costs) and the switch from gasoline to ethanol to hydrogen over time. Of course, the curves and fuel choices are simplified for the purposes of exposition. In this story, the rising marginal cost of gasoline eventually makes renewable ethanol and hydrogen fuels relatively inexpensive. Technological advancements, including the development of cost-effective ways to separate hydrogen atoms from oxygen atoms in water, make hydrogen the least expensive of these fuels in the long run. When the marginal cost of gasoline reaches the marginal cost of ethanol at T_1, drivers switch to ethanol. In terms of fuel costs, the transition will be smooth because the marginal costs of the two fuels will be equivalent at the switch point. In reality, the switch will not be immediate or costless, due in part to changes that may be necessary in the automobiles we drive. At T_2, hydrogen has become relatively inexpensive, and drivers can be expected to switch fuels once again.

As the marginal cost of a resource increases, market forces lead to the development and use of relatively less expensive alternative resources.

In the case of oil, depending on the type of use, substitutes might include any of the alternative energy sources discussed in Chapter 7. Another alternative to virgin resource use is recycled forms of the resource. Although the oil burned in combustion engines cannot be recycled, motor oil is readily recycled, as are plastics and related petroleum products. The American Petroleum Institute says that one gallon of recycled oil can generate enough energy to power the average home for about 24 hours. Depletable resources range in their ability to be recycled, from single-use resources like natural gas, to durable resources that lose little or nothing in reuse, such as gold and gemstones.

Water

Droughts of increasing length and severity are stressing ecosystems and threatening large and growing segments of the world population with water scarcity. Audrey Azoulay, the director-general of the United Nations Educational, Scientific, and Cultural Organization, gave the following warning about water resources:

> *There is an urgent need to establish strong international mechanisms to prevent the global water crisis from spiraling out of control. Water is our common future and it is essential to act together to share it equitably and manage it sustainably.*[10]

Chapter 6 discusses the value of water and water pollution levels. This section focuses on issues of water availability and allocation.

Many water distributors employ pricing structures that provide no incentive for water conservation.

The major cities were built around waterways when water was critical to transportation; it is still essential to industry, recreation, and the environment. Unlike most resources, for which substitutes are a possibility, switching to an alternative resource is not an option when it comes to water's role in sustaining life. There will be no switch points for water. Management issues involve the efficient allocation of surface water, the sustainable

10 See www.unesco.org/en/articles/imminent-risk-global-water-crisis-warns-un-world-water-development-report-2023.

extraction of replenishable groundwater, and the prudent mining of nonreplenishable groundwater.

Water is the most abundant of the Earth's resources, but having it where it is needed, when it is needed, with tolerable levels of pollution and salt, is a growing problem. Freshwater withdrawals for all purposes average over 1,000 gallons per U.S. resident per day. Elsewhere, 2.1 billion people do not have safe drinking water at home.[11] Ninety-seven percent of the Earth's water contains too much salt for drinking or irrigation. At present, it is not economically practical to remove salt from water for most purposes, although the U.S. Water Desalination Act of 1996 and similar efforts are promoting new technology to reduce desalination costs.[12] Most of the world's freshwater is in the ice caps or in inaccessible underground aquifers.

The *hydrologic cycle*, in which moisture falls to the Earth as precipitation and returns to the atmosphere via evaporation and transpiration from plants, replenishes some of our water supplies. *Surface water* in lakes, rivers, and oceans receives runoff from watersheds. *Groundwater* has accumulated over millennia in underground aquifers of sand, gravel, and fractured rock. About 2.5 percent of the extractable groundwater in the United States is replenishable by percolation (seepage) into aquifers. The remainder is depletable in the sense that after it is "mined," water will no longer be available from that groundwater source. After the water serves its purpose in agriculture, industry, or household use, the hydrologic cycle will distribute the withdrawn water across the mostly inaccessible havens for moisture on the Earth and in the atmosphere. Groundwater can be used sustainably if extraction rates do not exceed replenishment rates. The next section shows how the efficient allocation of depletable groundwater resources can be analyzed like that of oil resources.

Surface Water Allocation

The 100,000 cubic kilometers of surface water in rivers and lakes would be more than adequate to serve our current demands if it were all in the right places. The great variation in precipitation and freshwater storage volume across the world results in ample supplies in some places and shortages elsewhere. The wettest place on Earth, Mawsynram, Meghalaya, India, averages 467 inches of annual rainfall. The driest place on Earth, the Atacama Desert in Chile, averages 0.6 inches of annual rainfall. As is sometimes the case with food supplies, the high cost of transporting large volumes of water makes more equal

---

11 See www.who.int/news-room/detail/12-07-2017-2-1-billion-people-lack-safe-drinking-water-at-home-more-than-twice-as-many-lack-safe-sanitation.

12 See https://abcnews.go.com/US/scientists-find-new-desalinate-seawater-solar-power-study/story?id=105807454.

distribution a challenge aside from issues of ownership. Ambitious ideas include the movement of icebergs[13] and the seeding of clouds with silver iodide crystals or dry ice to stimulate precipitation over relatively dry areas—at the expense of having less rain downwind.[14]

The replenishable nature of surface water makes its management largely an issue of who has the right to what part of the annual flow. The efficient allocation of water between two groups can be studied with a static model similar to the dynamic model used for allocations between two periods. Consider the drought in Taiwan that forced officials to allocate water between farmers for their rice paddies and high-tech computer chip makers for washing chips after they are etched. The width of Figure 14.2 represents the water supply available for allocation within a given period. The purple line represents the hypothetical marginal value curve (or equivalently, the demand curve) for rice farmers. The blue line represents the marginal value of water to the chip industry. Because the quantity of water for industry begins at zero on the far right and increases to the left, the industry marginal

Figure 14.2
The Efficient Allocation of a Fixed Water Supply

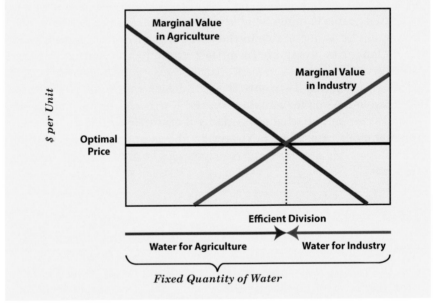

A drought in Taiwan forced difficult water-allocation decisions between the computer chip industry and rice farmers. The graph illustrates hypothetical marginal value (demand) curves for each sector. The efficient division occurs where the marginal value is equal for agriculture and industry. At any other allocation, the movement of one unit of water from the sector with the lower marginal value to the sector with the higher marginal value would yield a net gain for society.

13 See www.theguardian.com/environment/2017/may/05/could-towing-icebergs-to-hot-places-solve-the-worlds-water-shortage.

14 See www.independent.co.uk/news/long_reads/cloud-seeding-weather-control-manipulate-effects-chemicals-climate-change-a8160146.html.

value curve is correspondingly reversed to indicate decreasing marginal value as the quantity of water increases from right to left. The marginal extraction cost of water in this case is assumed to be zero.

The efficient division equates the marginal value of water for agriculture and industry. This occurs at the optimal price level found at the intersection of the marginal value curves as shown in Figure 14.2. If the price were lower, the demand for water would exceed the fixed supply. Higher prices would result in a surplus of water despite the drought. Thus, the market should settle into equilibrium at the optimal price. In reality, a free market for water would also involve private consumers, many of whom could not pay substantial prices for the water they need to survive. Thus, nonprice rationing is a consideration, as was the case in Taiwan. The government took 34,580 acres of rice paddies out of production, set aside sufficient water supplies for household use, and reallocated the remainder to industry. Notice that, if this were not the efficient division of water, the movement of one unit of water from the sector with the lower marginal value to the sector with the higher marginal value would have yielded a net gain for society.

Water Rights

Competing interests in water for agriculture, industry, cities, wildlife, and indigenous groups has intensified scrutiny of water rights. Systems of water rights vary across countries and even within countries such as England and the United States. A complex set of laws covers *use* rights, not ownership rights, and the diversion of water from its natural flow. **Riparian laws**, common in the eastern United States, protect the water use rights of landowners on riverbanks and lakeshores. In other words, the owners of land that adjoins water can use the water. These rights are generally retained regardless of whether the landowners make use of the water. Riparian water use is limited to reasonable, beneficial use. The water cannot be diverted for use elsewhere, and it cannot be stored for later use. When water supplies are inadequate for the demands of riparian users, it is typical for household use to take priority over commercial use, and for riparian use to take precedence over use by *appropriators* as described next.

Prior appropriation laws, common in the western United States, grant rights to those who first used the resource, provided that they are still using it. One such "first in time, first in right" law was upheld by the California State Supreme Court after cities in San Bernardino County tried to divert water away from farming operations for city use. These laws generally permit the trading or sale of use rights and the diversion of unclaimed water for beneficial use.

There is no particular reason to expect either riparian laws or prior appropriation laws to allocate water efficiently. Those whose property abuts water and those who used the water first may value the right to use marginal units of water less than someone else. Consider Figure 14.2 once more. With prior appropriation rights and a negligible marginal cost, the farmers in Taiwan or San Bernardino County or elsewhere will use water until the marginal value in agriculture curve meets the horizontal axis—exceeding the efficient division between agriculture and industry. At that point, the marginal value in agriculture is zero and the marginal value in industry is positive, so a reallocation of some water to industry would benefit industry more than it would hurt farmers. Requirements of continuous use under prior appropriation laws might encourage water use even by those who place little or no current value on it, so as not to lose the right to future use.

Alternatives to rights-based allocation methods are available, for better or worse. Chapter 13 explained the loss of overall benefits to society that occurs when rights are not an issue and many people have unrestricted access to a resource such as fish or the water they swim in. In some countries, the government oversees water allocation, the efficiency of which rests on the government's information, skills, and intentions. Canada, the United States, New Zealand, and Australia have granted special water rights to indigenous populations. The salability of appropriation rights makes market-based solutions another possible solution. There is an online "water rights market" that helps to bring together buyers, sellers, and traders of water rights in the western

Water rights are essential to successful farming in the American West.

United States.[15] Like the markets for fishing rights and pollution rights discussed previously, the market for water rights could serve efficiency goals in at least two ways: It provides an incentive for efficient water use by holders of water rights because they can sell the rights they don't use, and it allocates rights to those who value them the most because those with a lower use value will sell their rights to others with a higher use value.

Domestic Water Use

More than 13 million households draw their water from private wells in the United States. Even so, households are the largest users of public water supplies. The same supplies serve many businesses, although major agricultural and industrial users such as farms and mines often establish their own water supplies.[16] Household use accounts for 7 percent of all freshwater withdrawals. The largest overall use of water in the United States is for power plants, which draw surface water for their cooling systems. The second largest overall use of water is for irrigation, 62 percent of which is *consumptive*, meaning that the water is not returned to its source. This is compared with about 26 percent consumption for household use, 15 percent for commercial and industrial use, and 3 percent for thermoelectric (power plant) use.

Households receive priority in water allocation, which does not mean they enjoy unrestricted use. Municipalities often impose constraints during dry periods, including limits on watering lawns and washing cars. While these restrictions may limit some unnecessary use, they do not allocate water to those who value it the most or in other ways lead to the distributive efficiency economists seek. Water use will be efficient if the price users pay equals the marginal cost to society. Users will then withdraw water only as long as their marginal benefit exceeds the societal marginal cost. Some pricing mechanisms can lead to efficiency, yet the most common pricing structures are not the most efficient. Here are several price structure options:

Flat rate: A fixed amount paid per unit of water used

Flat fee: A fixed amount paid per period regardless of water use

Decreasing block: A per-unit price that decreases as water use increases

Increasing block: A per-unit price that increases with water use

Average cost: A flat rate equal to the average cost of providing water

15 See https://venturewell.org/water-market-mammoth-trading/.

16 See www.usgs.gov/mission-areas/water-resources/science/water-use-united-states.

Peak-load: The price increases at times of shortage

Marginal cost: The price reflects the marginal cost of providing water

College residence halls typically charge a flat fee for utilities including water, as do some cities. A flat fee does not encourage efficient use because, after paying the fee necessary to obtain the first unit, the marginal cost of water to the customer is zero and not the positive marginal cost to society. Paying a flat fee for water is like going to an all-you-can-eat restaurant: There is no incentive to stop consuming until the marginal benefit of another unit is zero.[17] Although they are among the most common price structures, flat rate, decreasing block, and average cost prices also do not reflect the increasing marginal cost of providing water.

Increasing block, peak-load, and marginal cost pricing systems provide improved incentives for conservation and efficiency. The price of water increases as more is used or less is available, more closely mirroring the social marginal cost. Inefficiencies still exist because users who are inexpensive to serve effectively subsidize those who are relatively expensive to serve: Nearby users subsidize distant users and large users subsidize small users. Such complexities make it difficult to equate price and marginal cost. Ironically, Luby et al. (2018) found that many cities with problematic water scarcity have relatively low water prices. Charles Howe (2005) found that increasing block structures can create prices that resemble the true marginal cost of water and thereby promote efficiency.

The sustainability of water use can also be improved through water recycling. Modern wastewater treatment plants release water that is clean enough to drink. Although consumers may find the thought of drinking recycled water unsavory, it is commonly used for industrial cooling, landscape irrigation, and groundwater replenishment. These

Figure 14.3

Water Pollution Limits the Usable Supply

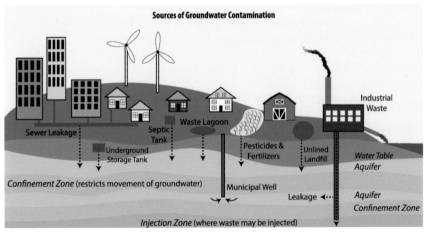

Sources of Groundwater Contamination

Industrial Waste

Sewer Leakage

Septic Tank

Waste Lagoon

Underground Storage Tank

Pesticides & Fertilizers

Unlined Landfill

Water Table Aquifer

Confinement Zone (restricts movement of groundwater)

Municipal Well

Leakage

Aquifer Confinement Zone

Injection Zone (where waste may be injected)

17 Flat fees are sometimes justified by the cost of installing meters to measure household water use.

efforts are important because existing water supplies are vulnerable. Waterways and aquifers are threatened by point and nonpoint source pollution worldwide. Pollution can jeopardize water supplies even where the total volume of water is large. Figure 14.3 illustrates some of the many sources of groundwater contamination that can make existing freshwater supplies unusable.

The Hidden Trade of Water

The increasing scarcity of water in many parts of the world makes water exports unpopular and, in many cases, illegal. It is more acceptable to export goods, even if they are produced using a lot of water. Trade in water-intensive goods is also more efficient. For example, it is easier to transport an orange than to transport the 92 liters of water needed to produce an orange. And instead of importing the 1,000,000 liters of water a dairy farm needs to produce enough milk for 200 kilograms of cheese, a country can simply import 200 kilograms of cheese.

Trade in water-intensive goods, rather than in the water needed to produce those goods, obscures the water loss such exports embody. This includes the 3,800 liters of water needed for the cotton to produce each pair of jeans, the 1,950 liters of water needed for each 4 ounces of chocolate, and the 1,800 liters of water needed to grow 4 ounces of almonds.

Laws meant to protect water are insufficient if they neglect trade in water-intensive goods. Consider the state of Arizona, which has problematic water shortages and prohibits water exports. Most areas of the state do allow farms to use all the water they need, so farms in Arizona grow alfalfa, a water-intensive crop, and export it to other countries as cattle feed.

The scarce natural resources embedded in exports extend beyond water. Examples include the petroleum in England's fertilizer exports, the lithium in Australia's battery exports, and the iron ore in Japan's steel exports. Governments face the dilemma of how to limit the repercussions of exports. One approach would be to broaden limits on water exports to include water-intensive goods. That way, a cap on water exports would apply both to water in tankers and water in, say, hazelnuts, which require 1,200 liters of water per 4 ounces.

Another approach would be a severance tax, meaning a tax on the extraction of natural resources such as water, petroleum, lithium, and

(continued)

Summary

Oil, among other fossil fuels and mineral resources, is classified as depletable because its time frame for renewal is beyond the scope of practical consideration. There is no positive sustainable yield for depletable resources, making their management a matter of when to exhaust economically viable supplies. A two-period model shows that the dynamically efficient use of these resources is accelerated by higher current net benefits, higher discount rates, lower future net benefits, and the availability of reasonable alternatives. Harold Hotelling predicted that producers will extract stocks such that the equilibrium economic rent from a depletable resource increases over time at a rate equal to the discount rate. This implies that the value of a mineral reserve is equal to the current net price multiplied by the volume of extractable reserves.

Water is a life-sustaining replenishable resource, meaning that although stocks do not grow over time, some can be replenished within a reasonable time period. Water for consumptive uses is not returned directly to its accessible source but returns to the Earth through the hydrologic cycle in largely inaccessible forms—primarily as seawater. Fresh surface water is replenishable, but it is often not where it is needed, it is vulnerable to pollution, and it is a small fraction of the total water supply. Only 2.5 percent of extractable groundwater is replenishable; the remainder accumulates in aquifers created over millions of years and is depletable.

The efficient allocation of water occurs when the net marginal value of water is equivalent across users. Some water allocation is under the purview of governments. Some regions, including the eastern United States, have riparian laws that protect the water use rights of landowners along waterways. In other regions, prior appropriation laws dominate, meaning that those who first used water from a source and continue to do so will retain the right to that use. Appropriation rights are now marketable in the western United States and elsewhere, which allows existing users who place a relatively low value on the rights to sell them to those who would gain the most from them. Efficiency in domestic water use can be encouraged with pricing schedules that resemble the marginal cost of use, and peak-load pricing that discourages use during periods of excess demand.

Problems for Review

1. Suppose the current price of iron ore is $100 per dry metric ton unit and the extraction cost is $55 per dry metric ton unit. Given Hotelling's valuation principle and constant extraction costs, what is the total value of a stock of 1 million dry metric ton units of extractable iron ore?

2. Classify the following as depletable, renewable, or replenishable resources:

   a) sheep

   b) water

   c) coal

   d) bamboo

3. Water is scarce in many parts of the world.

   a) Do you anticipate future wars over water? Why or why not?

   b) What do you see as the most realistic policy approach to minimizing future conflicts over water?

   c) Would you be in favor of water rationing in your area? Why or why not?

4. True or false and explain your answer: Riparian water rights are more efficient than prior appropriation rights.

5. Figure 14.2 shows the marginal value of water in industry and agriculture, net of any extraction costs, which are assumed to be zero in the example. Explain how the following changes would affect the curves in Figure 14.2 and the optimal allocation of water between the two uses.

   a) Water has a positive marginal extraction cost

   b) Computer chips demand increases

   c) The supply of water increases

6. Draw a graph with the price per unit of water on the vertical axis and the quantity of units purchased on the horizontal axis. On the same graph, draw the relationship between the price per unit (the marginal cost to the customer) and the quantity for the following plans:

   a) A flat fee

   b) A flat rate

7. Draw a new graph with the price per unit of water on the vertical axis and the quantity of units purchased on the horizontal axis. On the same graph, draw the relationship between the price per unit (the marginal cost to the customer) and the quantity for the following plans:

   a) A decreasing-block plan under which the price decreases after each 10-unit increase in usage.

   b) An increasing-block plan under which the price increases after each 10-unit increase in usage.

8. Consider the plans for which you drew graphs in Problems 6 and 7.

   a) Rank these plans in order from best to worst for promoting socially efficient resource use.

b) Which of these plans most closely resembles the pricing plan you face for water?

c) Which of these plans most closely resembles the pricing plan you face for oil?

d) Why do you suppose a more socially efficient pricing plan is not in place for water or oil?

9. True or false and explain your answer: If all water use were 100 percent nonconsumptive, the only major problems regarding the world water supply would involve allocation and distribution.

10. The chapter explains how some countries are effectively exporting water without sending any liquid overseas.

a) Explain how that is done.

b) Identify two policy approaches to this problem.

The last problem draws on information from the Appendix.

11. Indicate whether the following statement is true, false, or uncertain, and illustrate your answer using a set of graphs similar to those in Figure 14.4: If the marginal cost of extraction increases over time and the marginal benefit curve is the same for each period, then the undiscounted net marginal benefit decreases over time and the resource price must rise in order for marginal rents to rise.

websurfer's challenge

1. Find a website for a firm or association in a depletable resource industry that has some mention of the scarcity of its resource. Evaluate the even-handedness of the statement.

2. Find a website that provides water resource data for your state or country. Summarize the condition of the water supply in a few sentences.

The Lucky Peak Power Plant in Idaho collects energy from the Boise River as it flows through a dam.

Key Terms

Choke price
Depletable resources
Hotelling's rule
Hotelling's valuation principle

Hubbert's peak
Prior appropriation laws
Replenishable resources
Riparian laws

Internet Resources

American Water Resources Association:
www.awra.org

American Water Works Association:
www.awwa.org

Canadian Water Resources Association:
www.cwra.org

Chartered Institution of Water and
Environmental Management, London:
www.ciwem.org

Department of Water Resources, India:
https://jalshakti-dowr.gov.in/

Energy Information Administration:
www.eia.doe.gov

IMF Publication on Peak Oil:
*www.imf.org/external/pubs/ft/
fandd/2021/06/the-future-of-oil-arezki-and-
nysveen.htm*

Ministry of Water Resources,
P.R. China:
www.mwr.gov.cn/english/

State of the Environment Report,
South Africa:
https://soer.environment.gov.za/soer/

U.S. Geological Survey Water
Resources Division:
http://water.usgs.gov

Further Reading

Anderson, Soren T., Ryan Kellogg, and Stephen W. Salant.
"Hotelling Under Pressure." *Journal of Political Economy 126*, no. *3*
(2018): 984–1026. https://doi.org/10.1086/697203. Examines a modification of Hotelling's rule applied to net revenues from oil in Texas.

Chermak, Janie M., and Robert H. Patrick. "A Microeconometric
Test of the Theory of Exhaustible Resources." *Journal of Environmental Economics and Management 42*, no. *1* (2001): 82–103. An
examination of depletable resource theory which finds, contrary to
many studies, that the theory may apply to reality.

Debaere, Peter. "The Global Economics of Water: Is Water a Source of
Comparative Advantage?." *American Economic Journal: Applied Economics 6*, no. *2* (2014): 32–48. Finds that water provides a comparative
advantage and that countries with more water export relatively water-
intensive products.

Ferreira da Cunha, Roberto, and Antoine Missemer. "The Hotelling Rule in Non-Renewable Resource Economics: A Reassessment."
Canadian Journal of Economics 53, no. *2* (2020): 800–820. https://doi.
org/10.1111/caje.12444. Discusses appropriate applications of, and mod-
ifications in, Hotelling's contributions in light of empirical findings.

Hotelling, Harold. "The Economics of Exhaustible Resources."
Journal of Political Economy 39, no. *2* (1931): 137–175. The seminal
article on resource depletion that sets forth Hotelling's rule.

Howe, Charles W. "The Functions, Impacts, and Effectiveness of
Water Pricing: Evidence from the United States and Canada." *Interna-
tional Journal of Water Resources Development 21*, no. *1* (2005): 43–54.
An authoritative overview of the influence of water pricing structures.

Hubbert, M. King. "The Energy Resources of the Earth." *Scientific
American 225*, no. *3* (September 1971): 60–70. A discussion of Hubbert's
peak, with forecasts of the depletion rates of oil, coal, and other energy
sources.

Luby, Ian H., Stephen Polasky, and Deborah L. Swackhamer.
"U.S. Urban Water Prices: Cheaper When Drier." *Water Resources
Research 54*, no. *9* (2018): 6126–6132. https://doi.org/10.1029/
2018WR023258. A study of water pricing and perverse incentives,
finding that cities with more water scarcity often have relatively low
water prices.

Miller, Merton H., and Charles W. Upton. "A Test of the Hotelling Valuation Principle." *Journal of Political Economy 93*, no. *1* (1985): 1–25. Introduces a method of valuing depletable resource stocks based on Hotelling's rule and studies the method's empirical validity.

Osmundsen, Petter. "Dynamic Taxation of Non-Renewable Natural Resources Under Asymmetric Information About Reserves." *Canadian Journal of Economics 31*, no. *4* (1998): 933–951. A rigorous analysis of optimal regulation of resource extraction that considers more complexities than the simpler model in the text.

Renzetti, Steven. "Evaluating the Welfare Effects of Reforming Municipal Water Prices." *Journal of Environmental Economics and Management 22*, no. *2* (1992): 147–163. Uses a simulation to estimate aggregate consumer surplus under various water-pricing plans.

Renzetti, Steven. "Municipal Water Supply and Sewage Treatment: Costs, Prices, and Distortions." *Canadian Journal of Economics 32*, no. *3* (1999): 688–704. Compares average prices with marginal cost as a measure of efficiency and estimates the welfare loss from overconsumption.

Tsur, Yacov. "Optimal Water Pricing: Accounting for Environmental Externalities." *Ecological Economics 170* (April 2020): 106429. https://doi.org/10.1016/j.ecolecon.2019.106429. Suggests an improvement in water allocation via prices that reflect the value of water that is left in waterways and the scarcity of recycled water.

Appendix

Intertemporal Allocation and Hotelling's Rule

Allocation Between Periods

The model introduced in Chapter 8 to describe the efficient allocation of depletable resources between two periods can be applied to the allocation of oil supplies between the present and the future. The object is to decide how much oil to use in the first period and how much to conserve for use in the second period. With reality necessitating a series of decisions between use now and use later, the two-period model is a reasonable simplification of dynamic allocation.

The solid lines in the top graph of Figure 14.4 represent the marginal extraction cost and marginal benefit in a hypothetical oil market. The vertical distance between these two lines is the net marginal benefit for period 1, shown in the bottom graph decreasing from left to right as consumption increases. If the first period were the only period as in a static model, or the last period that anyone cared about (implying an infinite discount rate and discounted period 2 benefits equal to zero), then consumption would continue up to Q_A, at which point the net marginal benefit (NMB) in period 1 is zero. In a static model, Q_A is the competitive equilibrium quantity at which marginal extraction cost equals marginal benefit and period 1 welfare is maximized.

The bottom graph depicts the two-period model. The width of the horizontal axis represents the fixed supply of oil. Consumption in period 1 starts at zero on the left and increases to the right; consumption in period 2 starts at zero on the right and increases to the left. The NMB curves for periods 1 and 2 represent the decreasing net marginal benefit of consuming oil. NMB_1 is the difference between marginal benefit and marginal extraction cost from the top graph, and NMB_2 is the difference between period 2 marginal benefit and marginal extraction cost.

The two-period model incorporates the opportunity cost of using resources now, which is that they can't be used later. The discounted NMB_2 is the present value *in period 1* of period 2 benefits, calculated at each consumption level as $(NMB_2)/(1 + r)$, with r being the discount rate. The solid discounted NMB_2 line ascending from left to right is sometimes called the *in situ value*, meaning the value of the resource if left in its original place. As the value forgone by consumption in period 1, the in situ value of each incremental unit is the marginal opportunity cost of present use. After the intersection of NMB_1 and discounted NMB_2 at quantity Q_B, the opportunity cost of using another unit of oil in period 1 exceeds the present value of using it in period 2, and consumption

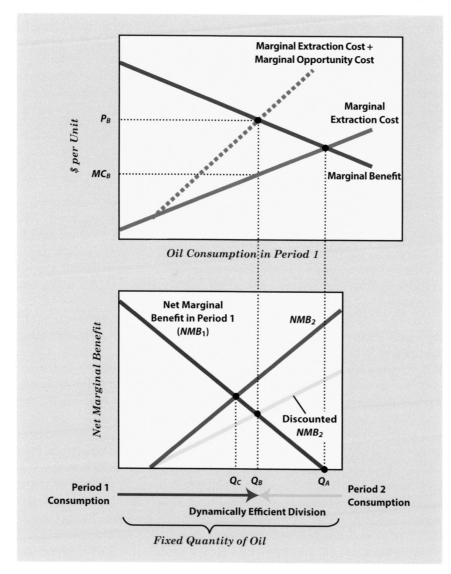

Figure 14.4

Oil Now or Then

should cease. Quantity Q_B thus represents the dynamically efficient allocation of oil to the present; the rest should be saved for future use.

The solid lines in the top graph show the period 1 marginal benefit and marginal extraction cost. The difference between these two values is the net marginal benefit in period 1, as shown in the bottom graph. In a static model with perfect competition, the equilibrium would occur at the intersection of marginal extraction cost and marginal benefit. The corresponding quantity Q_A maximizes the net marginal benefit in period 1. When two periods are considered, every unit consumed in period 1 represents a unit that cannot be consumed in period 2. Period 2 consumption increases from right to left on the bottom graph as period 1 consumption decreases. The present value of the opportunity cost of period 1 consumption is the discounted net marginal benefit in period 2. The dynamically efficient outcome equates the present value of net marginal benefits in each period, which occurs when Q_B is consumed in period 1 and the remainder in period 2. This corresponds with the intersection of marginal benefit and marginal extraction cost plus marginal opportunity cost, as shown by the dotted red line in the top graph.

Hotelling's Rule Explained

It is apparent from the two-period model that, despite the assumption of identical net marginal benefit curves for each period, progressively less of a depletable resource should be used over time if the discount rate is positive. The higher the discount rate, the lower the present discounted value of future use, and therefore the more that is used in period 1. Let's consider the logic behind this important relationship and how it fits into our model.

Firms want to maximize the present value of the economic rent they gain from control of a scarce resource. They do this by equating the present value of marginal rent (rent from the last unit sold) in each period. Suppose that with a given extraction schedule, the marginal rent this period is $Rent_1$, and the marginal rent next period is $Rent_2$. With a discount rate of r, the present value of next period's marginal rent is $Rent_2/(1 + r)$. If the present value of leaving another unit until next period is greater than that unit's use value this period, that is, if $Rent_2/(1 + r) > Rent_1$, then that unit should be left for extraction next period. The logic is the same looking across any number of periods: The units should be allocated to equate the present value of marginal rent from each period. If this value is larger in one period than in some other period, resources should be allocated from the low-rent period to the high-rent period until there is no longer gain from reallocation.

Going back to the simplified two-period model, the rent-maximizing allocation occurs when

$$Rent_1 = \frac{Rent_2}{1+r}.$$

or equivalently,

$$Rent_1 (1+r) = Rent_2.$$

This is Hotelling's result—that marginal rent in each subsequent period (in this case $Rent_2$) should equal marginal rent in the previous period ($Rent_1$) plus growth at the discount rate. This condition holds in the model in Figure 14.4.

Looking at the top graph, we know that, despite the assumption of competition in this market, firms extracting scarce oil will produce less than the quantity that equates the marginal extraction cost and the marginal benefit because of the added opportunity cost of forgone period 2 sales. The competitive industry supply curve is the vertical sum of marginal extraction cost and marginal opportunity cost, represented by the thick dotted line. The equilibrium quantity is found where this aggregation of marginal costs equals marginal benefit, at Q_B.

At equilibrium, period 1 marginal rent, $Rent_1$, equals the difference between P_B and MC_B, as shown in the top graph. These levels of price and marginal cost differ by the discounted NMB for period 2, which defines the marginal opportunity cost of use in period 1. In other words, in equilibrium, $Rent_1 = NMB_2/(1 + r)$. Note also that in each period, $NMB = MB - MC_{extraction}$ and $Rent = P - MC_{extraction}$, and because $MB = P$ at equilibrium, NMB will equal $Rent$ at equilibrium as well. Thus, $NMB_2/(1 + r)$ is equivalent to $Rent_2/(1 + r)$, and at equilibrium, the equality of $Rent_1$ and $NMB_2/(1 + r)$ corresponds with the equality of $Rent_1$ and $Rent_2/(1 + r)$ as theorized by Hotelling.

With a constant extraction cost, the rent from every unit within a period will be the same, and the present value of rent from all periods will be the same, according to Hotelling's rule. A useful application is that the value of a resource is simply the current net price multiplied by the total extractable volume.

For a resource with no extraction cost, the marginal rent equals the price, and Hotelling's rule implies that price will increase at the discount rate. A positive extraction cost implies price increases at less than the discount rate. For example, with an initial price of $30, an extraction cost of $10, and a discount rate of 5 percent, marginal rent in the first period is $20, and we would expect a 5 percent ($1) increase in marginal rent in the second period. With a stable extraction cost, this $1 marginal rent increase represents a $1/$30 = 3.33 percent price increase. If the demand curve does not change over time, increasing prices are achieved only by decreasing supply (extraction) over time.

The dynamically efficient decrease in exploitation described in the two-period model and implied by Hotelling's rule is accelerated by increases in the extraction cost or the discount rate. A higher extraction cost in period 2 decreases NMB_2, and a higher discount rate decreases $NMB_2/(1 + r)$, thereby lowering the opportunity cost of consumption in period 1. It is reasonable to expect the extraction cost to rise over time as the most readily available supplies are tapped and smaller, deeper, more remote sources must be sought. For coal, ores, and similarly heterogeneous resources, materials of the highest quality will be removed first, followed by materials that are more costly to remove and refine. On the other hand, new discoveries and improved extraction technology may cause costs to decrease. In the long run if not sooner, however, easily accessible supplies will be consumed, and costs can be expected to rise.

Environmental Dispute Resolution

Environmental initiatives may be socially efficient, but they often face inertia because progress isn't cheap. It takes a financial commitment to clean up dumpsites, limit commercial fishing, forego development, and purchase emissions-reduction equipment. To complicate matters, these environmental efforts benefit many at the expense of a few, and the few have a monetary incentive to resist the efforts. When the few cry foul, environmental disputes are born. More broadly, disputes arise over

- *What should or should not be done*

- *Who should pay for it*

- *Who owns or controls what resources*

- *What constitutes compensable environmental damage*

- *How quickly progress should be made*

- *What steps are necessary to ensure endangered species viability*

and the list goes on. It is common for disputes to arise over compliance with the environmental legislation discussed in this text, including the Clean Air Act, the Clean Water Act, CERCLA, and the Endangered Species Act. The importance of environmental dispute resolution is clear from the large expenditures made to resolve related disputes, the law firms and institutes devoted to

393

DOI: 10.4324/9781003428732-15

environmental and natural resource disputes, and the vast litera-
ture on the topic.[1]

Recent responses to the burgeoning cost of environmental dis-
pute resolution include legislation and heightened interest in alter-
native dispute resolution (ADR). For example, the Small Business
Liability Relief and Brownfields Revitalization Act limits corpo-
rate liability for abandoned commercial sites that threaten envi-
ronmental and human health.[2] On the ADR side, RESOLVE, Inc.
exemplifies the nonprofit organizations specializing in environ-
mental dispute resolution. RESOLVE uses alternative dispute res-
olution techniques to mediate disputes over environmental cleanup
costs, facilitate agreements over incentives for sustainable fishing,
build consensus on estuary protection, assist conflict resolution
over endangered salmon, and promote policy dialogue about wind
energy.[3] New and improved methods of dispute resolution may
offer faster and cheaper remedies. In this chapter you will learn
more about these and related techniques for the dispute resolution
that is critical to the advancement of environmental goals.

Litigation

Litigation is the civilized remedy of last resort for environmental
disputes. The litigation process is often costly, slow, and potentially
injurious to the reputations of individuals and firms. For instance, litigation over the 1989 Exxon Valdez oil spill was not resolved until 2015.
What may be the worst environmental disaster in U.S. history, the 2010
British Petroleum Deepwater Horizon oil spill, lead to a record civil settlement of $20.8 billion and related litigation is ongoing.[4]

The U.S. Comprehensive Environmental Response, Compensation,
and Liability Act (CERCLA) spawned a wave of litigation over liability
for hazardous waste cleanups. Global environmental concerns over acid
deposition, deforestation, and global warming threaten international
conflict, with fewer formal avenues of last resort. Regional examples

1   Examples of the literature appear in the Further Reading section. The Diepenbrock Law Firm is among the
    many specializing in environmental and natural resources law (www.diepenbrock.com). Institutes include
    the U.S. Institute for Environmental Conflict Resolution (www.ecr.gov). Expenditures on Superfund litigation
    alone exceed $10 billion. Schools such as Brown University teach entire courses on environmental conflict
    resolution.

2   See www.epa.gov/brownfields/summary-small-business-liability-relief-and-brownfields-revitalization-act.

3   See www.resolv.org.

4   See www.pbs.org/newshour/nation/cleanup-workers-who-became-ill-after-deepwater-horizon-oil-spill-are-
    suing-bp-for-compensation.

include a suit filed by the U.S. Department of Justice over emissions from a Japanese garbage incinerator near the Atsugi naval air base. A legal dispute involving Ecuadorian tribes and Texaco Petroleum (now owned by Chevron) over the cleanup of alleged toxic contamination in Ecuador lasted for decades, stymied by contention over which courts should hear the case—those of Ecuador, the United States, or a third party. And a 2023 derailment of train cars carrying hazardous chemicals in Ohio led to multiple class-action lawsuits, a suit filed by Ohio authorities, and a federal civil suit brought by the Justice Department and the EPA.

Environmental law flourished in the United States as a direct result of a series of environmental statutes adopted in the 1970s and 1980s, including the National Environmental Policy Act, the Resource Conservation and Recovery Act, the Clean Water Act, the Clean Air Act, the Toxic Substances Control Act, and CERCLA, which gave rise to the *Superfund*, a trust fund used to clean up hazardous waste sites. Upon using the Superfund, the EPA attempts to recoup cleanup costs by taking legal action, when necessary, against past polluters as explained further in the upcoming Reality Check.

Superfund litigation increases concerns about **brownfields**, which are former industrial sites with the potential to be contaminated by hazardous substances. Brownfields are not typically as dangerous as Superfund sites, but nonetheless they make developers wary because of the potential for environmental lawsuits over harm from pollution at the sites. **Greenfields** are fields and forests that have not yet been developed, which provide attractive alternatives to brownfields and the potentially costly repercussions of past environmental harm. Risk-averse manufacturers are especially dissuaded by the threat of trial by jury because juries are unpredictable and create a wide range of potential legal costs. If improved dispute resolution techniques—the focus of this chapter—make solutions less costly and outcomes more predictable, the expected liability cost of using brownfields will decrease, and their use for manufacturing will increase relative to the use of pristine and environmentally sensitive greenfields.

A Simple Bargaining Model

Pretrial settlement can limit the losses of money, time, and reputation resulting from litigation. Settlement occurs often, but not often enough. This section presents a simple bargaining model and defines some sufficient conditions for settlement. The **plaintiff (p)** is the party filing the claim. The **defendant (d)** is the party accused of wrongdoing. The parties' **threat points** are the best outcomes they can achieve by going to trial rather than negotiating. The defendant's threat point is the amount the defendant expects to pay if they proceed to trial, including the jury award and attorney fees. The plaintiff's threat point is the amount the plaintiff expects to receive at trial minus the plaintiff's attorney fees. A risk-neutral party will not accept an offer inferior to their threat point. The **settlement range** is the set of values between the two parties' threat points. Both parties are better off settling for an amount in the settlement range than going to trial, which is why most cases settle. The **bargaining rent** is the total amount of money saved by the parties if they settle now rather than going to trial, which is the sum of the attorney fees that would be paid only if the case proceeds to trial. The *jury award* is the amount of money the jury decides the defendant must pay to the plaintiff at trial. Several abbreviations are useful:

$$J_d, J_p = \text{Defendant's and plaintiff's respective expected jury awards}$$

$$F_d, F_p = \text{Expected future attorney fees (and any other unrecoverable litigation costs for each side if they proceed to trial)}$$

$$T_d, T_p = \text{The threat points of the defendant and plaintiff, respectively}$$

$$BR = \text{bargaining rent} = F_d + F_p$$

Any legal costs the parties have already paid and cannot get back are *sunk costs* and should be ignored. The defendant expects to pay the expected jury award (J_d) plus expected future attorney fees (F_d) if the case proceeds to trial. Her threat point is thus $J_d + F_d$ because an offer to settle for more than that would be inferior to (have her pay more than) the expected trial outcome. Likewise, the plaintiff expects to receive an award of J_p at trial, less attorney fees of F_p, so the plaintiff's threat point is $J_p - F_p$. An offer of below that would give the plaintiff less than she expects to receive at trial.

A litigant whose goal is to maximize financial gain (or minimize financial loss) from the case should not consider settlement offers inferior to her threat point. A litigant will also refuse offers preferable to her threat point if she believes she can negotiate an even better settlement. Self-perceived bargaining positions depend on subjective assessments of relative experience levels, optimism, persuasive ability, and monetary resources, as well as the political, precedential, and emotional repercussions of trial. These assessments can change as learning occurs during the bargaining process.

Settlement negotiations are essentially an exercise of deciding how to divide the bargaining rent. If the parties feel evenly matched, each will expect to acquire half of the bargaining rent, and settlement can occur for the amount of the expected jury award. Mutually understood inequities in bargaining skills or finances can cause both parties to expect the same unequal distribution of the bargaining rent. For example, both sides might expect the defendant to gain one-third of the bargaining rent and the plaintiff to gain two-thirds. Differing levels of information or optimism can lead to overlapping expectations, as if both sides expected to gain two-thirds of the bargaining rent.

Figure 15.1 illustrates the simple bargaining situation with number lines. In the top scenario, the plaintiff and the defendant expect the same jury award ($J_d = J_p$). Because there is no overlap between the two parties' expectations, the settlement range is as wide as the sum of the two parties' attorney fees. Both parties would prefer to settle for any value in the settlement range over going to trial. Values outside of the settlement range are inferior to the expected trial outcome for one party or the other.

In the middle scenario, the two parties have differing, optimistic expectations for the trial outcome. Optimism about a low jury award gives the defendant a lower threat point than if J_d equaled J_p, meaning the defendant would rule out some of the amounts in the high end of the settlement range of the top scenario. Optimism about a high jury award gives the plaintiff a relatively high threat point, meaning the plaintiff would rule out some of the amounts on the low end of the settlement range in the top scenario. The resulting settlement range is smaller than in the case of equal expectations. If the parties were relatively pessimistic rather than optimistic, the settlement range would increase relative to the case of equal expectations. Can you illustrate a scenario with relative pessimism on a number line like those in Figure 15.1?

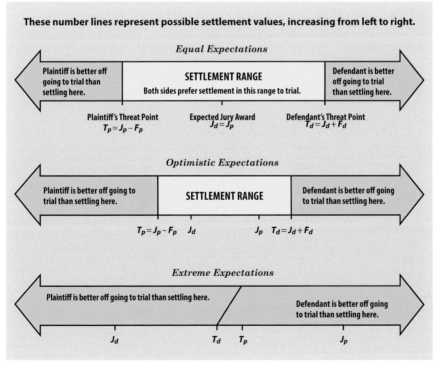

Figure 15.1

The Settlement Range

The last line represents an extreme case in which very different and relatively optimistic expectations result in the absence of a settlement range. Because the plaintiff expects more from going to trial than the defendant expects to pay, there is no out-of-court settlement that both sides would prefer over trial. The plaintiff will reject offers less than T_p, the defendant will reject demands greater than T_d, and neither party will settle for an amount between T_d and T_p.

Beyond formal legal disputes, this model applies to everyday disputes and negotiations. For example, large purchases often involve bargaining over price, employment involves wage negotiations, and international relations involve disagreements over property. In negotiations with someone you might hire to install solar panels on your roof, the installer's threat point will be the opportunity cost of her time—perhaps the $12,000 she could earn installing panels for someone down the road. Your threat point is what you could pay someone else to do the job, or the benefit you would receive from the panels, whichever is less. Suppose your benefit is $20,000 and no one else is available to install the panels. Because the installer's threat point is below yours ($12,000 < $20,000), a settlement range exists, and the bargaining rent is the $8,000 difference between your threat point and hers. Both sides would prefer any price between $12,000 and $20,000 to no deal. Disputes over prices are unlikely to find their way to trial, but the task is still one of deciding how the bargaining rent will be divided between the two parties. To settle on a price of $15,000, for instance, would be to give the installer $3,000 of the bargaining rent and you $5,000 of the bargaining rent.

Superfund Dispute Resolution

reality check

The U.S. generates 700,000 tons of hazardous waste every day as by-products of our consumption of gasoline, metals, chemicals, textiles, electronics, wood, food, and material possessions. The difficulty of monitoring the millions of U.S. business establishments, not to mention all levels of government, military installations, hospitals, and universities, means that dangerous amounts of waste find their way into unsafe resting places. The Comprehensive Environmental Response, Compensation, and Liability Act (CERCLA) set out to clean up hazardous waste sites with Superfund legislation that applies a "polluter pays" policy.

The Superfund act applies the doctrines of **strict liability** and *joint and several liability*. Under a strict liability standard, a party can be held liable whether or not it was negligent. For example, a firm can be responsible for cleaning up a hazardous dumpsite even if the firm's disposal practices broke no existing laws. Joint and several liability means that those contributing to the hazards of a dumpsite can be sued individually or collectively for the cleanup costs. Unlike a **negligence standard** of liability, which requires plaintiffs to demonstrate that defendants were negligent in their actions, strict liability typically simplifies the litigation process by avoiding the need to prove negligence. However, the large number of Superfund sites and the enormity of the associated cleanup expenses have resulted in a level of expenditures on environmental dispute resolution that irks everyone involved, save perhaps some lawyers on the receiving end. *Potentially responsible parties* (PRPs) are in litigation with the EPA over their involvement and appropriate contribution levels. PRPs are also suing each other in attempts to collect payment for "orphan shares"—portions of cleanup costs attributable to unidentified, defunct, or insolvent parties.

The EPA has a large legal staff, for whom the Superfund is the largest single project. Total legal costs on Superfund litigation have exceeded $10 billion, and more than $31 billion have been collected by "potentially responsible parties" to help fund the cleanup efforts. As of late-2023, 1,533 sites had been cleaned up, 144 were known to not be under control, and 166 more needed further investigation to determine whether people could still be exposed to contamination. More hazardous sites are discovered each year. All of this means that environmental dispute resolution will continue to be of great importance for some time to come.

Deadlock and trial can be averted by any one of the following:

- Brute force. *For better or worse, the successful application of force can supersede interests in negotiation or litigation.*

- A decision rule. *One of the disputants or a third party can put forth an acceptable or enforceable means of deciding the outcome.*

- Agreement. *There might be agreement over the expected trial outcome and the appropriate division of the bargaining rent. Alternatively, the parties may be relatively optimistic about the expected trial outcome but relatively pessimistic about the share of the bargaining rent they can obtain, making for compatible differences.*

- Ability to make a take-it-or-leave-it offer. *One party may be capable and willing to make a credible final offer within an existing settlement range, forcing the other party to either take the offer or leave it and receive an inferior outcome at trial.*

Efforts to avert or resolve environmental conflicts should, as a minimum requirement, seek to satisfy one of these four conditions. The remainder of this chapter describes attempts to achieve one of these conditions with varying trade-offs between ease of application, fairness, speed, and success.

Dispute Remedies

Brute Force

Various world powers, environmentalists, and industrialists have adopted brute force as a remedy of last resort. Too often, destructive force is applied out of frustration with the perceived ineffectiveness of alternative dispute remedies. Chapter 12 discussed the violent actions taken by ecoterrorists in the name of protecting the environment. Between 2005 and 2017, the anti-whaling group Sea Shepherd used acid, blockades, and other physical means to disrupt whale harvests, as shown on the television series *Whale Wars*. Ecoterrorism became a negotiating tactic in France, where workers laid off from a chemical plant won a severance package after dumping 790 gallons of sulfuric acid into a tributary of the Meuse River.

War is the most primitive and costly means of dividing natural capital. As seen in Ukraine, Palestine, Sudan, and elsewhere, war continues to be prominent in international land disputes when no authority can apply

Brute force is a primitive yet common form of dispute resolution.

or enforce civilized dispute resolution processes. Around 100,000 lives are lost to wars around the world in a typical year,[5] and global military expenditures exceed $1 trillion annually. Beyond the human casualties of violence and the financial losses to violent solutions, war destroys scarce natural capital. During the Vietnam War, forests and wildlife were destroyed by over 11 million gallons of dioxin-laden Agent Orange.[6] Nuclear proliferation increases the environmental stakes for violent conflict resolution. Joshua Pearce and David Denkenberger (2018) explain the need to limit the size of nuclear arsenals given the threat that smoke and dust from a large nuclear war could cause a life-decimating nuclear winter as discussed by Carl Sagan and Richard Turco (1990).

Like the threat of trials, the prospect of violence may moderate optimism that can separate the expectations and demands of parties to a dispute. When the threat of war does not foster acceptance of settlement offers, the force of combat can end a dispute. Unfortunately, sometimes the battle is fought to lend credibility to future threats. Worse, ego, pride, emotion, and greed can distract parties' priorities away from the efficient use of environmental resources.

The costly consequences of both legal and military battles can be effective motivators for more cordial settlement when such alternatives are available. The use of violence has waned with the advancement of legal authority and alternative conflict resolution techniques. Trial costs deter some needless claims and encourage a large majority of claimants to settle before trial. One might argue that trials should be lengthier and more expensive, thus inhibiting more lawsuits. Although high litigation costs do promote tolerance and discourage frivolous suits, they also inhibit underfunded plaintiffs with meritorious suits and may lead to vigilante justice if alternatives are not provided. If higher court costs are deemed desirable, an alternative to permitting inefficiency in the civil justice system would be to tax disputes and give the receipts to a worthy cause.

Decision Rules

Particularly in disputes with small stakes, parties sometimes acquiesce in decision rules as an efficient alternative to violence. The catalyst might be a third-party mediator or arbitrator whose decision each side has agreed to accept. Decision rules that are sometimes favorable include

- *Tradition*
- *Rules of thumb[7]*
- *Strict adherence to religious teachings*

5　See https://ourworldindata.org/grapher/deaths-in-armed-conflicts-by-type.

6　See Boffey (1998).

7　For example: leave things better than you found them; protect children first; first come first served.

- *Precedent*

- *Arbitration*

- *Flipping a coin*

- *Drawing straws and similar games*[8]

All of these methods address a question of eligibility to receive certain benefits.[9] Sporting events, jousting, duels, fist fights, wars, and related tests of strength or bravery have been used for the same purpose. None of these methods offers a panacea; the more harmless solutions often lack enforceability, while the more violent solutions lack popular appeal.

The incentive to comply with unforced decision rules is inversely related to the size of the stakes. In small-stakes cases like a dispute over the removal of a tree straddling two lots, the transaction costs associated with trying to override a decision are likely to exceed the benefits. Large stakes motivate parties who are disadvantaged by traditional rules to forego the rules' convenience in favor of a more involved battle for privilege. In a dispute over national boundaries, it is likely that arduous settlement negotiations or a more authoritative determination will be needed to supplant unforced decision rules. The ease of dispute resolution under authoritarian rule is a strength among the many weaknesses of dictators and other autocrats.

Fair Division

Dividing a forest or other natural capital can resemble the cutting of a cake. The cake might represent the Middle East, a mineral-laden continental shelf, or the Arctic National Wildlife Refuge. Interests in natural resources drive many disputes over borders. Consider some examples from Latin America: Honduras and Nicaragua are feuding over fishing and oil rights in the Caribbean and share a dispute with El Salvador over division of the shrimp-rich Gulf of Fonseca. Venezuela and Guyana have long disputed rights to land in the Essequibo River region of Guyana, which holds plentiful mineral and oil reserves. Neighboring Caribbean countries disagree over the division of Suriname, Belize, Aves Island, the Gulf of Venezuela, and the waters separating Venezuela, Trinidad, and Tobago. There are many cakes to be divided.

Divide and Choose Simple solutions exist for simple conflicts. This is the case when equal division is the goal and the adverse parties have similar preferences. For natural capital that can be measured

---

8 You may remember "grab the bat," "eeny-meeny-miny-moe," and "rock-paper-scissors."

9 See Brams and Taylor (1996) for a comprehensive review of decision rules.

accurately, as with minerals or uniform land, fair division is straight-forward. If the resource is not uniform or a reliable measuring device is not available, a *divide-and-choose* method can sometimes render an agreed-upon division between two adversaries. The **divide-and-choose** solution allows one party to divide the resource into two parts and the other to choose between the two allotments. The divider has an incentive to divide the resource equally in order to maximize the value of the inferior (if not equal) part that will remain after the chooser makes a selection. The United Nations Division of Ocean Affairs and the Law of the Sea advocates the divide-and-choose method for dividing mineral-rich sections of the seafloor among neighboring nations.[10]

Some conflicts are not so simple. The equitable appeal of the divide-and-choose method diminishes when the preferences of the parties differ. Suppose that the Nature Conservancy (NC) and the American Petroleum Institute (API) are trying to divide part of the Arctic National Wildlife Refuge into a section to preserve and a section to tap for oil. The land in question contains mountains on one side, ocean on the other, and flat land in the middle. Suppose the NC prefers to preserve the mountains while the API prefers to drill on the flat land near the ocean. If the NC makes the division, it will include more than half of the land in the section with the mountains, knowing that the API is still likely to select the section near the ocean. Likewise, if the API divides, it will include a disproportionate amount of land in the section with the ocean, knowing that the NC will favor a smaller section with mountains over a

Disputes over development are common. This sign arose during a dispute over new hotels planned for wilderness areas on the island of Oahu in Hawai'i.

10 See www.un.org/depts/los/.

larger section on the sea. This dispute is not a good candidate for simple fair division methods.

Say Stop The **say-stop** method can divide uniform but hard-to-measure assets among parties wishing to maximize their shares. It can also divide nonuniform assets among parties with similar preferences. This method offers the advantage of being easier to administer than divide and choose when more than two parties are involved.[11] Consider the dispute between fishers and environmentalists over giant cuttlefish living along reefs near Australia. Suppose that environmentalists, Australian fishers, and international fishers want to divide the cuttlefish habitat into three equally desirable areas and that subjective estimates of cuttlefish densities at different points along the reefs prevent a simple division on the basis of kilometers.

A solution would be to have a boat travel the length of the reef, with a rule that the representative from any of the three parties can call out "stop" at any time, and the caller's share will be the stretch that the boat has already passed. After the first share has been claimed, the remaining two representatives will call out when they feel the boat divides the remaining reef into two equally favorable sections. If the boat had not reached what a party sees as a fair dividing point, calling out would leave the other party with the better side. If the boat were beyond what the parties see as the fair dividing point, both would want to claim the section behind the boat before the other captured the advantage. When the boat begins, the choice is thus between the section of reef behind the boat and half of the reef's natural capital ahead of the boat. If less than one-third of the reef's capital has been passed, then more than two-thirds (more than one-third for each of the last two callers) has not been passed, and it pays to wait. If more than one-third has been passed, less than two-thirds remains (less than one-third for each of the last two callers), and it pays to call out. So the incentives are for the reef to be divided into three sections that do not elicit envy from the perspective of any of the recipients.

The theoretical result of the say-stop method with n participating parties is that a representative will say stop whenever, to the best of any party's knowledge, $1/n$ th of the asset has been passed by the divider (in this example, the boat). The division may not be perfect, but when there is no satisfactory objective measure, this method divides an asset with minimal expense and to the satisfaction of everyone involved.

11 Divide and choose can indeed be applied with three parties. The first divides the asset into three pieces, the second trims what is in her perception the largest piece so that it is the same size as the second-largest piece, and the third gets the first choice of the three pieces. The trimmer gets the second choice and the divider gets the third choice. The result is a division that everyone is content with, but then the trimmings must be divided by the same method, and then the trimmings of the trimmings, and so on.

Like divide and choose, the say-stop method cannot assure an equitable outcome when dividing a nonuniform resource between parties with differing tastes. The advantage will go to the party who prefers the characteristics of the asset on the side that the divider starts on. In the Arctic National Wildlife Refuge example, if the divider starts on the mountainous side of the refuge the Nature Conservancy favors, the NC will wait to say stop until the divider has reached the point where the American Petroleum Institute is just short of indifference between the two sides, at which point the side with the mountains will have a larger portion of the middle ground than the side with the ocean. Likewise, the API would gain the advantage if the divider started on the coast.

Strict Alternation   With **strict alternation**, separable assets can sometimes be divided fairly by allowing the parties to take turns choosing them. This works particularly well when there is an even number of items with particularly high or low value and when the parties' knowledge of each other's preferences cannot lead to an unfair advantage. Consider the dispute between Venezuela and Guyana over the environmentally rich region of the Essequibo River. There are many islands at the mouth of the river. We will examine a simplified version of the allocation of these islands with a goal of fair division.

Assume Venezuela and Guyana would each like to control as many islands as possible, and that they each favor control of the larger islands. If the islands in question are similar in size and location and are even in number, strict alternation of island selection could yield a fair resolution. If, instead, there are three large islands and one small island, the first chooser will gain the advantage. The first chooser will select the first large island, the second chooser will select the second large island, the first chooser will select the third large island, and then the second chooser will select the small island. Figure 15.2 illustrates the final tally in this case. The first chooser would control two large islands and the second chooser would control one large and one small island.

To examine the role of known preferences, imagine there is one small, one medium-sized, and one large unpopulated island, and one large island populated by citizens of Guyana. Guyana is the first chooser and prefers to have larger islands. Understandably, it places the highest priority on the island already populated by its citizens. Venezuela is the second chooser. It likes the large unpopulated island the best and the small island the least but would choose the medium island over the large island on which it would have to contend with citizens of its adversary. If these preferences are hidden and Guyana assumes that size is the primary selection criterion for Venezuela, Guyana will first select the populated island (because otherwise it has no assurance of receiving it). Venezuela will choose the other large island, Guyana will choose the medium-sized island, and then Venezuela will choose the small island. This provides each country with control over its most favored island, plus either a small or a medium island. The first chooser has a small advantage.

If Guyana knows Venezuela's preferences, Guyana will begin by selecting the large, unpopulated island. Then Venezuela will select the medium island because, as Guyana knows, this takes precedence over the larger island populated by Guyanese. In the second round, Guyana will select the populated island and Venezuela will select the small island. This leaves Venezuela with the small and medium-sized islands and Guyana with the two large islands. The first-chooser advantage for Guyana is significant. Such an advantage is generally unfavorable from a policy standpoint, although it may be appropriate as a means of balancing other inequities, such as those between developed and developing countries. Strict alternation is applicable to divisions among more than two parties when knowledge of preferences and heterogeneity of assets are not a problem, although the opportunities for strategic manipulation increase considerably.

Balanced Alternation   In some cases, the inequities of strict alternation can be resolved with the **balanced alternation** method of taking turns at taking turns. That is, in the island-choosing example, rather than selecting in the order Guyana Venezuela Guyana Venezuela, Guyana could choose first in the first round and Venezuela could choose first in the second round, making the order Guyana Venezuela Venezuela Guyana. This would prevent Guyana from garnering both of the large islands, but it does not guarantee the most equitable solution. If there were one large island and three small islands such that control over all three of the small islands would give either party the same satisfaction as control over the one large island, it would then be more equitable to give the first chooser one selection and allow the second chooser to make three selections in a row. So the order might be Guyana Venezuela Venezuela Venezuela. In similar ways, balanced alternation can be customized to yield a fair solution in some situations when strict alternation will not.

Balanced alternation cannot provide balance when the parties value one asset by more than the other assets combined. It can also result in an inefficient outcome when preferences are unknown by those deciding on the balancing scheme. Suppose Guyana cared only about the number of islands saved and had unknown indifference between the large island and any one of three small islands, whereas Venezuela equated the one large island with all three of the small islands combined. The ordering of Guyana Venezuela Venezuela Venezuela that worked well in the previous example could result in Guyana choosing the large island and Venezuela choosing three small islands. Yet both sides would be better off with a strict alternation of Venezuela Guyana Venezuela Guyana that would allow Venezuela to control a large island (subjectively equivalent to three small islands) plus a small island, and allow Guyana to control two small islands (subjectively equivalent to two large islands).

Figure 15.2
*Allocating
Islands in
the Essequibo
Estuary: Order
and Information
Matter*

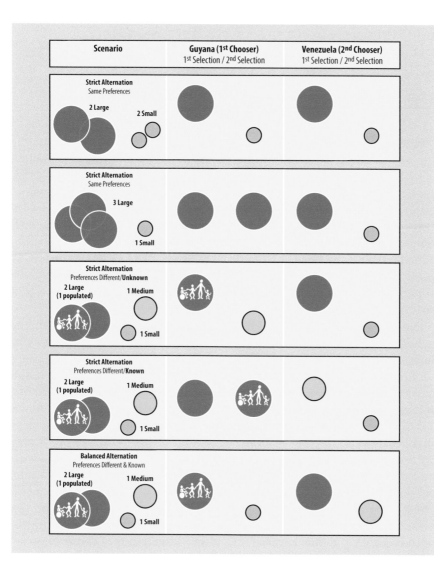

Adjusted Winner The **adjusted-winner** method can provide a balanced allocation of resources in some cases when alternation cannot.[12] Take, for example, the disputes between Greece and Türkiye (formerly known as Turkey) over fish stocks in their contested territorial waters, land on the island of Imia, and rights to oil below the continental shelf of the Aegean Sea. An application of the adjusted-winner method would have each country assign 100 points among the three assets as weights to reflect the countries' preferences. Hypothetically, the outcome might be as shown in Table 15.1.

12 See Brams and Taylor (1999).

Table 15.1

Asset	Greece	Türkiye
Land	25	25
Oil	55	50
Fish	20	25

In the first phase, each asset is allocated to the country that assigns the largest number of points to it. If both countries assign the same number of points to some of the assets, these assets are distributed, one by one in any order, to the party with the fewest points worth of resources at the time when the distribution is being made. Thus, Greece starts out with rights to the oil for 55 points, and Türkiye starts with rights to the fish stocks for 25 points. Both countries assigned 25 points to the land, and the tie goes to Türkiye because it has received the fewest points worth of resources. With the land and the fish, Türkiye receives 50 points worth of resources, and Greece receives 55 points worth of resources, making Greece the winner of this phase of the process.

The first "winner" phase is followed by an "adjustment" phase. For each item initially allocated to the winner, the mediator or authority in charge calculates the ratio of the winner's point allocation to the loser's point allocation for that asset. Since Greece won and received the oil, the relevant ratio is 55/50 = 1.10. Similar calculations would be carried out for any other items the winner received. Lower ratios indicate that fewer points will be lost by the winner relative to the points gained by the loser when an item is transferred from the winner to the loser. For efficiency, any transfers necessary to reach equality between the parties begin with the items with the lowest ratios of winner's to loser's point allocations and end with those with the highest ratios. If the transfer of the entire lowest-ratio asset would not give the initial loser more points than the initial winner, that transfer is made, and then the adjustment phase continues with the new lowest-ratio asset in the hands of the winner.

If, as in the story of Greece and Türkiye, a transfer of the entire lowest-ratio asset would overcompensate the loser, a fraction of the lowest-ratio asset is transferred. Of course, indivisible assets cannot be selected for this purpose unless they can be sold or otherwise converted into divisible items. To determine the transfer of oil that would equalize the total point allocations received by Greece and Türkiye, the points received by each side can be set equal to each other, with x representing the fraction of oil that Türkiye receives:

$$55 - 55x = 25 + 25 + 50x.$$

The left side of the equation represents Greece's 55 points from oil minus the fraction of those points that will be transferred to Türkiye. The right side represents Türkiye's 25 points from land, 25 points from fish, and fraction of the 50 points from oil. Solving for x determines that if Türkiye receives 4.762 percent of the oil, each side will earn $55 - 55(0.04762) = 25 + 25 + 50(0.04762) = 52.38$ points.

Strategic behavior can rear its ugly head in the midst of the adjusted-winner procedure. There is an incentive for parties to understate the extent to which they favor assets so long as they do not lose control of the assets they value more highly than the other party. By attributing fewer points to items they receive, parties are granted more compensation as the initial loser or debited less as the initial winner. The best possible outcome for a party occurs when the party allocates just one point more than the adverse party for each item received. The worst possible outcome occurs when such misrepresentation leads to the parties receiving their least-favored assets. The strong incentives to lie at least a little, and the problems created when strategic behavior goes awry, make this method most useful when parties can be expected to be honest.

The techniques in this section are designed to yield equal distributions of benefits. Sometimes environmental concerns, historical precedent, current possession, or broader social welfare objectives make an unequal division appropriate. When equality is not the goal, the allocation process becomes more complex. If either the appropriate allocation or the measurement of the asset is in question,[13] subjective solutions may be necessary. Neutral third parties become valuable as facilitators and decision-makers under remedies that include arbitration, mediation, negotiated settlement, and judgment at trial. As disputes intensify, attorneys are hired, claims are filed, and the costly process of discovery, demands, threats, and counterclaims may be forthcoming.

Agreement

Agreement mends disputes. Unfortunately, there can be much to agree on. Consider a dispute between the EPA and the fictional MESSCO Corporation over the payment of cleanup costs for a Superfund site. Initially, both sides expect that a jury would award $200 million from MESSCO to the EPA for cleanup. If trial would necessitate $40 million in additional attorney fees for the EPA and $60 million in additional fees for MESSCO,[14] there is $100 million to be saved by resolving the dispute beforehand.

Even though the parties agree on the expected trial outcome, they may disagree over how to divide the $100 million of bargaining rent. The EPA expects to receive $200 million at trial minus $40 million in fees, so any settlement amount over $160 million would be better for the EPA than going to trial. MESSCO expects to pay $200 million plus $60 million in fees at trial, so any settlement amount below $260 million

---

13 Although measurement is seldom a problem with land, disputes can arise over the measurement of less uniform assets such as land with varying topography, buildings, or biodiversity.

14 Attorney fees in the vicinity of 30 percent of an award are common.

is preferred over going to trial by MESSCO. Their relative bargaining strengths will determine where in the settlement range between $160 million and $260 million they might settle. Strategic behavior, misperceptions of bargaining strength, and undue optimism can prevent agreement over how to divide the $100 million savings from an out-of-court settlement.

The uncertainty of jury awards impedes the alignment of parties' expectations. The wide range of possible outcomes also threatens risk-averse parties who cannot afford even a remote downside risk of paying an enormous award. In the past, multimillion-dollar awards were newsworthy. Exxon, General Motors, and Ford are among the growing numbers of U.S. companies facing damage awards in the billions of dollars. The risk burden of variations in judgments could be decreased with caps on damages, decisions made by judges rather than juries, or a standardization of awards for particular offenses or injuries.

Alternative Dispute Resolution

The involvement of a neutral third party can foster settlement by removing barriers including differing information, biased expectations for trial, unrealistic perceptions of bargaining strength, and strategic behavior. **Alternative dispute resolution** (ADR) techniques couple assistance with these barriers with an abbreviated process that is faster and less expensive than trial. Under **decisional approaches**, the neutral third party has the authority to impose a solution on the disputants. Under **facilitative approaches**, the neutral party helps the parties achieve their own solution. Under **advisory approaches**, the neutral party renders a decision that is suggestive but nonbinding.

Decisional ADR Techniques

Conventional arbitration brings disputants before a neutral third party or panel to present evidence. The neutral(s) review the arguments and then render a final, binding decision that is not subject to court approval or appeal. The possibility that a neutral might offer a solution that splits the difference between the two parties' proposals provides incentives for polarized offers and discourages concessions. **Final offer arbitration** (FOA) is meant to counter these incentives.[15] Under FOA, each side submits a last, best offer and the arbitrator selects one of the two offers as the final outcome. In contrast to the tendency for extreme offers under conventional arbitration, the incentive under FOA is to submit an offer that the arbitrator will deem fairer than the adversary's

15 A number of economists, including Bazerman and Farber (1985), have found that the decisions of arbitrators are largely independent of the demands of the parties, lessening concerns about conventional arbitration.

offer. FOA is sometimes called baseball arbitration because it is popular for settling baseball salary disputes. A variant of FOA under which the arbitrator selects an outcome prior to hearing the parties' offers has likewise been dubbed *night-time baseball arbitration*. The final outcome is the offer that comes closest to the arbitrator's selection.

Med-arb is a hybrid of mediation and arbitration. In an initial phase, the mediator attempts to bring the two parties to agreement on an outcome, but if the impasse is not resolved, the dispute is referred to binding arbitration.

Facilitative ADR Techniques

Mediation involves no third-party judgment, but one or more mediators work with the parties to help them reach a settlement. The mediators help the parties clarify their differences and try to dovetail interests to the satisfaction of both sides. For example, for almost 300 years, Spain and Britain have disputed ownership of the environmentally sensitive Rock of Gibraltar—the headland on the southern coast of Spain that overlooks the entrance to the Mediterranean Sea. The Rock is the European home for Barbary apes and a rest stop for hundreds of thousands of birds migrating between their breeding grounds in Europe and their wintering areas in Africa. If Britain cared most about preventing further development on the Rock, and Spain cared most about ownership rights, a mediator might bring these mutually agreeable interests to light and foster a settlement that placed the Rock in Spanish hands with the stipulation that wilderness areas would be protected.

Early neutral evaluation brings representatives of the two sides together with a neutral party shortly after a claim is filed to talk through the strengths and weaknesses of their arguments and clarify realistic outcomes. A *judicial settlement conference* brings lawyers representing the disputants together with a judge or magistrate to try to resolve the case short of trial. This is useful when settlement is deterred by at least one party having unrealistic expectations for trial because it helps bring everyone down to Earth.

Advisory ADR Techniques

Advisory efforts to bring parties closer together include explicit opinions from one or more neutral parties. In a **mini-trial**, a neutral panel presides over a hearing in which representatives of each side present the highlights of their cases. The presentations generally occur in a private forum and not in a courtroom. The neutrals themselves, not the parties, then try to negotiate an appropriate settlement based on the strengths and weaknesses of each side. The parties are typically not bound to the decision, but they may accept it as a resolution to the dispute. A **summary jury trial** is a more formal and adversarial abbreviated trial that gives litigants a snapshot of what would occur at trial and guidance for settlement. Each side's arguments are heard by a jury of peers resembling that in a traditional trial. That jury provides an advisory verdict that the parties and their attorneys can use to inform their settlement negotiations, if not as their final settlement.

Court-annexed arbitration is an attempt to bridge the gap between disputants who have reached court by diverting them to a brief, non-binding hearing before one or more attorneys or retired judges. After the hearing, the arbitrator writes an opinion and determines an award that is filed with the court. If a litigant is not satisfied with the award, the litigant can appeal the decision and reenter the traditional court system. However, in most cases, the court-annexed arbitration award is not appealed.

With **neutral fact-finding**, a neutral party undertakes an independent investigation of a contested factual matter, such as whether a proposed road would endanger wildlife habitat. The issues in question are chosen by the parties themselves, and the decision is usually non-binding. Alternative dispute resolution techniques such as these are required in some classes of civil suits and could be better utilized in others.

Compatible Perspectives

Even when parties to a dispute lack the information necessary to share the same expectation, parties can reach settlement with differing but compatible expectations. Relative optimism regarding the expected judgment might be accompanied by relative pessimism regarding the appropriate division of the savings from avoiding trial, thus making a range of settlement offers acceptable to both sides. This opportunity arises because the judgment depends on the merits of each party's case, while the division of the savings depends on the parties' relative bargaining strengths. Either party might over- or underestimate either aspect of their situation. The task, then, is to bring parties close enough for compatibility.

Civil justice reform is in its infancy, with fledgling attempts at new rules of civil procedure that lower demands, increase offers, or otherwise satisfy one of the four conditions for settlement. Specific rules, including the American rule, the English rule, Federal Rule of Civil Procedure 68, and many parallel state court rules have met with limited success in encouraging parties to reach fair settlements expeditiously.[16] These rules and some of their pitfalls are described here. The section that follows explains two alternative rules that could equitably resolve

16 See Shelton (2007) and Anderson and Rowe (1995).

conflicts despite the parties' incompatible self-perceptions of their relative bargaining strengths.

The American Rule

As illustrated in Figure 15.3, the traditional "American rule" shifts court costs, excluding legal fees, to the "losing" party in civil cases. Under this rule, each party pays its own attorney fees regardless of the outcome at trial. The losing party—the defendant if there is a positive verdict in favor of the plaintiff, and the plaintiff otherwise—is assessed court costs other than attorney fees, including reasonable court fees, transcript costs, printing costs, and witness fees. This rule generally applies to litigation costs even when supplemental rules governing attorney fees are in effect. Court costs are typically negligible, making this rule a slap on the wrist for parties who fail to settle and fail in court.

The English Rule

The English rule places a greater burden on the losing party by shifting the payment of attorney fees, and not just court costs, to the losing party. Under the English rule, the loser pays both court costs and reasonable attorney fees for both sides. This larger penalty for losing a case serves to discourage frivolous lawsuits. The downside is that the rule also discourages justified claims by parties who cannot afford the risk of an unfavorable judgment despite the merits of their cases. Related concerns have prevented acceptance of the English rule in the United States, although loser-pays rules have received considerable attention.

Federal Rule of Civil Procedure 68

Settlement rules, also called *offer-of-settlement devices*, are intended to encourage parties in litigation to settle out of court. A typical settlement rule allows a party in litigation to formalize a settlement offer. If the opposing party refuses that offer and does not improve on it at trial, the refusing party suffers a consequence. Existing Federal Rule of Civil Procedure 68 (Rule 68) mandates that defendants collect post-offer court costs, but usually not attorney fees, from the plaintiff if a refused offer is not improved upon at trial. In theory, because Rule 68 creates an added threat that the plaintiff will have to pay the defendant's costs at trial, the plaintiff is more likely to accept any given offer. Unfortunately, rather than making the same offers she would make in the absence of Rule 68, the defendant is likely to offer to pay less as a result of her improved bargaining position. In effect, both the defendant's maximum offer and the plaintiff's minimum demand decrease by similar amounts in response to the rule, shifting but not closing the gap between the two

Figure 15.3

Cost and Fee Entitlement Under Existing Rules

The number lines represent possible values for the jury award, increasing from left to right.

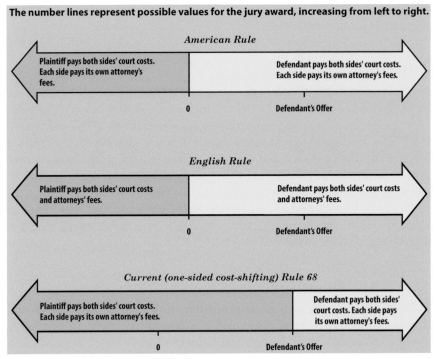

American Rule

Plaintiff pays both sides' court costs. Each side pays its own attorney's fees.

Defendant pays both sides' court costs. Each side pays its own attorney's fees.

0   Defendant's Offer

English Rule

Plaintiff pays both sides' court costs and attorneys' fees.

Defendant pays both sides' court costs and attorneys' fees.

0   Defendant's Offer

Current (one-sided cost-shifting) Rule 68

Plaintiff pays both sides' court costs. Each side pays its own attorney's fees.

Defendant pays both sides' court costs. Each side pays its own attorney's fees.

0   Defendant's Offer

sides. For this reason, the existing Rule 68 is underused and ineffective in encouraging settlement.[17] Proposed variants have been unable to satisfy the conditions for settlement and reconcile the offers and demands of litigants who would otherwise go to trial. In other words, there is plenty of room for improvement in settlement-encouraging rules by you and your generation.

Credible Take-It-or-Leave-It Offers

The fourth condition is satisfied when one party can extend a take-it-or-leave-it offer, the alternative being trial. Although the assumption that such an offer can be made is common in the law and economics literature, it is easier said than done. If a party had such power, assuming the parties were rational and seeking to maximize the expected value of their net gains, the offeror would make an offer slightly superior to the offeree's expected net outcome from trial. In the example of *EPA* vs. *MESSCO*, if the EPA could make a take-it-or-leave-it offer, they would offer to accept $259.99 million. If, instead, MESSCO could make the final offer, they would offer to pay $160.01 million. Since bargaining ends after the offer by definition of the final offer, the party receiving

17 See Anderson (1994) for expanded theoretical explanations for Rule 68's inadequacies and Anderson and Rowe (1995) for empirical evidence.

the offer would choose to accept it rather than continuing to trial for lesser expected gain and greater uncertainty.

Fortunately for those who might receive them, purported take-it-or-leave-it offers are seldom credible because it is not in the offeror's best interest to carry them out. After the rejection of any offer that would make the offeror better off than going to trial, further bargaining over settlement values that fall between the rejected offer and the net trial outcome for the offeror would be beneficial to both parties. From the standpoint of fairness, it is also undesirable for parties to be able to make take-it-or-leave-it offers because they permit the offeror to gain most of the benefits from settlement.

The Sincerity Rule

If enacted, the **sincerity rule** would allow either party to end negotiations with a final offer, while incentivizing an offer near the adversary's expected jury award. The idea is to provide a more equitable, relatively simple solution to disputes that holds up despite bargaining inequities. Under this rule

1 Either party may designate an offer as a sincerity offer.

2 If the offeree rejects the offer, the parties proceed to trial and the *offeror* pays the *offeree's* post-offer fees.

Those unable to stomach the high cost of a trial, or those facing unreasonable offers from well-positioned opponents, would have an incentive to make an acceptable sincerity offer. If the offeror offers an amount slightly better than the offeree's expected jury award, the offeree (unless risk-loving) will accept it. If the offeror offers an amount inferior to the offeree's expected jury award, the offeree can choose to proceed to trial at no additional cost, and the offeror must pay post-offer attorney fees for both sides. Acceptable offers are thus expected near the offeree's expected jury award and can be augmented to account for any possible risk-loving disposition on the part of the offeree.

The sincerity rule could provide equitable solutions to unfair demands among adversaries. Consider the hypothetical case in which the EPA and MESSCO dispute the payment of cleanup costs for a Superfund site. Knowing that the EPA has been criticized for its large legal expenses in the past and will not want to advance to trial, MESSCO may insist on a settlement near the EPA's threat point of $160 million. Rather than submitting to this unfair offer, the EPA could make a sincerity offer to settle for $199 million. This offer approximates the expected jury award, and if MESSCO seeks to minimize its payment, MESSCO will favor the $199 million payment over the expected $200 million payment at trial (with fees paid by the EPA). Sincerity offers could be used in a variety of situations to avoid trial, an inequitable settlement, or prolonged strategic bargaining.

Final Offer Auctions

Final offer auctions are another proposed method of avoiding negotiation deadlock by permitting a credible final offer, again adding a mechanism to avoid the potential inequities of unchecked final offers. A **final offer auction** allows the two parties to bid for the right to make the final offer. The amount of the winning bid is granted to the party who lost in the bidding phase. If the final offer is rejected, the parties must proceed to trial with no further bargaining, and each side is responsible for its own attorney fees.

In our Superfund example, each side expects a $200 million award from MESSCO to the EPA at trial. The savings on attorney fees by settling out of court would be $40 million for the EPA and $60 million for MESSCO. Given the opportunity to make one final offer, each would offer the other an amount just better than the adversary's threat point—their expected trial outcome net of fees. The EPA would make a final offer to settle for about $259.99 million and MESSCO would make a final offer to settle for about $160.01 million. Notice that the difference between what the EPA receives if it can make the final offer and what it receives if it cannot is about $100 million. The right to make a final offer thus conveys the ability to capture virtually all of the $100 million in bargaining rent.

Each side would bid up to half of the bargaining rent, or $50 million, to gain $100 million, because the alternative is to receive the other party's bid. For example, after an EPA bid of $49 million, MESSCO chooses between (1) not increasing its bid and receiving $49 million from the EPA or (2) bidding, say, $49.1 and either receiving a higher bid from the EPA or winning and receiving $100 − $49.1 = $50.9 million. It is rational for MESSCO to keep bidding until the EPA has bid $50 million. At that point, MESSCO should stop and take the $50 million rather than bidding $50.1 million and receiving $100 − $50.1 = $49.9 million.

If the EPA makes the winning bid of $50 million, it will offer to settle for $259.99 million and MESSCO should accept, rather than paying more at trial with added uncertainty and the tribulations of the court system. The net gain for the EPA would be $259.99 − $50 = $209.99 million. The net payment for MESSCO is the same: The $259.99 million settlement amount minus the $50 million bid received from the EPA. Thus, if both parties act rationally to maximize their net gains or minimize their net losses, the outcome will be halfway between the two parties' threat points. This is true whether or not the parties share the same expectation of the jury award. As with all of the existing and proposed rules, settlement is hindered when the parties' expectations for trial are unknown or differ by more than the total savings from avoiding trial.

Summary

The growing scarcity of natural capital exacerbates disputes over borders and use rights. Global industrialization increases the creation of toxic waste while environmental consciousness intensifies the calls for abatement. Litigation is a costly cure. This chapter highlights the causes of conflict, the conditions for settlement, and the variety of solutions available to avoid costly and prolonged disputes. Fair division techniques, including balanced alternation and the say-stop method, can apportion natural assets among parties when preferences are similar or the assets are homogeneous. The adjusted-winner technique can accommodate both differing preferences and heterogeneous assets, but it is subject to strategic manipulation. Alternative dispute resolution involves third-party neutrals who are able to handle complex and subjective issues. However, the trust of neutral parties, the adherence to nonbinding decisions, and the ability to enforce binding decisions are all more likely when conflicts are minor and domestic than when they are major and global.

In their intended role of resolving environmental disputes that would otherwise end in trial or violence, existing settlement rules are underused and ineffective. Stalemated and disadvantaged parties could benefit from the sincerity rule and final offer auctions, which show promise in theory and empirical testing. Use of the legal system to settle environmental disputes is not unlike use of fossil fuels to power automobiles and electric utilities. We lumber on with coal and oil for two-thirds of our energy needs despite the availability of techniques that, with wide acceptance, would provide inexpensive, renewable power and a lower environmental burden. Similarly, we often turn to the traditional approaches of trial and violence to settle conflicts over environmental assets and obligations when superior alternatives exist. The realm of environmental dispute resolution is one in which we stand to make great strides in economic efficiency.

Disputes are only natural. Blades of grass abut each other in a competition for space, food, and water. Animals compete for all of the same resources. With boundaries established and stomachs swollen, animals further engage their adversaries in quests for mates and status. Wildlife has its own forms of conflict resolution. When neighboring ferns compete for sunlight, a race for height and breadth determines fate. Polar bears resolve conflicts over mates with contests of brute force. As parasites like the torsalo fly exceed the carrying capacities of their hosts, death makes space for those with endurance. Conflicts between water and rock are always won by rock in the short run and water in the long run.

Homo sapiens have a civil litigation system unequaled in nature, although we frequently revert to alternatives that resemble those in the wild. Height and breadth are favored in a fist fight. Brute force wins turf wars. Those with endurance live to see the end of conflict-laden regimes. Humans may yield temporarily to stubbornness in a standoff, but over time, progressive flows of intellectual capital have generally dominated rigid political and socioeconomic systems. A difference between the resolution of disputes over natural resources and those between natural resources, one hopes, is the role of intelligence in the former. Thought contributes not only to the outcome of dispute resolution methods involving strategy, but also to the understanding and refinement of dispute resolution methods themselves. I hope this chapter, and this text as a whole, have enriched your thoughts about efficient solutions to our critical environmental and natural resource challenges.

• • • • • • • • • • • • • • • •

Problems for Review

1. Suppose two superpowers are vying for control of the moon. Superland wants to build a new settlement on the moon and prefers the side facing Earth because the residents will enjoy views of this planet. Supermania wants to store nuclear waste on the moon and prefers the side facing away from Earth to keep the mess out of sight. Both would like as much land on the moon as possible, and each knows the other's preferences.

   a) *Would the divide-and-choose method lead to a fair division of the land on the moon? Why or why not?*

   b) *How would your answer change if, instead of Supermania, the second party were Supervista, a country that wants to protect the Earth-facing side of the moon from development? (That is,*

   both parties prefer the Earth-facing side.)

2. Suppose a major oil spill has led to litigation between an oil company and residents of an oil-blackened coast. Both sides expect a jury award of $2 billion at trail. If the parties settle now, the savings on additional attorney fees will be $100 million for the defendant (the oil company) and $50 million for the plaintiffs (the residents). On a number line like those in Figure 15.1, indicate the expected judgment (J), each side's threat point (T_p and T_d), and the settlement range.

3. Answer the following questions in the context of the case explained in Problem 2.

   a) *What is the bargaining rent?*

   b) *What is the most either side should pay for the right to make a credible final offer?*

c) If the defendant can make a credible (non-sincerity rule) final offer, what should it be?

d) If the plaintiff makes a sincerity rule offer, what should it be?

4. The chapter explains four ways for disputants to avert deadlock and trial. Identify which of those ways was applied in each of the following cases:

a) R-land invaded U-land to capture disputed reserves of iron ore, lithium, and oil.

b) Aamon and Una flipped a coin to decide who would take the recycling bin to the curb.

c) Eight South American nations signed a pledge to end Amazon deforestation.

5. Why might a strict liability standard reduce the cost of environmental litigation?

6. What dispute resolution technique would you advise for Spain and Britain in their dispute over the Rock of Gibraltar? Explain your answer.

7. Which dispute resolution technique would you advise for the residents and developers in the dispute over new hotels planned for wilderness areas in Hawai'i? Explain your answer.

8. Some law firms use billboards and TV ads to seek plaintiffs with claims about exposure to environmental hazards.

a) What is a potential benefit of this advertising?

b) What is a potential risk of this advertising?

c) In your opinion, can most attorneys bring justice to those who need it while stopping short of encouraging unnecessary litigation? Explain your answer.

9. Disputants who negotiate ruthlessly and are intent on maximizing their personal gain are sometimes called "hawks.""Doves" give in easily, favoring peaceful accord over the potential gains from conflict. The presence of a "dove" makes settlement more likely.

a) In your view, what could be a drawback of dovish behavior?

b) Identify one decision rule that can lead to a fair resolution despite hawkish behavior on the part of one or both parties.

10. What potential problem with conventional arbitration is resolved by final offer arbitration?

websurfer's challenge

1. Find a description of an environmental dispute that was resolved using one of the techniques discussed in this chapter.

2. Find an argument for and against the English (fee-shifting) rule.

3. Find a website that advocates a specific type of legal reform and critique its main argument.

Key Terms

Adjusted winner
Advisory approaches
Alternative dispute resolution
Balanced alternation
Bargaining rent
Brownfields
Conventional arbitration
Court-annexed arbitration
Decisional approaches
Defendant
Divide and choose
Early neutral evaluation
Facilitative approaches
Final offer arbitration

Final offer auction
Greenfields
Mediation
Mini-trial
Negligence standard
Neutral fact-finding
Plaintiff
Say Stop
Settlement range
Sincerity rule
Summary jury trial
Strict Alternation
Strict liability
Threat points

Internet Resources

American Arbitration Association:
www.adr.org/

EPA Superfund site:
www.epa.gov/superfund

Association for International Arbitration:
www.arbitration-adr.org

Institute for Legal Reform:
www.instituteforlegalreform.com

Further Reading

Anderson, David A. "Improving Settlement Devices: Rule 68 and Beyond." *Journal of Legal Studies 23* (1994): 225–246. A theoretical investigation of settlement devices and the conditions for settlement.

Anderson, David A. "The Fair Division of Natural Resources." *Journal of Natural Resources and Environmental Law 15*, no. 2 (2001): 227–245. An overview of decision rules applicable to the division of natural resources.

Anderson, David A., and Thomas D. Rowe, Jr. "Empirical Evidence on Settlement Devices: Does Rule 68 Encourage Settlement?." *Chicago-Kent Law Review 71* (1995): 519–545. An empirical test of Rule 68 and several other settlement-encouraging legal rules.

Bazerman, M., and H. Farber. "Arbitrator Decision-Making: When Are Final Offers Important?." *Industrial and Labor Relations Review 39* (1985): 76–89.

Boffey, Philip M. "Agent Orange in Vietnam, 30 Years Later." *New York Times* (September 8, 1998). An overview of the use of and subsequent problems with Agent Orange.

Brams, Steven J., and Alan D. Taylor. *Fair Division: From Cake-Cutting to Dispute Resolution.* Cambridge: Cambridge University Press, 1996. A comprehensive look at fair division techniques.

Brams, Steven J., and Alan D. Taylor. *The Win-Win Solution.* New York: Norton, 1999. A sequel to their previous book, with an emphasis on the adjusted-winner division technique.

Jacques, Kristi. "Texaco's Oil Production in the Ecuadorian Rainforest." *University of Michigan Environmental Justice Case Studies.* Accessed November 22, 2002. www.umich.edu/~snre492/Jones/texaco.htm. Describes a sticky environmental dispute between a multinational corporation and indigenous tribes.

Pearce, Joshua M., and David C. Denkenberger. "A National Pragmatic Safety Limit for Nuclear Weapon Quantities." *Safety 4*, no. 2 (2018): 1–25. Given the desire for nuclear weapons, this article estimates a maximum "safe" size for nuclear arsenals.

Rowe, Thomas D., Jr., and David A. Anderson. "One-Way Fee Shifting Statutes and Offer of Judgment Rules: An Empirical

Experiment." *Jurimetrics Journal 36* (1996): 255–273. An empirical test of three alternative fee-shifting rules that resolve the overlapping influences of pro-defendant rules and pro-civil rights rules.

Sagan, Carl, and Richard Turco. *A Path Where No Man Thought: Nuclear Winter and the End of the Arms Race.* New York: Random House, 1990. Explains the stark realities of nuclear war.

Shelton, Danielle M. "Rewriting Rule 68: Realizing the Benefits of the Federal Settlement Rule by Injecting Certainty into Offers of Judgment." *Minnesota Law Revue 91* (2007): 865–937. A discussion of how offer-of-settlement rules can mitigate uncertainty in legal cases.

Studer, Tensin. "Environmental Accountability: A Case for International Conciliation?." *Journal of International Dispute Settlement 14, no. 3* (2023): 350–374. https://doi.org/10.1093/jnlids/idad008. Advocates dispute resolution with the assistance of third parties that help the disputants tell their story, listen, understand each other, and reach mutual consensus.

U.S. Department of Justice. *Justice Department Files Suit Against Japanese Company Over Air Pollution at U.S. Naval Base Near Tokyo.* Tokyo: U.S. Embassy Press Release, 2002. http://usembassy.state.gov/tokyo/wwwhp022.html. An example of an international environmental dispute.

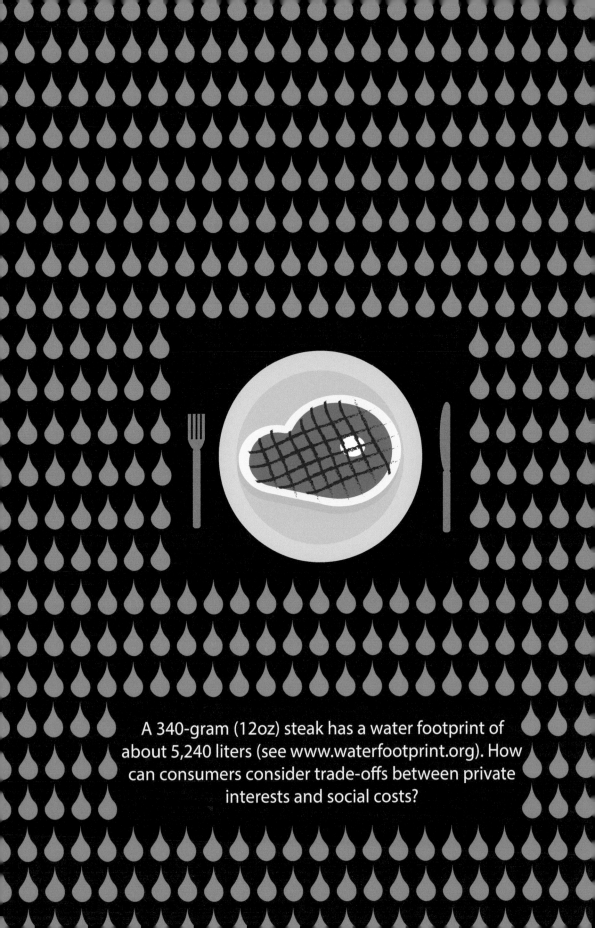

A 340-gram (12oz) steak has a water footprint of about 5,240 liters (see www.waterfootprint.org). How can consumers consider trade-offs between private interests and social costs?

Morals and Motivation

*S*ome constraints come from the laws of science. To split hydro-gen atoms from oxygen atoms in water, the water must be heated to 2,800°C. Other constraints are imposed by our own will. We set public and personal policies and determine what is socially acceptable. We choose our goals and our paths toward them. We decide where to draw the line when our lifestyle imposes burdens on others. Although humans can be stubborn about change, the last two decades of the twentieth century brought the fall of the Berlin Wall which divided East and West Germany, the formal end of racially oppressive apartheid in South Africa, and "pere-stroika" in Asia that restructured the Soviet political and economic systems. Sea changes are possible, even when they involve teaching old humans new tricks.

American philosopher and naturalist Henry David Thoreau (1817–1862) wrote, "I know of no more encouraging fact than the unquestionable ability of man to elevate his life by conscious endeavor." Given that society's treatment of the environment and natural resources is similarly subject to human discretion, this chapter explains approaches to the elemental ethical dimension of our choices.

Morals are standards for right and wrong. **Ethics** is the study of morals. In some contexts, these terms can be used interchange-ably. Ethical theories offer principles for the assessment of human behavior and provide the guidance needed in many instances of decision-making and policymaking. The modifier "normative" in the title of the next section indicates that these theories refer to the way things should be, not necessarily the way things are.

425

DOI: 10.4324/9781003428732-16

A protester in Hawaiʻi alleges unethical behavior by McDonald's.

How are ethical decisions made? Is any behavior unselfish? And what are the implications of selfishness on appropriate environmental policy? The allocation of environmental assets entails struggles between profits and preservation and between people and wildlife. In the end, critical decisions about consumption levels, policy compliance, discount rates, and conservation come down to moral judgments. The trade-offs discussed in Chapter 5 and the sustainability questions of Chapter 8 force similar moral dilemmas. What sacrifices should be made for future generations? Do our grandchildren deserve the same level of utility that we enjoy? Should they be left with the same natural capital that we have to work with? These questions are inseparably economic and ethical.

As we consider the selfishness of individuals, we must also consider the obligations of firms. Business ethics has much to do with environmental economics and natural resource management. The Monsanto Chemical Company and its subsidiaries decided to produce PCBs, Agent Orange, Bovine somatotropin (BST), and dioxin. Fast-food chains are often harangued for excessive packaging and alleged animal cruelty. Automakers must decide whether to oppose or surpass emissions and fuel economy standards.

If firms have no interest in ethics beyond visible efforts rewarded by customer loyalty, regulations may be necessary to promote ethical behavior behind the scenes. McDonald's objection to rainforest destruction for its beef supply,[1] Toyota's aggressive development of hybrid vehicles, and Starbucks' composting of coffee grounds may simply be responses to the social conscience of the consumer. The leaders of these corporations may or may not weigh the costs of these changes against anything beyond profits. This chapter discusses models for personal and firm behavior and their implications for environmental and natural resource economics.

---

1   See https://corporate.mcdonalds.com/content/dam/sites/corp/nfl/pdf/McDonalds_PurposeImpact_ ProgressReport_2022_2023.pdf.

Normative Ethical Theories

Human behavior is motivated by our objectives. Economists generally assume that rational individuals maximize a subjective utility function with weights on everything from the number of apples the individuals consume to the welfare of other people, the environment, wildlife, and future generations. For example, consider the hypothetical preferences of two individuals named Mark and Joan. Mark's utility function might be

$$\text{Utility} = 1,000 + 3(\text{apples}) + 17(\text{utility of friends}) +$$
$$2(\text{hours of television per day})(\text{utility of next generation})^2$$

while Joan's utility function is

$$\text{Utility} = 300 + 2(\text{number of animal species}) + 7(\text{utility of friends})$$
$$+ 47(\text{square feet in house}) / (\text{arsenic level in drinking water}).$$

Every time Mark receives another apple, his utility increases by 3 utils. Mark places no weight on the number of animal species, and he places a greater emphasis on the utility of friends than Joan. Each individual's utility function influences his or her decisions regarding the allocation of time, money, and other resources. The makeup of one's utility function is subjective, meaning that individuals make personal choices regarding their underlying objectives. The function that firms try to maximize has similarly subjective weights on such variables as profits, market share, service to the community, and loyalty to employees.

Some decisions rest on a sense of duty or obligation. These may include choices not to litter, smoke cigarettes around children, or hunt bald eagles. The extent to which upholding perceived duties and obligations corresponds with utility maximization is subjective and debatable. For some people, it might be a great burden to carry trash to the nearest bin, refrain from smoking, or forego a tempting kill. For them, utility maximization would lead interested parties to litter, smoke, and hunt when they can get away with it. For other people, guilt might cause headaches and lost sleep for having carried out these activities. For them, compliance with the perceived duties would be perfectly rational within a model of utility maximization. The role of moral duties in our behavior and in our utility functions depends on our own ethical stance. The ethical theories that follow offer alternative criteria for acceptable behavior that span from an emphasis on one's selfish gratification to a focus on social policies that are beneficial to everyone.

The **teleological**, or *consequentialist*, theories of ethical egoism, utilitarianism, and the common good focus on the consequences of actions and the achievement of a desired end, such as utility maximization. The **deontological**, or *nonconsequentialist*, ethical theories of rights, justice, and virtue focus on the duties and intentions of the decision maker. It is up to individuals and firms to decide which of the underlying principles to adopt.

Ethical Egoism

Ethical egoism, also known as *individualism*, is about "looking out for number one." The theory is that our moral obligation is to pursue personal interests, regardless of the effects of these pursuits on others. *Atlas Shrugged* author Ayn Rand embraced egoism as part of her larger philosophy of objectivism. Rand summarized her view this way: "My philosophy, in essence, is the concept of man as a heroic being, with his own happiness as the moral purpose of his life, with productive achievement as his noblest activity, and reason as his only absolute."[2] *Wealth of Nations* author Adam Smith described how selfishness could lead to efficiency in a free market under the right assumptions. Critics of ethical egoism argue that excessive inward focus can result in needless interpersonal conflict, neglect of others and the environment, and outcomes that are inferior to cooperative outcomes for everyone involved.

The suboptimal prisoner's dilemma outcomes in Chapter 3 illustrate ways in which selfish behavior can cause each party to be worse off than it would be with a cooperative solution. Land-use issues provide additional examples. Suppose that several parcels of land lie along a road in a county with no zoning ordinance. Should a go-cart track be built on one of the parcels, the property values of the other parcels could fall precipitously due to objectionable noise and air pollution. As stated by a neighboring landowner in such a case, "[the go-cart] race engine is very, very loud and very, very annoying."[3] Acting selfishly, an ethical egoist might be inclined to make the socially *in*efficient decision to build a track even if the neighbors' losses would exceed the track owner's personal gains.[4]

Although ethical egoists are not looking out for society, they must be careful not to let their interests in personal freedoms cause personal losses. Sometimes that which is best for society is also best for the individual. The freedom to pollute can backfire, allowing many polluters who cause illness and environmental degradation that harms the polluters themselves. In the context of the go-cart track, opposition

---

2 See www.aynrand.org.

3 "Hasty Grants Permit for Go-Cart Track," www.amnews.com/. Accessed March 2002. See also www.arkansasonline. com/news/2023/jun/01/planning-panel-denies-go-kart-track-proposal/.

4 This assumes the absence of successful Coasian bargaining, an assumption supported by the real-world case example.

to zoning laws is a mistake if cooperative support for stricter zoning would make everyone's property more valuable. It may be the case that a go-cart track will bring in more money than the sale of an unzoned lot. However, since big-spending commercial and residential land shoppers avoid property that could end up next to scores of unmuffled race engines, the security of zoning restrictions might make each of the parcels more valuable than a go-cart track. For similar reasons, planning and zoning boards often place limits on billboards, swine farms, junk cars, home sizes, and commercial development.

Go-cart tracks are fun, but please, not in my back yard!

Higher standards can benefit everyone involved.

Critics of ethical egoism point out that a narrow focus on self-interest can lead to greater damage than that from go-cart tracks. A standard example is the opportunity to kill one's rich relatives to obtain their wealth. Indeed, many an act of violence and destruction is carried out for personal advancement, neglecting the impact on others. Children really have been known to kill their families for wealth,[5] pharmacists have watered down chemotherapy drugs to save money at the expense of human lives,[6] the habitat of endangered species has been preemptively destroyed to avoid the need to comply with environmental regulations,[7] and deadly chemicals have been dumped into the air and waterways to avoid the cost of safer disposal. These acts would not be justified by personal gain if foreseeable retribution would cause greater harm to those carrying them out. Nonetheless, there is much that individuals can (or think they can) get away with if they choose to. Ethical egoists could disavow hideous self-serving acts by operating within a broader ethical framework that provides boundaries for behavior. The challenge,

---

5 A famous example is the Menendez brothers, who were convicted for killing their parents in 1996 in the family's Beverly Hills home. In Orange County, California, at least seven children have been accused of killing their parents since 2000.

6 See, for example, https://nypost.com/2018/07/06/pharmacist-sentenced-to-12-years-for-diluting-cancer-drugs/.

7 This is called the *scorched earth* technique for avoiding regulations.

in that case, is to define and defend this outer set of boundaries, which would likely resemble some of the following ethical theories.

Utilitarianism

Under **utilitarianism**, it is a moral obligation to steward the greatest good for the greatest number of people. Promoted by nineteenth-century economic philosophers Jeremy Bentham and John Stuart Mill, this criterion brings into the picture everyone who would be affected by a contemplated action. Rather than downplaying externalities as advocated by ethical egoism, Mill wrote that "the liberty of the individual must be thus far limited; he must not make himself a nuisance to other people." Bentham's utilitarian calculus called for a maximization of aggregate utility for all of society. Although difficulties with interpersonal utility comparisons make an actual summation of utility unrealistic, the goal of maximizing social welfare can provide guidance for many individual and public decisions. According to Bentham, we are to let individuals define what "good" means to them, examine policy options with regard to their effect on every person, and subscribe to those policies that provide the greatest balance of what the public perceives as good and as evil.

Utilitarianism is implicit in everyday deliberations. When a plan for a new wilderness area is evaluated in terms of its benefits to nature lovers, its boost to nearby property values, and the resulting losses to those who would otherwise develop the protected area, this is essentially a utilitarian exercise. In regard to the case study in the previous section, utilitarianism would not permit the construction of a go-cart track that caused more harm from noise and air pollution than good from profits, whether or not the track owner would benefit personally. The utilitarian approach gives equal consideration to the interests of each person and is appealing to some because it does not rely on tradition, superstition, prejudice, or religious doctrine. It does not follow a strict egalitarian tenet that each individual should receive the same allocation of goods or utility.

Detractors point out difficulties with the classical interpretation of good and evil as pleasure and pain. The satisfaction of selfish and sadistic preferences could be consistent with a goal of utility maximization even if the preferences involved were, say, the torture of animals or the burning of forests. For this reason, utilitarians often speak of a range of acceptable preferences or define good and evil in broader terms with less room for immoral pleasures. As an example, good can be defined as the production of *agent-neutral* or *intrinsic* goods that every rational person values, such as health, beauty, or knowledge.

Utilitarianism is also criticized for permitting the unjust distribution of resources. Discrimination, exploitation, and the concentration of wealth among a small number of individuals can all be justified under this theory, so long as they lead to the maximization of aggregate utility. If rich, healthy, and young people were found to receive more utility per dollar spent cleaning up nearby toxic dumpsites than poor, sick, or older people, no dumpsites would be cleaned up near the homes of the latter groups until every site had been cleaned up near the former groups. Philosopher John Rawls advocated an alternative **maximin** approach that targets utility improvements for those who are the worst off. The Rawlsian approach would have us devote resources to better the lives of the poor and downtrodden first and continue to do so until they are as well-off as the rich and healthy.[8]

There are many situations in which one person's utility is related to another person's utility. For example, suppose the use of a forest is divided between a hunter and a hiker. Each person's utility increases with the share of the forest allocated for their purpose, and it is unsafe for the same land to be used for hiking and hunting. In that case, as the hiker's utility increases because more land is allocated for hiking, the hunter's utility decreases because less land is left for hunting. On a graph with each person's utility level on an axis, we would illustrate the possible distributions of utility for the hiker and hunter as a downward-sloping line because as one person's utility increases, the other's decreases. In the case of two lovers, if one's utility increases when the other's utility increases because they are happy to see each other happy, the utility distribution curve would be upward sloping.

Figure 16.1 illustrates hypothetical utility distribution curves and the favored allocation under each of the theories mentioned thus far. Each curve represents the set of achievable combinations of utility for Artemis and Brutus in a fictional scenario.[9] The point labels indicate the associated theory as follows:

8 See Rawls (1999) for more on this and related approaches.

9 In Greek mythology, the hunting goddess Artemis is the daughter of Leto and Zeus, and the twin sister of Apollo. She is known as the "Mistress of Animals" and is often found frolicking in the forest. Brutus was a Roman politician and general who conspired to assassinate Julius Caesar, and in another incarnation, he is the nemesis of Popeye.

Figure 16.1
Utility
Distribution
Curves

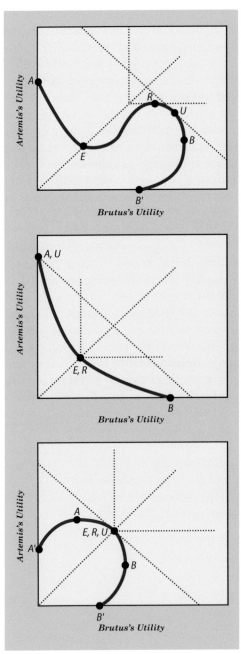

Brutus's Utility

Brutus's Utility

Brutus's Utility

E = Egalitarian

R = Rawlsian

U = Utilitarian

A = Cooperative ethical egoism—Artemis

A' = Noncoop. ethical egoism—Artemis

B = Cooperative ethical egoism—Brutus

B' = Noncoop. ethical egoism—Brutus

These graphs show achievable combinations of utility levels for Artemis and Brutus. As indicated by the points labeled *E* in each graph, an egalitarian allocation divides utility equally, so it lies on the 45° line representing points of equal utility. The points labeled *R* satisfy Rawls's maximin criterion, which yields an outcome on the highest point on an L-shaped line that has its kink on the 45° line and touches the utility distribution curve. The point that achieves a utilitarian maximization of utility, labeled *U*, is found at the point on the highest line with a slope of –1 that touches the curve. The points labeled *A* and *B* indicate Artemis's and Brutus's preferred outcomes as ethical egoists. Points *A'* and *B'* indicate the noncooperative egoist outcomes for Artemis and Brutus, respectively.

The curve in the top scenario does not necessarily resemble a real-world set of possibilities; it is constructed to differentiate the preferred outcomes among the various theories. An egalitarian allocation divides utility equally, so it is represented by a point on the 45° line representing points of equal utility for Artemis and Brutus. Points to the right of that line give Brutus more utility than Artemis; points to the left favor Artemis.

The point that satisfies the Rawlsian maximin criterion is found by imagining an L-shaped line with its "kink" on the 45° line and extending it out away from the origin until it touches the utility distribution curve at just one point. That point maximizes the smaller of the two utility levels. How do we know? Because you will notice that at any other point on the utility distribution curve, the worst-off person has a lower utility level than at point R.

The point that achieves a utilitarian maximization of utility, labeled U, is found by drawing a line with a slope of -1, and extending it out away from the origin until it touches the utility distribution curve at just one point. The significance of the line with a slope of -1 is that movements along it represent one-for-one trade-offs between Artemis's and Brutus's utility levels. Remember that slope is "rise over run." Along this line, if Brutus gained, say, 12 utils (a "run" of 12), Artemis would lose 12 utils (a "rise" of -12), to correspond with the slope of

$$\frac{rise}{run} = \frac{-12}{12} = -1.$$

If there were possible utility distribution points *above* that line, then a trade-off of more than one-for-one would be available, and the sum of the two utility levels could be increased. Thus, the furthest-out point that touches that line represents the utility-maximizing distribution.

The points labeled A and B indicate Artemis's and Brutus's preferred outcomes as ethical egoists. The players would choose those points if they cared only about themselves and not about equality, the worst-off person, or the sum of societal utility. In the middle scenario, point A provides zero utility for Brutus, and point B provides zero utility for Artemis. Selfish and deadly acts of pollution and violence, including those described in the section on ethical egoism and in the Reality Check in this chapter, are real-world examples of such outcomes.

Points A' and B' are noncooperative ethical egoist outcomes achieved if the player sacrifices utility by deliberately or naively preventing the other from receiving any utility. For example, if Brutus does not recognize that he gains by helping Artemis up to point B in the top and bottom scenarios, he will suffer the same fate as those caught up in a prisoner's dilemma or mutual free riding. If Artemis or Brutus ignore the benefits of cooperation in the bottom scenario, everyone will be worse off than they would be at points above and to the left of points A' and B', such as at points R and U.

The middle utility distribution curve could tell a story of spite, jealousy, or war. A belligerent Brutus doesn't want Artemis to receive any benefits, and if she does receive benefits, Brutus's utility will fall considerably. As Artemis receives more and more utility, Brutus's utility continues to fall, but at a decreasing rate. The more utility Artemis receives, the less an additional increase in Artemis's utility hurts Brutus. The same story is true for Artemis: She does not want Brutus to receive any utility. In reality, fire and chemical defoliants including Agent Orange have been used expressly to destroy the forest environments of enemies and, thereby, their utility if not their lives. Notice that the utilitarian outcome is on the vertical axis. The greatest sum of utility levels occurs when Artemis wins and receives all of the utility. The egalitarian and Rawlsian solutions provide each party with the same level of utility.

The bottom curve could resemble the utility distributions between environmentalists and developers. Let Artemis be a developer and Brutus be an environmentalist on an otherwise deserted island. If Brutus prevents Artemis from developing any of the island at all, Artemis receives no utility, and Brutus is without any shelter. This places them at point B'. The first few developments improve both parties' utility levels, providing Artemis with an occupation and both residents with shelter, and leading to point B. Additional developments make Artemis happy but cut into wilderness that Brutus would like to protect. The negative slope of the curve between points A and B represents these trade-offs. For Artemis to restrict Brutus's utility below the level he receives at point A would decrease Artemis's utility as well. Such decreases in environmental interests would hinder the viability of hunting and fishing stocks and detract from the natural beauty both castaways enjoy. Thus, a movement from A to A' would cause both Artemis and Brutus to suffer. The egalitarian, Rawlsian, and utilitarian outcome in this case would be to provide each party with the same level of utility. A scenario in which the Rawlsian and the egalitarian allocations are identical, but the utilitarian allocation is different, is left for you to discover in the problem set.

The Common Good

Originated in the ancient writings of Plato, Aristotle, and Cicero, the notion of the common good is that society is a community whose members share the pursuit of common goals. The welfare of individuals is inextricably bound to the good of the community. John Rawls wrote, "The **common good** I think of as certain general conditions that are in an appropriate sense equally to everyone's advantage."[10] Likewise, common-good policies are focused on outcomes that are beneficial to all, examples being clean air and water, health care, public safety, and stable global temperatures. This approach respects and encourages individuals' freedom to pursue private goals but asks that we recognize and advance those goals we share in common.

Like many of the ethical theories, this approach leaves some ambiguities. Which particular outcomes are for the common good? When may individual freedoms be exercised at the expense of others? How do we assess activities that benefit some people and harm others? While it may involve few absolutes, this approach clearly entails a community mindset in which we recognize the interconnectedness of our utility and act collectively and

10 See Rawls (1999, p. 217).

cooperatively in the interest of social welfare. Plato himself emphasized the social benefits of education in this regard. In the environmental realm, the maintenance of soils and climates suitable for agriculture and the protection of wildlife species that provide widespread benefits from their mere existence are further examples of allocations for the common good.

Virtue

Deontological ethicists argue that we should focus on duties and intentions rather than on moral principles and outcomes. With roots in the writings of Plato, Homer, and Sophocles, the normative theory of virtue is the oldest ethical theory in Western philosophy. Adherents to the **theory of virtue** make decisions about their behavior by reflecting on the types of people they intend, or have the duty, to be. According to Aristotle, "That which is the prize and end of virtue seems to be the best thing in the world." He opined that a moral virtue is the mean between two vices. For example, courage is a moral virtue between the vices of cowardice and rashness. Modesty falls between the vices of arrogance and low self-esteem. Aristotle's theory of virtue enjoyed great popularity during the Middle Ages when it was endorsed by philosopher Thomas Aquinas. Despite the rise of competing approaches, virtue theory remains among the most prominent ethical theories of modern times.

Virtues are character traits, attitudes, and dispositions that influence our utility functions and actions. Examples include generosity, courage, compassion, temperance, fortitude, honesty, fairness, and self-control. Virtues must be learned and practiced, and virtue theorists place special emphasis on education and the exercise of self-discipline. The theory of virtue holds that once a character trait is obtained, it becomes characteristic of the acquirer. Those who have formed the virtue of benevolence tend to practice benevolence often. Those who have developed many virtues become predisposed to act in accordance with moral principles. In that way, the virtuous person is the ethical person.

Critics argue that virtue ethics fails to address several issues relevant to the quest for right and wrong.[11] The lack of specific behavioral guidelines leaves common ethical dilemmas such as whether to recycle or whether to contribute to habitat preservation open to interpretation. Seemingly virtuous people might engage in questionable behavior for lack of specified criteria or offenses. Is it moral to hunt? Is it moral to drive an SUV? There is little in virtue theory to assist with such determinations. There is also the possibility of moral backsliding. If virtues are maintained only with practice, individuals might fall out of practice or temporarily deviate from the standards of their character traits. A focus on traits rather than conduct might overlook activities that would pass few other tests of morality.

---

11 See Louden (1997).

Rights

Eighteenth-century German philosopher Immanuel Kant, like Aristotle, argued that the morality of an action is determined by whether its performance adheres to moral duties rather than by the action's repercussions. Our moral duties, in Kant's view, include respecting individuals' rights to dignity and respect.

Consider the U.S. Department of Health and Human Services' determination that larger-than-expected portions of the nation and world were blanketed with nuclear fallout from Cold War nuclear testing, causing an estimated 15,000 cancer deaths in the United States alone.[12] One could make an "ignorance is bliss" argument that the withholding of information on this irreversible environmental disaster would permit higher levels of utility and do more for the common good than its rev-

Is it moral to buy bottled baby sharks?

elation. Kant felt that individuals have the right to learn the truth and would have applauded the U.S. government's report of these findings as duly honest treatment, regardless of the consequences. Other fundamental moral rights in Kant's theory include the rights of individuals to choose freely what they will do with their lives, to have privacy, to be free of punishment, and to receive what has been promised in a contract or agreement.

Kant advocated two additional principles that are relevant to environmental and natural resource economics. According to the **Principle of Ends**, one should never treat humanity as a means to an end but always as an end in itself. This suggests that individuals should not be exploited in the pursuit of profit. Kant taught that we are obligated to act out of respect for the human worth of others. Many environmentalists advocate the same principles of respect for wildlife and natural resources.

12 See www.theguardian.com/world/2002/mar/01/research.medicalscience.

Kant's **categorical imperative** states that we should choose only those actions that we would put forth as universal laws of nature. If we would not want everyone to build go-cart tracks in residential areas, we should not choose that behavior for ourselves. If we would want everyone to recycle and ride their bicycle to class or work rather than driving, we should do so as well. This view of moral behavior as that which we would rationally recommend to others provides clarity on the morality of specific actions that may or may not be acceptable under other deontological theories.

Justice

Aristotle and Plato saw fairness and justice as compelling measures of morality. These criteria for behavior have a solid intuitive foundation. Even young children call attention to actions they do not perceive as fair, with acuity that the actions are therefore not right. Aristotle's teaching that "equals should be treated equally" supports modern antidiscrimination and comparable-worth movements.[13] In regard to the environment, **justice theory** calls for decisions that are fair to present and future generations of humans. In the same way, justice for animals and other wildlife is central to the ethical theory of People for the Ethical Treatment of Animals (PETA) and various environmental organizations. Perhaps the largest challenge in applying the theory of justice is to determine which actions are indeed fair and just. For example, the theory makes it clear that citizens should have equal access to national wilderness areas, but it is not clear how much pollution it is fair to impose on one's neighbors.

Environmental Ethics

Ethical theories are useful for guiding human decisions that affect other humans. Human decisions that affect the environment or natural resources, and in turn affect humans, are also within the purview of these theories. Apart from some modern adaptations, however, the traditional theories are anthropocentric. The approaches discussed in this section shed more light on ecocentric morality.

Deep Ecology

Norwegian philosopher Arne Naess (1973) defined **shallow ecology** as the fight against pollution and resource depletion in order to improve the health and affluence of people in developed nations. **Deep ecology**, he said, adds to this fight the elements of ecocentrism and sustainability, and the intrinsic value of nonhuman nature. Naess and other deep

---

13 The comparable-worth doctrine states that those whose jobs are deemed equally important should receive equal compensation.

Chickens, some alive and some dead, on the way to market. Is our treatment of animals moral?

ecologists believe that moral evaluations should rest on ecological principles, with scientific insight into the interrelatedness of all systems of life. They believe that greater respect for the environment is morally right, in part because environmental peril could be the end of us all. In other words, that which is good for the environment is good for humans and all living things.

In contrast to ethical egoism, which is about individuals focusing on themselves, Naess suggests that we should identify with the ecosphere and the plants and animals therein. Such a focus would encourage behavior that is consistent with what he feels science tells us is necessary for the well-being of life on Earth.

Social Ecology

Murray Bookchin noted that "by so radically separating humanity and society from nature, or naively reducing them to mere zoological entities, we can no longer see how human nature is derived from nonhuman nature and social evolution from natural evolution."[14] Bookchin's eco-anarchist theory of **social ecology** has inspired many environmental ethicists to connect environmental interests with socialistic ideals. Social ecologists promote social equality and ecological interests within a framework of revolutionary libertarian socialism. A common view among people with these leanings is that the concentration of economic and political power, the homogenization of culture, and the strengthening of social hierarchies are barriers to freedom and are the principal causes of what social ecologists see as an ongoing ecological crisis. Social

14 See www.spunk.org/library/writers/bookchin/sp000514.txt.

ecologists ask people to play an active role in social evolution that will remedy these perceived imbalances of power, diversity, and equality. In the extreme, this can lead to vigilante justice as discussed in Chapter 12. The annexation of radical political views with environmentalism, valid or not, brings with it opposition from those with moderate and opposing political views. Critiques of the underlying political philosophies, aside from the brief coverage in the chapter on government, are beyond the scope of this text.

Ecofeminism

Ecofeminism, a term coined by French feminist François d'Eaubonne, associates the unethical domination of women with the domination of nature. Ecofeminists see a common thread of immorality in varying levels of disrespect for women and for wilderness. Disregard for the feelings of other human beings and disinterest in the state of the environment may be rationalized in a similar manner, and conversely, the virtue of compassion may carry over from compassion for humans to compassion for animals. With a focus on the interconnected spheres of feminism, development, and community, ecofeminism has become a popular grassroots activist movement over the last three decades.

Resolving Ethical Dilemmas

Would you be able to sleep at night after illegally dumping hazardous waste where it might reach drinking water supplies? Is it conscionable to wash your car during a water use moratorium, to pitch recyclables, to dump trash out your car window, or to buy unneeded material goods? Our daily ethical dilemmas as individuals, business owners, and policymakers can be simplified by the application of decision rules. Almost any behavior can be rationalized with arguments like, "other people are doing it," "it's legal," "it provides a product that people demand," or "if I didn't do it, someone else would." However, these statements take no account of the harm caused by the activity, possible alternatives for the activity, the likelihood that others will mimic the behavior, or the effect of the behavior on one's conscience.

Economics provides the guidance that behaviors should continue until the marginal cost exceeds the marginal benefit. Unfortunately, dilemmas still arise regarding which costs and benefits to consider. For example, ethical egoism suggests consideration only of one's own costs and benefits, while utilitarianism seeks aggregate utility maximization. As extensions of moral theories, the following criteria are available to assist with decisions governing environmental and natural resources:

Ethical Egoism: *Is this action good for me?*

Utilitarianism: *Does it bring the greatest good to the greatest number of people?*

The Common Good: *Is it good for society as a whole?*

Virtue: *Does this action reflect balance between vices?*

Rights: *Does this action respect the moral rights of everyone?*

Kantianism: *Would I want everyone to do it?*

Justice: *Is this action fair and just? Does it treat equals equally?*

Deep Ecology: *Is this action sustainable and ecocentric?*

Social Ecology: *Is it consistent with social equality and ecological interests?*

Ecofeminism: *Does this action show due respect for living things?*

The following are three additional questions that some people like to consider when facing ethical dilemmas:

1   **Would you like to see it in a headline?**
    The **front-page-of-the-newspaper test** promotes consideration of what other members of society would think about an action. The criterion is: Would you carry out the behavior in question if you knew that a description of it would appear in a newspaper headline? If you envision that the headline would be incriminating or make you feel embarrassed, that's a sign that the contemplated act is morally unsound.

    How would you feel if one of the following headlines appeared in the newspaper?

    > SMOKER RELEASES CARBON MONOXIDE INTO PUBLIC RESTAURANT

    > STUDENT DRIVES CAR TWO BLOCKS RATHER THAN WALKING

    > STRIP-MINING CONTINUES DUE TO LACK OF RECYCLING

    Your response could help you decide whether these actions are acceptable. Here are some actual newspaper headlines:

    > ENDANGERED WHALES LIVE IN AREA EARMARKED FOR GAS EXPLORATION INCO LTD. FACES LAWSUIT ALLEGING NICKEL REFINERY POSES A BIG HEALTH RISK

    > EXPERT SAYS ALABAMA PLANT SHOULD PAY $8.6 MILLION FOR PCB DAMAGE

Do you think the decision-makers would have behaved differently had they subjected themselves to the front-page-of-the-newspaper test in advance?

2 **How does this decision make you feel?**
   Intuitionism is the doctrine that, rather than applying an explicit formula to reach moral decisions, we should follow our own intuition. Intuitionists feel that while we should adhere to certain moral principles, the principles express self-evident propositions. Thus, as Jiminy Cricket said to Pinocchio, "Just let your conscience be your guide." This is a common approach, and it can work well for those who are in close touch with their conscience. Pinocchio, on the other hand, said, "What's a conscience?"

3 **What would my role model do?**
   Many people have heroes, spiritual leaders, or role models whose behavior they admire. When faced with an ethical dilemma, they imagine what that other person would do if faced with the same situation. Those who admire their parents, community leaders, clergy, or teachers, for example, may try to allocate resources the way those people would. Friends and neighbors influence each other in the same way. When a few people in a neighborhood start putting materials out for recycling, that decision can be contagious.

441

MORALS AND MOTIVATION

The Arsenic Pond

Nicola Hanna, the chief federal law enforcement officer for the Central District of California, says that "our nation's environmental laws are specifically designed to ensure that hazardous wastes are properly handled . . . from the point of generation to the point of disposal."[15] It doesn't always work out that way.

To make Crystal Geyser bottled water in Olancha, California, CG Roxane LLC draws water from natural sources and filters it to remove toxins. Unfortunately, that filtering isn't just for show: Naturally occurring arsenic is removed in the process. A 2018 indictment alleges that CG Roxane flushed the arsenic out of the filters and into a man-made pond that they called "the Arsenic Pond." After testing, the California Department of Toxic Substances Control identified the water in the pond as hazardous waste and told CG Roxane to transport it—labeled as hazardous waste—to an appropriate hazardous waste facility.

(continued)

15 See www.justice.gov/usao-cdca/pr/three-companies-including-crystal-geyser-charged-illegally-transporting-hazardous-waste.

According to the indictment, in May of 2015, the companies hired to transport the wastewater did not properly identify it as hazardous waste, they did not indicate that the wastewater had arsenic in it, and they took it to a facility that was not permitted to treat hazardous waste.

What could go wrong? The World Health Organization reports that arsenic can cause cancer and skin lesions, and arsenic has been associated with cardiovascular disease, impaired cognitive development, and diabetes.

Investigations by the EPA, the Department of Transportation, and the California Department of Toxic Substances Control followed. In 2020, the U.S. Department of Justice posted a headline that does not pass the front-page-of-the-newspaper test: "Bottler of Crystal Geyser Water Pleads Guilty to Illegally Storing and Transporting Hazardous Wastewater Contaminated with Arsenic." CG Roxane paid a $5 million fine after a plea agreement.

Who knew that making clean water could be a dirty business?

The answer to the question "Does it maximize profits?" provides one means of decision-making. The questions listed earlier provide alternative criteria for all those who choose to consider morality among the factors that determine resource allocations. If you question the ties between environmental economics and ethics, consider whether externalities, social discount rates, or most nonuse values for natural assets would matter in the absence of ethical considerations.

Summary

At the core of common debates over environmental and natural resource economics are moral dilemmas involving the appropriate treatment of flora, fauna, fellow humans, and future generations of the same. This chapter considers the motives behind our behavior and the composition of our utility functions. Ethical theories offer guidance in decision-making, including criteria for the way we use resources. These theories help individuals choose actions that are conscionable and provide firms with alternatives to simple profit maximization. Economists, too, must decide how assorted costs and benefits should weigh into measures of efficiency. The allocation of natural resources cannot be divested from ethical issues.

Suppose five organ-transplant candidates could be saved with the benefit of organs from one healthy human sacrifice. Should the sacrifice be made? If each individual gained the same utility from life, the strict utilitarian response would be yes, in order to maximize social welfare. Rights advocates would say no because the healthy person has the right to live. Would you sacrifice five lives in exchange for the right of individuals to earn profits from an enterprise that pollutes? Would you sacrifice five forests for the same purpose? How about five animal species? Unfortunately, economic decisions like these must be made routinely. Although ethical theories do not remove the pain from these trade-offs, they do provide structure for well-reasoned decisions. Perhaps the assistance of these theories is the most that we can hope for, and the least that those whose lives hang in the balance should expect.

• • • • • • • • • • • • • • • • • •

Problems for Review

1. Consider your utility function.

   a) What are three things that weigh heavily in that function, meaning that more of them makes you much happier? These can be any sort of activities, goods, or services.

   b) Research finds that relationships and activities such as helping others typically provide more happiness than material things. Does your utility function resemble this environmentally friendly finding?

2. Suppose you face an ethical dilemma such as whether to pay extra for organic cotton clothing, pick up litter, drive less, or purchase carbon offsets for your travel.

   a) What question or criteria do you use to decide what to do?

   b) Do you think any of the new ideas you picked up from this chapter will influence your behavior? If so, in what way?

3. The giant Asian pond turtle is endangered. It is also highly sought after for medicinal and food purposes.

   a) Would you hunt and kill one of these turtles for $50?

   b) Would you hunt and kill one of these turtles for $10 million?

   c) At what price would you become willing to hunt a turtle?

   d) Suppose exorbitant prices lead otherwise moral people to poach endangered species. Based on your understanding of how market prices are determined, if the goal is to avoid immoral behavior in hidden markets encouraged by high prices, would it be better for a government policy to decrease the supply of a species or to decrease the demand?

4. Consider farmers who raise cage-free chickens, restaurants that serve local food to reduce transportation externalities, and manufacturers that use biodegradable packaging

materials. In your estimation, what portion of their motivation to do these things comes from their own morals, and what portion is driven by the interests of the customers these businesses are trying to attract?

5. In your experience, are the government policies in your country generally aligned with the common good? Do you trust politicians to act on the behalf of society? Comment in one or two paragraphs.

6. A National Academy of Sciences report estimated that 60,000 women in the United States put their fetuses "at risk" of brain damage from mercury in the fish they ate. U.S. Food and Drug Administration (FDA) scientists warn that a woman should eat only one can of tuna per week and that "The action levels we have in place are not protective enough for this—the fetuses." Nonetheless, after meetings with the U.S. Tuna Foundation and other seafood industry representatives, the FDA decided only to suggest that pregnant women eat fish in moderation. Which ethical theories would support this decision?

7. Suppose Muchland and Littleland are neighboring countries. Littleland has relatively few natural resources, although new discoveries of lithium are adding to residents' utility levels. The leader of Muchland, President Much, gains utility when the leader of Littleland, President Little, gains utility from discoveries *up to a point* but then becomes jealous. If President Little has 0 utils, President Much feels bad and has 0 utils. For the first 6 utils President Little gains, President Much gains 1 util for every 2 utils gained by President Little. After that, more utils for President Little make President Much less happy: President Much losses 1 util for every 2 utils gained by President Much beyond 6. On a graph with President Little's utility on the vertical axis and President Much's utility on the horizontal axis, draw a utility distribution curve for this situation. Also draw a dotted 45° line to show where the leaders' utility levels would be equal. Label the points representing the Rawlsian, utilitarian, and egalitarian solutions R, U, and E, respectively.

8. Draw a utility distribution curve on which the Rawlsian and the egalitarian allocations are identical, but the utilitarian allocation is different. Label the points representing each solution with the first letter of the associated theory.

9. Utilitarian Jeremy Bentham argued that if utility is good, then it is good irrespective of whose utility it is, and thus the sum of societal utility should be maximized. Do you agree? Why or why not?

10. The tenets of ecofeminism suggest that those who seek to dominate or exploit women and those who

seek to dominate or exploit the wilderness have a similar mindset. Are the people you know who do one of these things the types of people that you believe would do both? Are the people you know who would never do one of these things also the types who would be unlikely to do the other? Explain your observations.

websurfer's challenge

1. Find a website that explains a general ethical theory *not* described in this chapter and summarize it in one paragraph.

2. Find a website that explains the view of a prominent ethicist and describe how her or his view could be applied to environmental economics.

Key Terms

Categorical imperative
Common good
Deep ecology
Deontological
Ecofeminism
Ethics
Ethical egoism
Front-page-of-the-newspaper test
Intuitionism

Justice theory
Maximin
Morals
Principle of ends
Shallow ecology
Social ecology
Teleological
Utilitarianism

Internet Resources

Women and Life on Earth:
www.wloe.org/what-is-ecofeminism.76.0.html

Institute for Social Ecology:
www.social-ecology.org

International Society for
Environmental Ethics:
https://enviroethics.org

Opensecrets.org:
www.opensecrets.org

People for the Ethical Treatment of
Animals:
www.peta.com

On Environmental Ethics:
https://plato.stanford.edu/entries/ethics-environmental/

Further Reading

Anderson, David A. *Treading Lightly: The Joy of Conservation, Moderation, and Simple Living*. Danville, KY: Pensive Press, 2009. This book discusses environmental ethics and provides examples of relevant practices and lifestyles.

Aristotle. *Nicomachean Ethics*. New York: Oxford University Press, 1998. Aristotle's classic treatise on ethics and virtue.

Bookchin, Murray. *Toward an Ecological Society*. Montreal: Black Rose Books, 1988. This book outlines the social ecology perspective of its greatest proponent.

d'Eaubonne, François. "The Time for Ecofeminism." In *Key Concepts in Critical Theory: Ecology*, edited by Carolyn Merchant and translated by Ruth Hottell. Atlantic Highlands, NJ: Humanities Press, 1994. An edited volume on the themes of deep ecology, ecofeminism, and environmental justice.

Geirsson, Heimir, and Margaret R. Holmgren (eds.). *Ethical Theory: A Concise Anthology*. 3rd ed. Peterborough, ON: Broadview Press, 2018. A collection of readings on the predominant ethical theories.

Kant, Immanuel. *Critique of the Power of Judgment*. Cambridge: Cambridge University Press, 2002. A taste of classical Kantian philosophy.

Katz, Eric, Andrew Light, and David Rothenberg (eds). *Beneath the Surface: Critical Essays in the Philosophy of Deep Ecology*. Cambridge, MA: MIT Press, 2000. Compares deep ecology's philosophical ideas with other schools of thought, including social ecology, ecofeminism, and moral pluralism.

Leopold, Aldo. *A Sand County Almanac*. New York: Oxford University Press, 2001. A collection of essays on conservation in which it is claimed that the source of the ecological crisis is philosophical.

Louden, Robert B. "On Some Vices of Virtue Ethics." In *Virtue Ethics*, edited by Roger Crisp and Michael Slote. New York: Oxford University Press, 1997. A collection of essays on virtue ethics by prominent modern philosophers from several Western countries.

Naess, Arne. "The Shallow and the Deep, Long-Range Ecology Movements: A Summary." *Inquiry 16* (1973): 95–100. The article with which this Norwegian philosopher started the deep ecology movement.

Nash, Roderick Frazier. *The Rights of Nature: A History of Environmental Ethics.* Madison, WI: University of Wisconsin Press, 1989. A well-researched history of environmental thought and politics in the United States

Niebuhr, Reinhold. *Moral Man and Immoral Society: A Study in Ethics and Politics.* Louisville, KY: Westminster John Knox Press, 2002. Argues that corporations cannot be ethical because they have an essential vested interest in their own survival.

Rawls, John. *A Theory of Justice.* Cambridge, MA: Belknap Press, 1999. This remarkably influential contemporary philosopher explains his doctrine of "justice as fairness."

Washington, Haydn, and Michelle Maloney. "The Need for Ecological Ethics in a New Ecological Economics." *Ecological Economics 169* (March 2020): 106478. https://doi.org/10.1016/j.ecolecon.2019.106478. An argument for more attention to ethics in discussions of natural resources, a decommodification of nature, and a shift to an ecocentric worldview.

Index

Note: Page numbers in *italics* indicate figures, and page numbers in **bold** indicate tables in the text.

INDEX

World Commission on
    Environment and
    Development 208
World Medical Association 160
world population growth *244*
World Trade organization (WTO)
    292–293
World Travel and Tourism
    Council 306

Yangtze River tragedy 353, 357
Yucca Mountain 184

zero emission vehicles (ZEVs) 185
zero-rent level of effort 349
zero waste initiatives 6
Zia, A. 249
Zoellick, R. 181
zoning ordinances 428